The Urban Book Series

The Urban Book Series is a resource for urban studies and geography research worldwide. It provides a unique and innovative resource for the latest developments in the field, nurturing a comprehensive and encompassing publication venue for urban studies, urban geography, planning and regional development.

The series publishes peer-reviewed volumes related to urbanization, sustainability, urban environments, sustainable urbanism, governance, globalization, urban and sustainable development, spatial and area studies, urban management, transport systems, urban infrastructure, urban dynamics, green cities and urban landscapes. It also invites research which documents urbanization processes and urban dynamics on a national, regional and local level, welcoming case studies, as well as comparative and applied research.

The series will appeal to urbanists, geographers, planners, engineers, architects, policy makers, and to all of those interested in a wide-ranging overview of contemporary urban studies and innovations in the field. It accepts monographs, edited volumes and textbooks.

Indexed by Scopus.

Graciela H. Tonon
Editor

Urban Inequalities

A Multidimensional and International Perspective

Springer

Editor
Graciela H. Tonon
UNICOM, Faculty of Social Sciences
National University of Lomas de Zamora
Lomas de Zamora, Argentina

ISSN 2365-757X ISSN 2365-7588 (electronic)
The Urban Book Series
ISBN 978-3-031-59745-9 ISBN 978-3-031-59746-6 (eBook)
https://doi.org/10.1007/978-3-031-59746-6

This Springer imprint is published by the registered company Springer Nature Switzerland AG
The registered company address is: Gewerbestrasse 11, 6330 Cham, Switzerland

For my children Pedro and Erica who always give me their love and energy to continue learning.
To my husband, Walter, who supports me every day with his love and care.

Preface

This book is dedicated to the study and analysis of urban inequalities from a multidimensional and international perspective. The idea of the book is to present a new look at the issue of urban inequalities, which is reflected in some chapters dedicated to theory and others dedicated to the different methodological approaches used to measure inequalities. Likewise, the text includes chapters that contain proposals for concrete action to prevent and resolve inequalities.

The book begins with a chapter of my own in which urban inequalities are theoretically defined and analyzed; to do this the theoretical proposals of the capability approach, the quality of life, and the community well-being were used; the chapter also reflects on public policies to overcome inequalities.

The following chapters are dedicated to urban inequalities: economic, health, education, housing, diversities, well-being, and capabilities, in countries from different continents: Bangladesh, Peru, Hong Kong, and Germany.

Below are the chapters dedicated to the environment and climate change, one with a world global vision and the other that presents the case of Argentina as well as a chapter that analyzes the role of inequality for digital learning in pandemic times in Mexico City.

The block of the following chapters shows different types of children's inequalities. Some are dedicated to children in Africa; health, access to services, and education of children living in cities of the Sub-Saharan region; urban deprivation and inequalities on children and adolescents in the two largest cities of Nigeria; and socio-economic status and children's subjective and psychological well-being in South Africa. Others focused on the inequalities of children in Latin America: children's vital inequalities in urban contexts of Argentina's Great North, considering child mortality rate and health vulnerability index and children in street situations in Brazil as unequal subject of rights.

The subsequent chapters focus on other types of inequalities such as those of the Taiwanese urban indigenous people, and another one that studies the impact of globalization on citizen's political autonomy in the cities in Europe, such as the case of London.

Finally, the book presents two chapters to try to overcome inequalities: the case of Wellbeing Advances Equity and Civic Leadership in Santa Monica, USA, and the proposal to prevent urban inequalities using POZE and AI to optimize quality of life in urban settings.

Authors from different countries, different thematic fields, different universities, international organizations, and NGOs have collaborated with their chapters. I want to thank them for their generous participation which massively enriches this book.

Lomas de Zamora, Argentina
July 2023

Graciela H. Tonon

Acknowledgements With my great thanks to Dr. Prof. Richard Estes, a great teacher and social worker, pioneer in the field of development, inequalities, and vulnerabilities studies, and an extraordinary human being.

Contents

Chapter 1
Re-defining Urban Inequalities and Its Policies: Through the Lens of the Capability Approach, Quality of Life, and Community Well-Being

Graciela H. Tonon

Abstract To study the redefinition of urban inequalities, it is necessary first to review the concept of inequalities. The Nobel Prize Amartya Sen (Desarrollo y libertad, Editorial Siglo, Bogotá, 2000) argues that the analysis of inequality must consider the choice of space, that is, the variable in terms of which inequality is assessed. In addition, Sen (Desarrollo y libertad, Editorial Siglo XXI, Bogotá, 2000, pp. 137–138) proposed the capability approach and explained that inequality had been traditionally assessed, in terms of income, to the neglect of other types of inequality, such as poor health, lack of education, and/or social exclusion. However, he argued that income inequality can differ significantly from inequality in other spaces and the contrast between the different income perspectives. In this line of reflection, we will recognize that inequalities matter for the sake of the well-being and prosperity of a society. In this chapter, we shall revise the concept of inequality and its different classifications, the peculiarities of inequalities in the urban space in relation to people's quality of life and community well-being, concluding with the analysis of public policies in response to inequalities.

Keywords Inequalities · Urban space · Human capabilities · Quality of life · Community well-being · Public policies

1.1 Introduction to the Study of Inequalities

The study of inequalities acquired importance during the 1990s and expanded during the twenty-first century, first developing in the field of economics. In this regard, we can quote the general definition: "inequality is the size of the relative differences in resource endowments between units belonging to a particular environment" (Pettinato 2002, p. 24).

G. H. Tonon (✉)
UNICOM-Institute of Social Studies, School of Social Sciences, Universidad Nacional de Lomas de Zamora, Lomas de Zamora, Argentina

G. H. Tonon (ed.), *Urban Inequalities*, The Urban Book Series,
https://doi.org/10.1007/978-3-031-59746-6_1

The Nobel Prize Amartya Sen (2000) argues that the analysis of inequality must consider the choice of space; that is, the variable in terms of which inequality is assessed. In addition, Sen (2000, pp. 137–138) explains that inequality has been traditionally assessed in terms of income, neglecting other types of inequality, such as poor health, lack of education, and/or social exclusion. However, he argued that income inequality can differ significantly from inequality in other spaces and with the contrast between the different perspectives of income. According to Kanbur (2021, p. 3), "the study of the economic dimension of inequality has a long history and is not only important per se, but also linked to many other topics in the social sciences."

There is a degree of consensus regarding the two major premises in the study of inequality: one of the points of agreement is the acknowledgment of a multiplicity of inequalities, namely gender, age, levels of education, access to the labor market, geographic location (type of neighborhood), access to basic services, individual traits, as well as the characteristics of the public policies of the country in which a person resides.

The second premise puts forth the need to study inequality taking into consideration different scales of analysis, i.e., not only taking into account the individual scale, but also institutional and structural issues, the social media people subscribe as well as their practice and social interaction (Dammert Guardia et al. 2019, pp. 8–9).

On the other hand, Pardo and Prato (2021, p. 6) pointed out that, as inequality was re-defined in different areas of action and interchange, the associated power relationships were likewise re-defined.

In the twenty-first century, when dealing with the different types of inequalities, we can specifically point out the inequalities by age and by gender, which are considered among the determining axes of the distribution of well-being and power in society, because they are one of the bases of the social organization around which responsibilities and roles are assigned (ECLAC 2016).

In this respect, and in the particular case of Latin America, ECLAC (2016) has identified dimensions which determine the structuring of patterns of social inequality—aside from the income and socio-economic level—namely (a) gender inequality; (b) ethnic and racial inequalities; (c) inequalities related to the life cycle (especially childhood, youth, and old age); and (d) territorial inequalities (between rural and urban zones, and between the different regions of each country), pointing out that those dimensions intersect and potentiate each other, particularly affecting certain population groups, such as indigenous and afro-descendant women.

It is also important to remember Arendt (1998) who expressed that much inequality is generated through ideology, corruption, and privilege, compounding differences in nature and culture, and recognizing the inequalities that exist in the field of politics,

> Political inequality means more than exclusion from formal participation in the political process, as even when people do have the vote, it can be controlled through vertical relationships that trade access to benefits from those in power in return for electoral mobilization (Bartlett et al. 2015, p. 7).

We can identify inequalities related to the places where people live, whether urban or rural spaces. In this sense, Tonkiss (2018) expressed that the groups, which simultaneously experience spatial and intersecting inequalities, are the ones that tend to be most persistently and left furthest behind.

Likewise, the convergence of multiple inequalities characterizes the hard cores of poverty, vulnerability, and social exclusion and makes them persist and reproduce (ECLAC 2016), so in this sense, inequalities matter to the well-being and prosperity of a society (Kabeer 2014, p. 91).

1.2 The Concept of Inequality in the Twenty-First Century

Since the late 1990s and throughout the twenty-first century, different authors have studied and classified inequalities. Next, we will review the most outstanding contributions: inter-category and intra-category inequalities, inequalities and capabilities, horizontal inequalities, categorical inequalities, intersecting inequalities, and relational inequalities.

1.2.1 Inter-categorical and Intra-categorical Inequalities

In 1996, Fitoussi and Rosanvallon identified new dimensions of analysis of inequalities and proposed the study of two types of inequalities: inter-category and intra-category inequalities. The former are the inequalities that have been traditionally studied (also called structural inequalities) which have been internalized by society, being fundamentally associated with the economic dimension. On the other hand, intra-category inequalities are considered new inequalities which involve other dimensions of analysis such as social, cultural, and religious dimensions. The authors noticed that intra-category inequalities could disturb the representation that people can have of themselves and can therefore affect identity; and (Fitoussi and Rosanvallon 2003, p. 76) also pointed out that when inequalities increase, they affect the structure of society. In this sense, the perception of the increase in inequalities in a society can be generated from: a weakening of the principles of equality in society, an increase in structural inequalities, and/or the emergence of new inequalities, resulting in a process in which the three categories are dependent on each other.

1.2.2 Inequalities and Capabilities

Also, the 1998 Nobel Prize winner, Amartya Sen, tackled the issue of inequalities. Sen (2000) argues that the analysis of inequality must consider the choice of space, that is, the variable in terms of which inequality is assessed. In addition, Sen

(2000, pp. 137–138) explains that inequality has been traditionally assessed in terms of income, thus neglecting other types of inequality, such as poor health, lack of education, and/or social exclusion. However, he argued that income inequality can differ significantly from inequality in other spaces, and that the contrast between the different perspectives of income and capability has a direct impact on the space in which inequality is to be examined.

Sen (2000, p. 16) proposed the capability approach that identifies freedom as a major principle of development and, in this respect, the capability approach represents a theoretical proposition for the assessment of life satisfaction, perception of social situations, design of public policies which affect economic development, social policies, and international development. The capability approach provides an overarching structure for the understanding and measuring of equality, which focuses on what matters to people, recognizes diversity in needs, lays emphasis on barriers, constraints, structures, and processes, while recognizing diversity as regards goals (Buchardt and Vizard 2007, p. 7 quoted in Tonon 2022a, b, p. 13).

Sen (2000, p. 121) stated that inequalities must be studied on the basis of choice of space as well as other relevant variables such as well-being, freedom, and the different aspects of the quality of life. In his theoretical proposal, Sen (2000, p. 120) expressed that the feeling of inequality can erode social cohesion and even make it difficult to achieve efficiency.

1.2.3 Horizontal Inequality

Another author, concerned about this topic, is Stewart (2013), who defines inequality as a major source of injustice and a cause of poverty, and sometimes of conflict, thus adopting a plural approach to inequality by encompassing several important dimensions. The "vertical" model of inequality is based on ranking individuals/households by their income, assets, or human capabilities (Stewart 2002). Inequality in income distribution is a summary measure of the dimension of income and employment which provides only a partial indicator of the other dimensions; besides being a vertical measure, since it considers the members of society from top to bottom, measuring their income and the consequent inequality (Stewart 2000, p. 13).

The Stewart (2013) proposed a horizontal measurement of inequality which would measure the inequality between groups, defined by region, ethnicity, class, and/or religion according to the most appropriate type of group identification in each particular society. These groups are socially defined by their members, or by others, often based on common cultural characteristics and behavior, appearance, or place of birth (Stewart 2009, p. 316). Horizontal inequalities greatly affect people's well-being directly, as well as other objectives, since people are affected not only by their individual circumstances but also by how well their group is doing since being part of a group is often an important aspect of a person's identity. Horizontal inequalities cut across the economic categories and are related to social discrimination.

One of Stewart's conclusions was that "group inequality tends to be more persistent, and more difficult to tackle than individual inequality because of certain group features, mainly asymmetries in social and cultural capital, as well as discrimination" (Stewart 2009, p. 333).

1.2.4 Intersecting Inequalities

Another classification is that of intersecting inequalities, which combine both horizontal and vertical characteristics. The two approaches (vertical-horizontal) focus on distinct axes of disadvantage: resources and identity (Kabeer 2014, p. 92). The former approach to inequalities (vertical) is focused on income. The latter is based on social discrimination, which takes into consideration the social identity of different groups.

Group-based inequalities are the product of cultural norms, values, and practices that enable the participation as equal citizens in the life of their community (Kabeer and Santos 2017, p. 6). Yet people may suffer social discrimination without necessarily facing material deprivation, and "it is the intersection between these different forms of inequality that serves to define the most severe and often the most enduring form of social exclusion in societies" (Kabeer and Santos 2017, p. 6).

The intersecting inequalities that give rise to social exclusion are deeply entrenched in the historical structures and everyday practices of societies, making them appear remarkably resistant to change (Kabeer 2014, p. 98). As Arauco et al. said:

> It is the intersection, rather than the simple addition of these different forms of inequality (economic, social, spatial, and political)—the fact that they reinforce and exacerbate each other—that explains the enduring nature of the exclusions experienced by certain groups within a society (Arauco et al. 2014, p. 11).

1.2.5 Categorical Inequalities

On the other hand, Tilly (2007) made reference to categorical inequalities by expressing that "inequality is a relation between persons or sets of persons in which interaction generates greater advantages for one than for the other" (Tilly 2007, p. 247). The author argues that inequalities situate people on one side or the other of the categorical boundaries. Categorical inequalities ought to be observed as related to gender, language, and religion. The categories are crucial because they shape both inequalities and identities. (Tilly 2007, p. 246). These classifications set categorical limits which people use to organize social life and reproduce the inequality between members of different categories (Tilly 2007, p. 248).

1.2.6 Relational Inequalities

An original proposal in the study of inequality is one that focuses on relationships, which has led us to review the research work developed by Ghosh (2019). The author believes that human beings are inherently social and that their individual capacities are socially regulated, and in turn, regulate their own social relationships. These interactions, likewise, undergo changes in time and are affected by the dynamics of broader economic and social processes. Though, traditionally, most debates on distributive inequality have generally been focused on material goods or services, today there is a recognition of what is known as relational equality, which involves egalitarianism, as well as the demands of contemporary social movements (Ghosh 2019). The aforementioned goes beyond distributive inequality, for it acknowledges different kinds of social and hierarchical inequality—particularly in the form of persistent group inequalities—which are sustained by laws, norms, or habits, that generate higher or lower levels of social interaction (Anderson 2014).

At this point, we must refer to the research work carried out by Nussbaum (2012) whose version of the focus on capabilities is part of the notion of human dignity, for her list of essential human capabilities recognizes the importance of humans as social beings, thus regarding such capabilities in terms of social interaction. This necessarily leads us to consider the fact that capabilities are likely to be regulated and circumscribed by social hierarchies.

In this respect, Ghosh (2019, pp. 182–183) identifies three types of social hierarchies, namely: dominance hierarchies, esteem hierarchies, and prestige-based hierarchies.

- The dominance or leadership hierarchies are the ones in which those who occupy lower social positions are subjected to the arbitrary authority of the higher social classes and are therefore defenseless—to the point that they are compelled to obey their orders and obtain their permission to a free exercise of their freedoms, such as gender relationships.
- Esteem needs hierarchies are the ones in which those who occupy inferior positions are stigmatized, on the basis that they are defined by certain group identities and are, therefore, subjected to segregation, discrimination, persecution, or even violence. These stigmatized identities may be based, among other things, on sexual orientation, religion, language, clothing, or name, and distinct ethnic characteristics.
- Prestige-based hierarchies are the ones which give special relevance to the opinion of others, as well as to social institutions, and the interests of those who occupy high social positions, thus enjoying more rights, privileges, opportunities, or benefits than the rest (Ghosh 2019).

These types of hierarchies may also juxtapose, thus creating extreme forms of relational inequality. (Ghosh 2019, p. 182).

In other words, the inequality studies research field has made contributions in conceptual and theoretical issues, on the one hand, and far-reaching investigations of empirical problems, on the other (Kanbur 2021, pp. 3–6).

1.3 Inequalities in the Urban Space

Urban inequality is a highly complex phenomenon which calls for a multidimensional approach.

In the twenty-first century, cities are characterized by demographic growth as well as a multiplicity of complex issues, namely: territorial segregation, a formal-informal dichotomy, resource depletion which affects cities, the difficulties experienced in human interactions, mobility, and management deficit (Hiernaux-Nicolas and González Gómez 2017, p. 93). Inequality is detrimental to cities in many ways since it results in labor inequality, educational inequality, mobility inequality, spatial inequality, environmental mismatch, digital divide, food deserts, and unequal access to basic services (Nijman and Wei 2020, p. 1). Urban inequalities affect people of different economic conditions, from the wealthy to the deprived.

Segura (2014, p. 14) points out that spatial inequality involves a double dimension, namely the unequal access to urban space; and urban space as a dimension that has an impact on the production and reproduction of inequality. The author observes an unequal access to the city as a place of residence, housing, infrastructure, and urban services, therefore affecting the basic dimensions, constitutive of social life, and conditioning the production and reproduction of social inequalities, since urban space is the result of social relationships and is, therefore, a dimension of social life that has specific effects on people.

On the other hand, Ciccolella and Baer (2011, p. 232) believe that, in the case of Latin American cities, the price of the land has historically been the main factor that determines territorial distribution, and in this sense the social structure of the city.

This is shown in the research work carried out by Haddad and Nedović-Budić (2006) who developed a spatial econometric analysis of interurban inequality in different districts of the municipality of Säo Pablo (Brazil), exploring the relationship between the dwelling index as compared to the public service provision and investment on social programs. The results they obtained showed that public intervention reinforced the preexistent differences between center and periphery since public services and social programs benefited the more densely populated districts. Moreover, in urban areas, residency in low-income settlements can be an instrument of social exclusion (Bartlett, Mitlin, and Satterthwait 2015, p. 6).

Kabeer (2010) noted that there were other dimensions to social exclusion that were not fully captured by the interplay between economic deprivation and social discrimination. Spatial inequalities are differences in development associated with place-based characteristics, such as remoteness, high density, or poor municipal infrastructure provision, and often coincide with intersecting inequalities (Kabeer and Santos 2017). Spatial inequalities often overlap with intersecting inequalities, i.e., socially excluded groups concentrated in the most disadvantaged locations and neighborhoods. These social, economic, and spatial inequalities in turn contributed to political disenfranchisement: such groups were generally denied voice and influence in collective decisions that affected their lives (Kabeer 2010).

During the last decades, urban expansion and the decentralization of employment policies have become the processes with the highest incidence on spatial mismatch, since these processes have changed people's home location options, as well as their job opportunities (Glaeser and Kahn 2001; Wei and Ewing 2018).

The new economy has boosted urban inequality in three different ways, namely a greater labor market bifurcation, resulting in a deepening income inequality; an increase in interurban inequality; and the resurgence of central cities which have gradually become more unequal, i.e., greater classification and inequality between the different suburban areas (Nijman and Wei 2020, p. 2).

Regarding the job mobility rate, Giuliano and Small (1993) studied the complexity of commuting from home to work in urban spaces characterized by their extension—which implies that it may take hours—thus relating this issue with labor opportunity and waste of time. In the case of Latin America, a research study indicates that the average time devoted to commuting in cities like Guadalajara, the Autonomous City of Buenos Aires, Bogotá, Santa Cruz de la Sierra, Río de Janeiro, and Caracas ranges between 60 and 70 min a day (Jordán et al. 2017, p. 257). In addition, public transport—including travel by bicycle or on foot—is the predominant form of mobility in the region; moreover, in the case of cities like the Autonomous City of Buenos Aires, Santiago, Brasilia, Cali, Montevideo, Panamá, San José, and Rosario (Argentina) pedestrian transportation predominates (Jordán et al. 2017, pp. 254–255).

Although the debates on urban inequality have given greater relevance to the study of residential segregation patterns, the different dimensions of urban inequality are interactively intertwined and interrelated. Raudenbush and Kasim (1998) had already pointed out the existence of the interrelations between different dimensions, or forms of urban inequality, which reinforces the need to expand studies in that field.

1.4 Urban Inequalities, Quality of Life, and Well-Being

When analyzing the subject of urban inequality in the twenty-first century, quality of life and well-being should also be taken into account—a focus which several authors have begun to adopt.

This is the case of Bartlett et al. (2015, p. 5) who refer to urban housing by pointing out that "urban inequalities are expressed through highly stratified spatial realities, structured both by the market and by government." Contrasts can be identified in urban housing quality, density, and levels of provision, with significant implications for health and quality of life.

If, on the other hand, we analyze the relationship between quality of life and urban mobility, we should consider the research work carried out by Nuvolati (2009) who regards the concept of accessibility of services and public transport as crucial in the assessment of quality of life in the city, further expressing that, though accessibility of services may be assessed on the basis of educational, cultural, and economic disparity, it may also be approached in terms of traffic, public transportation system, and other issues related to urban mobility.

It is likewise possible to observe that data-based economy is definitely concentrated in urban areas, thus generating new forms of spatial inequality, both within and between cities and urban populations. There has been a widening of the income gap, disparity in cities has increased, and the suburbs have been broadly reclassified according to social groups, ethnic, and cultural origins. During the last decades, new forms of production have emerged—digital, data-based, consumer-oriented—which point to unprecedented urbanization in the world economy, i.e., what (Scott 2017) calls the third wave (Nijman and Wei 2020, p. 1).

In the case of daily life in urban communities, community well-being is worth taking into account. According to Wiseman and Brasher (2008, p. 358) community well-being is the combination of social, economic, environmental, cultural, and political conditions, properly identified to flourish and reach their full potential. La Placa et al. (2013) believe that this concept has generally referred to the social, cultural, and psychological needs of the subjects, their families, and their communities. These authors (2013, p. 119) noticed that the definition of community well-being was influenced by health, poverty, public transportation, economic activity, as well as by environmental and ecological considerations. In this regard, the notion of well-being ought to be considered, not only from a personal perspective but also in social terms, since the opportunities are of equal importance, whether they are offered to a group or to an individual (Royuela et al. 2003, p. 52), which means that all definitions of well-being should be contextualized, taking into account both the population and the place in which its members develop their daily lives. (Wiseman and Brasher 2008, p. 357).

At this point, the concept of life satisfaction ought to be clarified, for it represents the report of the way in which individuals assess their lives as a whole (Diener 2005, p. 3) whether at a certain point in time, or the evaluation of their life development since they were born. Since the question about their general life satisfaction allows people to make a fundamental connection with the feelings they are experiencing at that moment (Cummins 2005, p. 1), the questions that should be asked must be related to the specific dimensions of these persons' lives to obtain a more concrete assessment of the situations they are going through. In the concrete case of community satisfaction, the latter implies satisfaction with a number of situations, namely: perception of safety on the streets and in public places, social services, interaction with their neighbors, equipment and infrastructure, public transportation, the possibility of working in the community, the enjoyment of leisure time with other members of the community, as well as the possibility of exchanging points of view and discussing worries and problems of mutual concern (Tonon 2012).

Another major indicator to be considered is neighborhood cohesion, defined by Buckner (1988) as a collective attribute composed of three elements: sense of community, attachment to the place, and interaction with the neighbors. In other words, community satisfaction is the result of people's self-assessment of how they regard their lives as individuals in their communities (Tonon 2017, p. 8).

Finally, if we consider the ages of the people who populate the cities, we ought to make reference to inequalities related to the lifecycle. In this respect, Atkinson (1970) and Paglin (1975) stated that it is important to understand the extent of the changes reflected in natural inequality (inequality related to the life cycles) due to the nations' demographic changes—as a result of changes in production technology and income distribution—since the data reveal that natural inequality is a substantial component of real inequality.

1.5 Public Policies to Overcome Inequalities

The analysis of the public policies to overcome inequalities leads us to the notion that, though social inequality is measured culturally and historically and varies from one moment or place to another, it can still be argued that contemporary inequalities are mostly determined by the prevailing forms of production and associated to labor relations which tend to operate at larger, even global scales (Nijman and Wei 2020, p. 2). The previous characteristics are vital to the analysis and design of public policies which have led us to the study of inequalities in relation to democracy.

In this respect, Dreze and Sen (2002) stated that the practice of democracy may be limited for at least three distinct reasons: firstly, democratic institutions may become dysfunctional in terms of corruption or inefficiency; secondly, there may be an inadequate use of functional democratic institutions, and thirdly the existence of social inequalities. Sen (2000, p. 230) believes that "the absence of democracy is in itself an inequality—in this case of rights and political power." A capability focus implies the acknowledgment of authentic human well-being, needs, and dignity as the final goal of social policy (Eiffe 2013, p. 39).

Considering the above-mentioned, we may say that an equal society protects and promotes equality of valuable capabilities—the central and important things that people are able to do and become—so that everyone may enjoy the substantial freedom to live in the ways that they value and choose. Moreover, an equal society recognizes the individuals' diverse needs, situations, and goals, and seeks to expand their capabilities by doing away with discrimination and prejudice while tackling the economic, political, legal, social, and physical conditions that constrain people's achievements and limit their substantial freedom (Buchardt and Vizard 2007, p. 3), that is, a society that identifies its existing inequalities and makes decisions to solve them.

The national government plays a fundamental role in the elimination of the inequalities of a country and, in this regard, Bartlett et al. (2015, p. 8) have pointed out that "local government is most often the official level through which inequalities in services and provision are addressed."

Drèze and Sen (1989) made a wide distinction between strategies aimed at expanding the capabilities focused on direct state provision, and market-oriented strategies which aim at moderating the market. With the passing of time, this idea has expanded, since there are, at present, publications devoted to the topic of capabilities and public action as means of approaching the subject of inequalities (Vizard 2016, p. 455).

In his paper "Sobre el pensamiento lateral" (On Lateral Thinking), Atkinson (2011) tackled these conceptual issues by pointing out that inequality excels the income and revised the question of inequality in other dimensions such as health and education. In this regard, Kanbur (2021, p. 8) states that the main conceptual point lies in the fact that the correlation between income and health is a natural source of inequality, detached from income inequality and health inequality per se. This ought to be considered when it comes to deciding on public policies on the matter.

Atkinson (2015, p. 3) identified five areas in the distribution of income to mitigate inequality: technology, employment, social security, the sharing of capital, and taxation. According to Kanbur (2021) what Atkinson (2011) proposes aims at a general improvement regarding social protection and security, achieved through the universal right to a basic income, in a new system of social security. Moreover, Atkinson (2011) proposes a set of public policy tools that may exert an influence over the nature and extent of the inequality generated by the market. He further points out the importance of public intervention aimed at increasing work performance, investing on education, boosting the role of the state in processes of technological change, minimum salaries, and employment guarantee, together with new forms of power compensation that may balance the labor market power—including the role of the trade unions—, and the existence of an Economic and Social Committee.

At this point, we must consider the concept of social protection system, which Sen (2000) defines as a stable social protection framework, i.e., the State's fundamental institutional mechanisms, as well as the citizens' extraordinary support in emergency situations. In this respect, Fitoussi and Rosanvallon (2003, p. 113) also proposed a social protection system that contemplates the subjects' biographical trajectories, to prevent the incidents that may occur in their daily lives, deriving in social determinants (Tonon 2022b, pp. 14–15).

We have further considered the research work on inequality developed by Stiglitz, which proposes an extension of individual rights, i.e., public policies to mitigate inequality. In this regard, Vizard (2016, p. 457, quoting Stiglitz 1999, 2002) reminds us that the author proposed an active participation of civil society as public assets that may function to control abuse of power and influence.

Stiglitz (2015) expressed

> "......inequality is not, for the most part, the result of economic forces. It is not, in this sense, the result of inexorable economic laws. Rather, it is the consequence of our policies, and in turn, a reflection of our politics. Yes, there are economic forces at play that are dividing the country, but even these forces are shaped by—and sometimes driven by—our policies. The strongest evidence in support of this perspective is that inequality is so different in countries with similar economic structures".

Inequalities have cumulative economic, social, and political effects, each of which are contributing factors that require separate and concerted attention (Doyle and Stiglitz 2014, p. 3).

Further analysis also aims at a more generalized extension of individual rights—including legal protection against discrimination, equality standards, and a human rights code—as public policies to counteract inequality. (Vizard 2016, p. 458).

It is vital to consider the citizens' participation in the processes of policy decision, to reduce inequality; for citizenship, as Tonon (2022b, p. 73, quoting Sommers 1997) expresses, cannot only be considered as a natural category since it depends on public and intersubjective relationships. In the last decades of the twentieth century, the notion of citizenship has undergone a process of redefinition that made recognition and diversity visible, giving rise to the issue of ethnic and gender identities, thus expanding the debate on the classic notion of citizenship from the discourse that promotes integration to one of recognition of diversity (Tonon 2022b, p. 75). Castillo García (2007) recognizes two outlooks on citizenship: one, considered traditional and which covers the citizens' relations with the State, society, and the rules of participation in public life; and another, defined as "new citizenships," which has emerged as a result of the process of globalization, the advance of technology, migration movements, intercultural processes, and social recoveries (Tonon 2022b, p. 75). Benedicto and Moran (s/f, p. 48) explain that the different processes of acquisition and transformation of citizenship take place around the processes that shape social life and the individuals' life trajectories. This is based on a dynamic and relational conception of citizenship in which social practices are the center of the argument, and in which citizenship, in its multidimensional character, is conceived as being composed by a few elements whose specific interrelation defines its social dynamics in each group or social context (Tonon 2022b, p. 78).

Espinoza (2009) argues that citizen participation is a society-state relationship circumscribed by the notions of democracy and citizenship, and understood as a space for interaction, communication, and differentiation between the state and the social system. The author defines citizen participation as a set of actions through which citizens become involved in the design, decision, and implementation of public affairs affecting or concerning them (Tonon 2022b, p. 79). When group inequality is reproduced in the exercise of political power and the access to public institutions, it undermines the confidence of the socially disfavored groups and further hinders the government's capacity to rule with justice (Kabeer 2014, pp. 102–103).

Kanbur (2021, p. 4) stated that the prominence of inequality in the global political discourse has gone hand in hand with an acceleration of research work on economic inequality, in the last two decades. This has raised the question as to the way in which research might be better oriented toward political issues and, at the same time, discovering whether research highlights elements which are either underestimated or absent in political discourse.

In response to this statement, Tonon (2022b) proposed the use of quality-of-life research and human capabilities research results, for the decision of public policies, which may enable governments to identify the needs of society and design public policies that may generate programs to boost people's quality of life and reduce inequalities.

1.6 Conclusion: Seeking to Overcome Inequalities

In this chapter, we have carried out a revision of the current classifications of the concept of inequality, focusing on urban inequality related to other concepts such as quality of life and well-being, as a prelude to the analysis of the public policies that approach the subject.

Having completed this tour we can conclude, in accordance with Kanbur (2021, p. 3) that, inequalities "are multidimensional in theory outcomes and causes, encompassing the economic, the social and the political; economic forces are not the only ones playing on inequality, but they are important by themselves and in conjunction with others."

Schokkaert (2016, p. 433) pointed out that a society with fewer inequalities requires a government that considers the current environmental challenge, migration, and the increasing healthcare costs.

An equal society protects and promotes the equality of valuable capabilities—the central and important things that people are able to do and to become—so that everyone may enjoy their freedom to live in the ways they choose and value (Tonon 2022a, b, p. 18).

An equal society acknowledges the individuals' diverse needs, situations, and goals, and seeks to expand their capabilities by removing discrimination and prejudice and tackling the economic, political, legal, social, and physical conditions that constrain people's achievements and limit their substantive freedom (Buchardt and Vizard 2007, p. 3).

"Equality is an important feature in a good society, but it is only one feature, and if the idea of a good society evolves, the way to think about inequality will also evolve" Schokkaert (2016, p. 432).

We thus consider it instrumental to continue to study and reflect, especially regarding the way(s) of reducing inequality in each society in general, and particularly in the urban ones, with the aim of improving everyone's quality of life.

References

Anderson R (2014) Human suffering and quality of life: conceptualizing stories and statistics. Briefs in well-being and quality of life research series. Springer

Arauco V, Gazdar H, Hevia-Pacheco P, Naila Kabeer N, Lenhardt A, Quratulain S, Naqvi S, Nayak N, Norton A, Sabharwal N, Scalise E, Shepherd A, Thapa D, Thorat S, Tran D, Vergara Camus L, Woldehanna T, Mariotti C (2014) Report strengthening social justice to address intersecting inequalities post-2015. UNDP

Arendt H (1998) The human condition, 2nd edn. University of Chicago Press

Atkinson AB (1970) On the measurement of inequality. J Econ Theory 2:244–263

Atkinson AB (2011) On lateral thinking. J Econ Inequal 9:319–328

Atkinson AB (2015) Inequality—what can be done? Working paper 2 November. London School of Economics and Political Science. https://eprints.lse.ac.uk/101810/1/Atkinson_inequality_what_can_be_done_wp2.pdf

Bartlett S, Mitlin D, Satterthwait D (2015) Urban inequalities. Addressing inequalities. The heart of the post-2015 development agenda and the future we want for all. Global Thematic Consultation. http://www.equityforchildren.org/wp-content/uploads/2013/11/sherry-Barlet-Urban-Inequalities.pdf
Benedicto J, Moran M (s/f) Los jóvenes ¿ciudadanos en proyecto? https://eprints.ucm.es/id/eprint/43789/1/Benedicto_y_Moran._Los_jovenes_ciudadano.pdf
Burchardt T, Vizard P (2007) Definition of equality and framework for measurement: final recommendations of the equalities review steering group on measurement. CASE/120. Centre for Analysis of Social Exclusion. London School of Economics. http://eprints.lse.ac.718uk/6218/1/Definition_of_equality_and_framework_for_measurement.pdf
Buckner J (1988) The development of an instrument to measure neighbourhood cohesion. Am J Commun Psychol 16(6):771–791. https://doi.org/10.1007/BF00930892
Castillo García J (2007) La configuración de las ciudadanías en estudiantes universitarios y universitarias de pregrado en Manizales, Colombia. Revista Latinoamericana De Ciencias Sociales, Niñez y Juventud, Manizales 5(2):755–810
Ciccolella P, Baer L (2011) Buenos Aires tras la crisis ¿Hacia una metrópoli más integradora o más excluyente? In Ciccolella P (ed) Metrópolis latinoamericanas: más allá de la globalización, OLACCHI, pp. 213–245
Cummins R (2005) Discussion document international well-being group. Australian Centre on Quality of Life.
Dammert Guardia M, Delgadillo V, Erazo J (2019) La ciudad, espacio de reproducción de las desigualdades. Andamios 16(ene./abr.):7–13. http://www.scielo.org.mx/pdf/anda/v16n39/1870-0063-anda-16-39-7.pdf
Doyle M, Stiglitz J (2014) Eliminating extreme inequality: a sustainable development goal, 2015–2030. Ethics Int Aff 28(1):5–13. https://doi.org/10.1017/S0892679414000021
Drèze J, Sen A (1989) Hunger and public action. Oxford University Press
Drèze J, Sen A (2002) Democratic practice and social inequality in India. J Asian Afr Stud 37(2):6–37. https://doi.org/10.1177/002190960203700202. http://jas.sagepub.com/cgi/content/abstract/37/2/6T
Diener E (2005) Guidelines for national indicators of subjective well-being and ill Being. Working paper. University of Illinois
ECLAC (2016) The social inequality matrix in Latin America. ECLAC-United Nations, Santiago
Eiffe F (2013). Implications for capability based social policy: a European perspective. In: Otto H, Ziegler H (eds) Enhancing capabilities. The role of social institutions. Verlag Barbara, pp 39–52
Espinoza M (2009). La participación ciudadana como una relación socio-estatal acotada por la concepción de democracia y ciudadanía. Andamios 4(10):71–109. http://www.scielo.org.mx/pdf/anda/v5n10/v5n10a4.pdf
Fitoussi JP, Rosanvallon P (2003) La nueva era de las desigualdades. Manantial
Ghosh J (2019) The uses and abuses of inequality. J Hum Dev Capabil 20(2):181–196. https://doi.org/10.1080/19452829.2019.1574282
Giuliano G, Small K (1993) Is the journey to work explained by urban structure? UC Berkeley. https://escholarship.org/uc/item/2ss7x5b1
Glaeser E, Kahn M (2001) Descentralized employment and the transformation of American city. Working Paper 8117. National Bureau of Economic Research. https://www.nber.org/system/files/working_papers/w8117/w8117.pdf
Haddad M, Zorica Nedović-Budić Z (2006) Using Spatial Statistics to Analyze Intra-urban Inequalities and Public Intervention in São Paulo, Brazil. J Hum Dev 7(1):85–109. https://doi.org/10.1080/14649880500502102
Hiernaux-Nicolàs D, González Gómez C (2017) La ciudad maquillada: las nuevas máscaras de la ciudad del siglo XXI. En Daniel Hiernaux-Nicolas y Carmen Imelda González-Gómez (Coordinadores) *La ciudad latinoamericana a debate. Perspectivas teóricas*. Universidad Autónoma de Querétaro Facultad de Ciencias Políticas y Sociales, pp 93–124

Jordán R, Riffo L, Prado A (2017). Desarrollo sostenible, urbanización y desigualdad en América Latina y El Caribe. Dinámicas y desafíos para el cambio estructural. CEPAL. https://repositorio.cepal.org/bitstream/handle/11362/42141/1/S1700701_es.pdf
Kabeer N (2010) Can the MDGs provide a pathway to social justice? The challenge of intersecting inequalities. Institute of Development Studies, Brighton/MDG Achievement Fund, New York
Kabeer N (2014) Social justice and the millennium development goals: the challenge of intersecting inequalities. In: The equal rights review, vol 13, pp 91–116. https://www.equalrightstrust.org/ertdocumentbank/ERR%2013.pdf
Kabeer N, Santos R (2017) Intersecting inequalities and the sustainable development goals: insights from Brazil. Working paper 14. August. International Inequalities Institute. The London School of Economics and Political Sciences. https://eprints.lse.ac.uk/101851/1/Kabeer_Santos_intersecting_inequalities_wp14.pdf
Kanbur R (2021) Sustainable development goals and the study of economic inequality. J Econ Ineq 19:3–11. https://doi.org/10.1007/s10888-020-09452-9
La Placa V, McNaught A, Knight A (2013) Discourse on well-being in research and 431 practice. Int J Well-Being 3(1):116–125. http://www.internationaljournalofwell-being.org/ijow/index.php/ijow/article/viewFile/177/315
Nijman J, Wei Y (2020) Urban inequalities in the 21st century economy. Appl Geogr 117:1–8. https://www.sciencedirect.com/journal/applied-geography/vol/117/suppl/C
Nuvolati G (2009) Quality of life in cities: a question of mobility and accessibility. In: Møller V, Huschka D (eds) Quality of life and the millennium challenge. Social Indicators Research Series, vol 35. Springer, Dordrecht, pp 177–191. https://doi.org/10.1007/978-1-4020-8569-7_12
Nussbaum M (2012) Crear capacidades: Propuesta para el desarrollo humano. Paidos
Paglin M (1975) The measurement and trend of inequality: a basic revision. Am Econ Rev 65(4):598–609. https://www.jstor.org/stable/1806537
Pardo I, Prato G (2021) Urban inequalities: ethnographically informed reflections. Palgrave studies in urban anthropology. Palgrave Mac Millan, Cham
Pettinato S (2002) A conceptual primer on the currents and trends in inequality. J Hum Dev 3(1):23–56. https://doi.org/10.1080/14649880120105371
Raudenbush S, Kasim R (1998) Cognitive skill and economic inequality: findings from the national adult literacy survey. Harvard Educ Rev 68(1):33–80. https://doi.org/10.17763/haer.68.1.1j47150021346123
Royuela V, Suriñach J, Reyes M (2003) Measuring quality of life in small areas over different periods of time. Soc Indic Res 64(1):51–74. 10.1023/A:1024789306734
Schokkaert E (2016) Putting inequality in context. J Hum Dev Capabil 17(3):429–433. https://doi.org/10.1080/19452829.2016.1203028
Scott A (2017) The constitution of the city. Palgrave McMillan
Segura R (2014) El espacio urbano y la (re)producción de desigualdades sociales. Desacoples entre distribución del ingreso y patrones de urbanización en ciudades latinoamericanas. Working Paper No. 65, desiguALdades.net Research Network on Interdependent Inequalities in Latin America. https://refubium.fu-berlin.de/bitstream/handle/fub188/22077/65-WP-Segura-Online.pdf;jsessionid=D667B59C3E6E66FA847F706CDDF12EE6?sequence=1
Sen A (2000) Desarrollo y libertad. Editorial Siglo XXI, Bogotá
Somers M (1997) ¿Qué hay de político o de cultural en la cultura política y en la esfera pública? Hacia una sociología histórica de la formación de conceptos. Zona Abierta no 77/78. Editorial Pablo Iglesias. https://dialnet.unirioja.es/ejemplar/5061
Stewart F (2000) Crisis prevention: tackling horizontal inequalities. QEH working paper series—QEHWPS33. Working Paper Number 33. https://ora.ox.ac.uk/catalog/uuid:c4d1275a-8ed9-4515-b2e0-44ecd63be570/download_file?file_format=application%2Fpdf&safe_filename=qehwps33.pdf
Stewart F (2002) Horizontal Inequality: a neglected dimension of development. WIDER Annual Development Lecture, Helsinki

Stewart F (2009) Horizontal inequality: two types of trap. J Hum Dev Capabil 10(3):315–340. https://doi.org/10.1080/19452820903041824
Stewart F (2013) Approaches towards inequality and inequity: concepts, measures and policies. Office of Research Discussion. Paper No. 2013–01, Florence: UNICEF Office of Research
Stiglitz J (1999) On liberty, the right to know, and public discourse: the role of transparency in public life. Oxford Amnesty Lecture, Oxford
Stigltz J (2002) Participation and development: perspectives from the comprehensive development paradigm. Rev Dev Econ 6(2):163–182. https://doi.org/10.1111/1467-9361.00148
Stiglitz, J. (2015) The origins of inequality, and policies to contain it. Natl Tax J 68(2):425–448. https://doi.org/10.17310/ntj.2015.2.09
Tilly C (2007) Unequal access to scientific knowledge. J Hum Dev 8(2):245–258. https://doi.org/10.1080/14649880701371133
Tonkiss F (2018) Unequal cities, divided spaces—the search for equality. Athena Lecture KTH Centre for the Future of Places: Stockholm, Sweden. https://www.kth.se/futureofplaces/events/athena-lecture-serie/unequal-cities-divided-spaces-the-search-for-equality-professor-fran-tonkiss-1.824528
Tonon G (2022a) Children's vulnerability and social justice in the south. In: Tonon G (ed) Social justice for children in the South. Evidence-based approaches to peace and conflict studies, vol 9. Springer. https://doi.org/10.1007/978-981-19-5045-2_1
Tonon G (2022b) (ed) Key actors in public policymaking for quality of life: facilitators and obstacles. Human well-being research and policy making series. Springer Nature. https://doi.org/10.1007/978-3-030-90467-8
Tonon G (2017) Rethinking community quality of life in Latin American Countries. In: Tonon G (ed) Quality of life in communities of Latin Countries. Springer, pp. 3–14. https://doi.org/10.1007/978-3-319-53183-0_1
Tonon G (2012) Young people's quality of life and construction of citizenship. Series Springer. Briefs in well-being and quality of life research. Springer. https://doi.org/10.1007/978-94-007-2996-4
Vizard P (2016) Inequality and public action. J Hum Dev Capabil 17(3):453–459. https://doi.org/10.1080/19452829.2016.1203032
Wei Y, Ewing R (2018) Urban expansion, sprawl and inequality. Landsc Urban Plan 177:259–265. https://doi.org/10.1016/j.landurbplan.2018.05.021
Wiseman J, Brasher K (2008) Community well-being in an unwell world: trends, 484 challenges, and possibilities. J Public Health Policy 29:353–366

Chapter 2
Urban Inequality in Bangladesh

Siddiqur R. Osmani and Binayak Sen

Abstract Bangladesh has experienced rapid economic development in the last three decades. During the same period, it has also experienced rapid growth in urban population (one of the fastest in the world), with average urban opulence rising faster than at the national level. There is clear evidence, however, that the poorer segment of the urban population has not always shared equitably in the fruits of economic growth. Inequality has widened in the distribution of household income and health outcomes but has narrowed in the distribution of educational opportunities. This chapter examines the processes underlying these diverse outcomes and argues for a stronger role of public action as a countervailing force against the inequalizing tendency of the market forces unleashed by the growth process. The paper demonstrates that where public action has been strong, inequality has narrowed, as in the case of education, and where public action has been weak, inequality has widened, as in the case of income and health.

Keywords Urban Bangladesh · Income inequality · Health inequality · Educational inequality · Social protection

2.1 Introduction

From being a predominantly rural society at the time of its independence in 1971, Bangladesh is fast becoming an urbanized one. On the eve of independence, only about one in twelve Bangladeshis lived in urban areas (in 1974); nearly half a century later, almost one in three did so (in 2022). Increasing urbanization is a consequence of economic growth and the structural transformation that comes with it, as the economic historian Kuznets (1966) had noted long ago; and, if properly managed, urbanization

S. R. Osmani
Ulster University, Belfast, UK
e-mail: sr.osmani@ulster.ac.uk

B. Sen (✉)
Bangladesh Institute of Development Studies (BIDS), Dhaka, Bangladesh
e-mail: binayak71@yahoo.com

G. H. Tonon (ed.), *Urban Inequalities*, The Urban Book Series,
https://doi.org/10.1007/978-3-031-59746-6_2

can in turn contribute to further economic growth through the economies of agglomeration.[1] This synergistic relationship between economic growth and urbanization must have played a role in ensuring that since about the early 1990s, rapid economic growth and rapid urbanization have gone hand in hand in Bangladesh. Indeed, much of the current discussion on urbanization in developing countries like Bangladesh focuses primarily on how best to exploit this synergy (e.g., Ellis and Roberts 2016).

As worthy as this goal is, it is equally important to trace the consequences of this synergy for the distribution of well-being among the urban population. Experience suggests that the combination of rapid economic growth and rapid urbanization has a tendency to create rising urban inequality. High and rising levels of inequality are not only undesirable from the point of view of social justice but can also unleash a whole range of disruptive forces that can potentially derail and distort economic growth itself. The policymakers should, therefore, keep a close eye on the evolution of urban inequality and take corrective measures, where necessary, as urbanization proceeds apace.

With this objective in mind, the present chapter undertakes an in-depth analysis of urban inequality in Bangladesh. Most of the past work on inequality in Bangladesh have focused on either the rural economy or the economy as a whole. A small number of studies have examined urban inequality, but they are mostly dated and are often very narrowly focused on some particular aspect of inequality.[2] The present study attempts to provide a more comprehensive picture, using more recent data and often undertaking novel analysis of raw data generated by various surveys. Section 2.2 traces the evolution of income inequality in urban Bangladesh, and the next two sections examine the determinants of this evolution—focusing on both market forces unleashed by economic growth (Sect. 2.3) and the role of the state in providing social protection (Sect. 2.4). Section 2.5 investigates the nature, trend, and determinants of non-income dimensions of inequality, paying special attention to the role of urban slums in shaping the dynamics of overall unban inequality. Section 2.6 concludes by bringing together the major findings and lessons.

2.2 Income Inequality in Urban Bangladesh

Bangladesh's transition to a higher growth path started from the early 1990s (Table 2.1). Over the three decades since then, Bangladesh has also become an increasingly unequal society. While both rural and urban parts of the economy have become more unequal to some extent, in the recent years it is urban inequality that is driving widening income inequality in the country as a whole.

---

[1] More than a century and a quarter ago, Marshall (1890) offered the first systematic analysis of the nature of the economies of agglomeration based on his observation of the unfolding of industrialization and urbanization of the nineteenth century England.

[2] Some of the more important studies include NIPORT/ICDDRB (2006, 2013), World Bank (2007), IFPRI/BBS/WFP (2008), UNICEF (2010), WFP (2015), and Ahmed et al. (2017).

Table 2.1 Decadal growth rates: 1970s–2010s (percent per annum)

	1970s	1980s	1990s	2000s	2010s
GDP growth	3.79	3.72	4.80	5.82	6.86
Population growth	2.40	2.01	1.70	1.32	1.27
Per capita GDP growth	1.39	1.68	3.05	4.44	5.59

Notes (1) Data refer to the average of annual growth rates for each period. (2) The 1970s refers to the period 1972/73–1979/80. (3) The 2010s refer to the period from 2010–11 to 2018–19, deliberatively leaving out 2019–20 so as to avoid the distortions caused by the onset of COVID pandemic
Source Calculated from time series data compiled by the authors from various statistical publications of the Government of Bangladesh and the World Bank, after making necessary adjustments for comparability, to the extent possible

As can be seen from Table 2.2, from 1991–92 to 2010, income inequality rose more or less in tandem in urban and rural areas, as measured by the Gini coefficient; if anything, rural inequality increased a little faster even though the level of inequality remained lower in the rural economy compared with the urban economy. Since 2010, however, rural inequality has remained more or less stable at around a Gini of 0.45, but urban inequality has surged ahead, with the Gini coefficient rising from 0.45 in 2010 to 0.50 in 2016 and further to 0.54 in 2022. With increasing urbanization of the economy, rising urban inequality is being translated into rising inequality in the country as a whole.

Consumption inequality also reflects a similar trend, although its level is understandably lower than that of income inequality. Thus, the Gini coefficient of consumption distribution has remained stable in rural areas at around 0.29 since 2010, but the urban Gini has gone up from 0.34 in 2010 to 0.36 in 2022, causing a slight increase in overall consumption inequality (Table 2.3).

The preceding discussion has relied exclusively on the Gini coefficient as the measure of inequality. Gini of course is the most popular and most widely used measure of inequality, partly because of its close association with the visually more intuitive Lorenz curve analysis of income distribution. But it is well-known that one of the limitations of the Gini coefficient is that by construction it attaches more

Table 2.2 Trend of income inequality (Gini coefficient)

Year	Urban	Rural	National
1991–92	0.40	0.36	0.39
1995–96	0.44	0.38	0.43
2000	0.50	0.39	0.45
2005	0.50	0.43	0.47
2010	0.45	0.43	0.46
2016	0.50	0.45	0.48
2022	0.54	0.45	0.50

Source Reports of *Household Income and Expenditure Surveys*, various years

Table 2.3 Trend of consumption inequality (Gini coefficient)

Year	Urban	Rural	National
1991–92	0.31	0.24	0.26
1995–96	0.36	0.27	0.30
2000	0.37	0.27	0.31
2005	0.35	0.28	0.31
2010	0.34	0.28	0.32
2016	0.33	0.30	0.32
2022	0.36	0.29	0.33

Sources Reports of *Household Income and Expenditure Surveys*, various years. The figures for 2016 are from the *Preliminary Report of HIES 2016*

weight to the middle of the distribution than to the tails. This would not be a problem for comparison between distributions if the middle of the distribution behaved in the same way as the tails—e.g., if any widening of the gap between the upper and lower tails was also accompanied by similar widening of the gap in the middle of the distribution. This is indeed the implicit assumption behind the widespread use of the Gini coefficient.

But the path-breaking work by Gabriel Palma has seriously questioned the empirical basis of this assumption. In an influential study on income distribution for a large number of countries around the globe, he has demonstrated that the middle of the distribution does not generally behave in the same way as the tails (Palma 2011). In fact, one of the stylized facts that emerges from his studies is that the middle class—representing 50% of the population belonging to the five deciles from the fifth to the ninth—manages to capture a fairly constant share of roughly half of national income in most countries most of the time. It is the changing division of the remaining half between the bottom 40% and the top 10% of the population that drives the change in overall income distribution. Thus, when income distribution worsens it is mainly because the share of the top 10% goes up at the expense of the bottom 40%, while the middle 50% more or less hold on to their share. The changing pattern of income distribution thus essentially represents a struggle between the two tails of the distribution for sharing the half of the national income that is not captured by the middle class.[3]

A simple way of capturing this struggle is to use what has come to be known as the Palma ratio of income inequality—it is the ratio of the income shares of the top 10% and the bottom 40% of the population. A higher value of this ratio obviously points to widening inequality. In Table 2.4, we present the trend of the Palma ratio

[3] This generalization has come to be known as the Palma hypothesis.

Table 2.4 An alternative measure of income inequality: the Palma ratio

	1980s	1990s	2000s	2016
Urban				
Income share of bottom 40%	17.80	15.12	13.67	13.32
Income share of middle 50%	53.06	48.95	48.42	45.24
Income share of top 10%	29.14	35.93	37.92	41.44
Total	100.00	100.00	100.00	100.00
Palma ratio	*1.64*	*2.44*	*2.78*	*3.11*
Rural				
Income share of bottom 40%	18.66	17.88	15.76	13.86
Income share of middle 50%	51.69	51.76	50.34	51.30
Income share of top 10%	29.65	30.36	33.91	34.84
Total	100.00	100.00	100.00	100.00
Palma ratio	*1.59*	*1.70*	*2.15*	*2.51*
National				
Income share of bottom 40%	18.22	16.30	14.34	13.01
Income share of middle 50%	51.53	49.72	48.92	48.83
Income share of top 10%	30.25	33.97	36.75	38.16
Total	100.00	100.00	100.00	100.00
Palma ratio	*1.66*	*2.10*	*2.56*	*2.93*

Notes and Sources

(1) Palma ratio is defined as the ratio of income shares of the top 10% and bottom 40% of the population in the income distribution

(2) Calculated by the authors from the data on income distribution by income deciles given in the reports of *Household Income and Expenditure Surveys*, various years

(3) The figure for 1980s is a simple average of the figures for 1983-84, 1985–86 and 1988–89. The 1990s refer to the average of 1991–92, 1995–96 and 2000; the 2000s refer to the average of 2005 and 2010. We have included the terminal year of a decade in the average of that decade since the situation in the terminal year is more likely to reflect the trend of what happened in the preceding decade than what is going to unfold in the following one

of income distribution in Bangladesh, where our main focus is on urban inequality, but for comparative purposes we have also included figures for rural areas and for the country as whole.[4]

The urban Palma ratio confirms what we have already learnt from the Gini ratio—that inequality has been widening in urban areas from the early days of Bangladesh's transition to higher growth path in the 1990s. In a slight deviation from the Palma hypothesis, however, this widening of inequality is not just a consequence of the

[4] Our table ends at the year 2016, because although there is *Household Income and Expenditure Survey* for 2022, the *Preliminary Report* that has been published so far does not contain data on detailed distribution of income (which is needed for calculating the Palma ratio). The Report does, however, provide estimates of poverty and Gini coefficients of income and consumption distribution which we have used in this chapter.

Table 2.5 Trend of poverty in Bangladesh: 1991–92 to 2022 (headcount ratio: percent)

	Urban		Rural		National	
	Poor	Extreme poor	Poor	Extreme poor	Poor	Extreme poor
1991–92	42.8	24.0	58.8	43.8	56.7	41.1
2000	35.2	20.0	52.3	27.9	48.9	34.3
2005	28.4	14.6	43.8	28.6	40.0	25.1
2010	21.3	7.7	35.2	21.1	31.5	17.6
2016	18.9	7.6	26.4	14.9	24.3	12.9
2022	14.7	3.8	20.5	6.5	18.7	5.6

Notes and Sources Reports of *Household Income and Expenditure Surveys (HIES)*, various years. The figures for 2022 are from the *Preliminary Report* of HIES 2022

bottom 40% of the population losing out to the top 10% in the struggle for income shares, because the middle half of the population has lost out too—their share of urban income has fallen from 53% in the 1980s to 45% in 2016.[5] This is clearly a case of the top 10% of the population pulling away from everyone else.[6]

Before embarking on an exploration of the possible causes behind this phenomenon, it is first necessary to look at one other feature of distribution. An important question is: while distribution was getting worse, what was happening to the absolute level of well-being at the bottom of the scale? Was the worsening of distribution a case of the poor getting poorer and the rich richer, or was it a case of both the poor and rich getting better off but the rich getting better off at a faster pace than the poor? To answer this question, we look at the trend of poverty as shown in Table 2.5. It is evident that poverty has gone down considerably in Bangladesh in the last three decades, including in urban areas. There was a discernible slowdown in the rate of urban poverty reduction after 2010, especially in the case of extreme poverty. Nonetheless, the decline in urban poverty from 43% to just under 15% in the space of three decades must count as remarkable progress in the absolute well-being of those at the bottom rungs of the distribution.[7]

It follows that the worsening of income inequality that is observed in urban Bangladesh since the onset of transition to a higher growth path is not a case of the poor getting poorer and the rich getting richer. The poor have become better off, considerably, in absolute terms, but evidently their economic conditions improved relatively slowly compared with the richest segment of the population, giving rise to the observed phenomenon of widening urban inequality.

[5] The falling share of the middle class may have potentially destabilizing consequences for both the politics and political economy of the country.

[6] The Palma hypothesis is more valid in rural Bangladesh where the middle half of the population has indeed held on to their share of half the rural income, while the bottom 40% has consistently lost out to the top 10 percent. The overall trend for Bangladesh as a whole also presents a picture roughly consistent with the Palma hypothesis, though not exactly because of the influence of urban trend.

[7] Similar progress is observed in rural areas too, and consequently in the country as a whole.

2.3 What Drives Widening Inequality in Urban Bangladesh?

If widening urban inequality is seen as a consequence of a struggle for income share in which the bottom rungs of the society are losing out to the richest, it is important to understand exactly how they are losing out and why. A good place to start is to examine the trend of real wages for unskilled workers, who would normally inhabit the bottom rungs of the income ladder.[8] A serious stumbling block, however, is the absence of separate wage data for urban workers. The Bangladesh Bureau of Statistics (BBS), the agency officially responsible for collecting and disseminating relevant statistical information about the country, does provide time series data on nominal wage indices—for unskilled workers in the country as a whole. It also provides time series data on consumer price indices (CPI)—separately for urban and rural areas, but the absence of separate urban data on nominal wage indices prevents a straightforward derivation of a time series of real wage indices for unskilled urban workers.

As the second best alternative, we followed a less straightforward procedure. We made use of the fact that although the BBS data on nominal wage indices are given at the national level, they are collected and reported separately for broad sectors of the economy—agriculture, industry and services.[9] We took a weighted average of the indices for unskilled workers in industry and services as a proxy for urban wage indices and the indices for agricultural workers as a proxy for rural wages indices.[10] By applying the indices of urban CPI to these proxy indices of urban nominal wages, we derived a series of real wage indices for unskilled urban workers; and similarly for rural real wage indices.

The evidence presented in Fig. 1 is striking. The average real wage of unskilled workers has fallen sharply in the 2010s; from a base of 100 in 2010–11, it fell to 92% in 1918–19. Most of the decline happened in the first half of the period, since then the real wage seems to have stagnated in urban areas. By contrast, while the rural real wage had also fallen by the middle of the decade, it recovered almost fully by the end of the decade.

---

[8] The bottom rungs are also inhabited by poor self-employed workers, but their real earnings can be expected to move roughly in line with the real wages of unskilled wage-employed workers, their fellow travelers at the lower end of income distribution, because a good deal of mobility across these two groups can be expected to exist in the large informal sector of the economy which is characterized by very little labor market restrictions. The trend of real wages of unskilled wage-employed workers can thus be expected to broadly reflect the trend of real earnings of all kinds of workers in the bottom rungs.

[9] Wage data are actually collected by BBS at highly disaggregated sub-sectoral levels and then aggregated for broad economic sectors by using labor force weights; see BBS (2015a).

[10] A partial justification for this rough assumption can be found in a recent study on rural–urban shares of GDP in Bangladesh (which we are going to draw upon more fully below), which shows that the urban sector accounts for some 62% of value-added in the industry sector, 54% of value-added in the service sector, and only about 3% of value-added in the agricultural sector; Osmani (2021).

Table 2.6 Growth of labor force: 2005 to 2016–17 (million)

	2005–2010	2010–2016/17
Urban	2.60	4.68
Rural	2.80	0.80
Bangladesh	2.75	1.76

Source BBS, *Labor Force Surveys*; various years

Why did the urban real wage fall so much in the 2010s? The short answer is that in the first instance the market forces unleashed by the process of economic growth led to this decline—through the route of rural-to-urban migration, which caused an excess supply of unskilled labor in the urban economy, pushing down urban wages. Furthermore, no countervailing effort was made by the government to prevent or at least moderate the decline. We reach this conclusion somewhat indirectly, however, since there is as yet no direct information on rural-to-urban migration in the decade of the 2010s.[11] We look instead at several independent pieces of evidence all of which point to the likelihood that there was a strong upsurge in rural-to-urban migration at least in the first part of the 2010s.

We first examine the labor force data. In Table 2.6, we compare the growth of labor force during the period 2005–2010 with the period 2010–2016/17.[12] The most salient aspect of this comparison is the striking contrast between urban and rural areas. During 2005–2010, both urban and rural labor force grew at roughly the same rate, but in the next period the urban growth rate experienced a huge upsurge—from 2.6 to 4.7% per annum, while the growth of rural labor force plummeted from 2.8% to a meager 0.8%. In consequence, urban labor force swelled from 13.3 ml in 2010 to 17.9 ml in 2016–17, while rural labor force rose only marginally (Table 2.7).

In principle, this swelling of urban labor force could come about in three, mutually non-exclusive, ways—(1) an upsurge in natural increase through higher fertility rate, (2) in upsurge in labor force participation rates, and (3) an upsurge in net rural-to-urban migration. The first possibility can be easily dismissed since there is clear evidence of sharp decline of fertility all over Bangladesh. The second possibility cannot be ruled out a priori, but the evidence from survey data does not support it.

[11] The recently completed *Population Census of 2022* would eventually provide the necessary data, but only a few summary statistics have been published so far, and it will take a while before the rate of migration can be worked out by analyzing raw data.

[12] There is a more recent labor force survey, for the year 2022, whose report has just been published. The Report shows a sharp decline in urban labor force and even sharper increase rural labor force, which may suggest that the post-2010 acceleration in rural-to-urban migration we have been discussing here may have subsequently been reversed. There is, however, a good deal of controversy over what actually happened. A complicating factor is that the 2022 survey also shows an exceptionally sharp rise in female labor force participation in rural areas and a less sharp but still substantial fall in female labor force participation in urban areas, which together might give a false impression of reversal of migration pattern if one looks simply at labor force data. Without controlling for changes in participation rates, it would be hazardous to draw any conclusion on migration from the rural–urban distribution of labor force alone. Since this issue requires further research, we have decided not to use the data for 2022 for the present study.

Table 2.7 Size of labor force: 2005 to 2016–17 (million)

	2005	2010	2016–17
Urban	11.7	13.3	17.9
Rural	37.8	43.4	45.7
Bangladesh	49.5	56.7	63.5

Source BBS, *Labor Force Surveys*; various years

BBS *Labor Force Surveys* show that the urban labor force participation rate declined slightly—from 57.3% in 2010 to 55.7% in 2016–17 (BBS 2017). This leaves an upsurge in rural-to-migration as the only plausible explanation of the explosion of urban labor force after 2010. This inference is strengthened by the fact that while the growth of urban labor force accelerated after 2010, the growth of rural labor force decelerated precipitously at the same time (Table 2.6).

Further support for the inference of accelerated migration can be found from two other independent sources, which show that poorer households were leaving their rural abode in droves. Table 2.8 presents the changing distribution of rural households by landowning status as found from successive rounds of *Household Income and Expenditure Surveys*. This table shows that the proportion of landless households—who constitute the poorest segment of the rural population – dropped precipitously from 51% in 2010 to 32% by 2016, after remaining virtually unchanged at around 50% in the preceding decade.

Of course, mere decline in the *proportion* of landless households does not by itself confirm that many of them actually left the rural areas, because their falling proportion could simply be a consequence of other categories of households growing strongly in numbers. But there is indeed evidence that not just the proportion, but the absolute number of landless households also declined sharply in the 2010s. This evidence comes from successive rounds of *Agricultural Census* conducted by BBS. As can be seen from Table 2.9, the absolute number of landless households was increasing steadily from 1983–84 to 2008 but then dropped sharply by 2019. The number of agricultural labor households, who come mainly from landless and near landless categories, also grew very little after 2008, reversing an earlier pattern of strong growth.

Table 2.8 Distribution of rural households by land ownership category (percentage of households)

Land ownership size	2000	2005	2010	2016
Landless (< 0.5 acre)	48.0	45.8	50.9	32.3
Functionally landless (0.5–0.50 acre)	13.0	15.9	15.9	42.1
Marginal (0.5–1.5 acre)	17.5	18.8	18.0	16.4
Small (1.5–2.5 acre)	9.2	8.8	6.8	4.9
Medium and large (> 2.5 acre)	12.4	10.7	8.4	4.2
All	100.0	100.0	100.0	100.0

Source Osmani (2018), Table VII, p. 14

Table 2.9 Number of landless and agricultural labor households in rural Bangladesh (thousand)

Household category	1983–84	1996	2008	2019
Landless households	1198	1815	3256	2323
Agricultural labor households	5495	6401	8849	8974
All rural households	13,818	17,828	25,355	29,625

Source BBS (2008, 2019a)

Where did all these land-poor people go? It is extremely unlikely that any significant number of them would have moved up the landownership ladder by acquiring more land.[13] Some of them may have moved abroad, as temporary overseas migration gathered steam in Bangladesh over the last few decades, but most sources of evidence suggest that very few landless poor households take part in this process simply because they can't afford the cost of overseas migration. That leaves the urban economy as the likely destination for most of them. There is thus a formidable case for the inference that there was an explosion of rural-to-urban migration after about 2010, which swelled the number of urban labor force at an unprecedented rate.

Before considering the implications of this exploding migration for the urban economy, it is necessary to discuss briefly what induced the upsurge in migration in the first place.[14] The process can be best understood in the framework of the Harris–Todaro model of rural-to-urban migration. The migration was induced in the first instance by both push and pull factors. A precipitous slowdown in the rate of agricultural growth would have exerted a strong push effect, forcing the rural poor to seek their livelihood elsewhere. At the same time, the perceived probability of doing better in urban areas would have acted as a pull factor. The perception would have been based on historical experience—the fact that not only was urban poverty considerably lower than rural poverty over the years, it also declined slightly faster than rural poverty until 2010.[15]

However, once migration started on a massive scale, the perception based on historical trend did not translate into reality for many a migrant. The essence of the argument is the rural poor may have been tempted to migrate to urban areas attracted by better prospects of earning a livelihood there, but when too many of them did so, a large number of them failed to improve their condition, resulting in an outcome where the average condition actually deteriorated. Another way of making the point is that a better *ex ante* probability of earning a livelihood in urban areas resulted in a worse *ex post* prospect of doing so, as migration occurred on a massive scale. What is essentially at work here is the fallacy of composition—what is good for an

[13] There is evidence that some landless households are moving up the ladder of *operational* holding by renting in land from others, but moving up the ladder of *landownership* is an entirely different, and difficult, proposition altogether.

[14] This question has been investigated in Osmani (2015, 2018). We present only a brief summary here.

[15] In the two decades from 1991/92 to 2010, urban poverty was nearly halved whereas rural poverty fell by about 40% (Table 2.5).

individual may not be good for a collection of individuals, especially if the collection happens to be very large.

This is precisely what happened in urban Bangladesh following the upsurge in migration after 2010. As the urban labor force was swelled at an unprecedented rate by massive migration from rural areas, the supply of unskilled urban workers increased at a much faster rate than ever before, outstripping any increase in the demand for labor, and thereby depressing urban real wages for unskilled workers.

This depression in real wages was in turn responsible for growing income inequality in rural areas. The connection between real wages and income distribution is best seen by considering the relationship between real wage and labor's share in total income and the latter's significance for income distribution. Since labor is the most important factor of production owned by the poor, evolution in the share of labor can reveal a great deal about the evolving share of the poor vis-à-vis the share of the rich in total income. The evolution in labor share can in turn be inferred by comparing the growth of real wage with the growth of labor productivity. If real wage and productivity grow at the same rate, the relative shares of labor and non-labor inputs (such as land and capital, including human capital) in total income will remain constant, and since labor input comes mostly from the poor and non-labor inputs mostly from the rich, the personal distribution of income will also remain relatively stable. If, however, real wage grows more slowly than productivity, this would lead to rising share of non-labor inputs, with the implication that the share of the rich is also perhaps rising i.e., personal income distribution is getting more unequal. The converse would be true if real wage grows faster than productivity. Comparison between the growth of real wage and the growth of labor productivity can thus illuminate the underlying forces shaping the evolution of income distribution.

The practical problem lies in obtaining reliable information on labor productivity, defined as total income divided by the size of the labor force. In the urban context, we do have fairly reliable data on labor force but lack of information on urban income is where the problem lies. There is no official data on the division of national GDP into its urban and rural components. A recent study tried to make an estimate for one particular year (2014–15) by analyzing BBS's estimates of gross value-added (GVA) at very detailed sub-sectoral levels and trying to decompose them into urban and rural components by using either direct information on urban and rural origin of value-added or by using some proxies of urban–rural ratios where such direct information did not exist (Osmani 2021). The detailed urban–rural ratios of sub-sectoral GVAs thus obtained for the year 2014–15 were then applied to the GVA data for other years to obtain a time series of rural and urban GDPs for the period from 2010–11 to 2018–19. An obvious limitation of this approach is that the time series thus constructed cannot be extended too far since the sub-sectoral urban–rural ratios may change over time. However, since such structural changes are likely to be slow, it may be permissible to compare over a short span of time. On that basis, we have used the urban GDPs of the years 2010–11 and 2016–17 to estimate changes in labor

Table 2.10 Productivity and real wage in urban Bangladesh

	2010–11	2016–17
GDP at constant prices (ml Taka)	4,029,693	5,564,070
Labor force (ml)	13.3	17.9
Average labor productivity (Taka)	302,984	310,842
Index of labor productivity (%)	100.0	102.6
Index of real wage (%)	100.0	92.0

Source Urban GDP figures are from Osmani (2021); labor force figures are from *Labor Force Surveys* of BBS, and the index of real wage was constructed by the authors as described in the context of Fig. 1

productivity between these two years and to compare it with changes in urban real wages during the same period. The results are presented in Table 2.10.

The table reveals a clear divergence in the evolution of real wage and labor productivity. While labor productivity grew in the 2010s (albeit modestly, presumably because of the influx of rural migrants discussed above), real wages of unskilled workers fell significantly—by about 8%. This means a decline in the share of urban income accruing to unskilled labor (both wage labor and poor self-employed workers who can potentially compete with wage labor in the informal labor market) and a corresponding increase in the share of income accruing to capital and skilled personnel. This is what is reflected in the divergent income shares of the bottom 40% and top 10% of the population, resulting in a rising Palma ratio (Table 2.4).[16]

2.4 Social Safety Net and Distributional Equity

While distributional equity is to a large extent an outcome of market forces, no society has to be entirely beholden to the market forces in this regard. The whole concept of the welfare state has emerged to achieve a certain degree of equity that the market on its may fail to deliver. Around the world, the use of social safety net programs (SSNP) has played a role in providing some countervailing force to modify the market outcome, with more success in some cases than in others. Unfortunately, in urban Bangladesh, and in the country as a whole, this countervailing force has been used to very little effect.[17]

[16] There remains the question as to how absolute poverty declined in the face of falling real wages of unskilled labor. This is not the place to go into the details of the explanation, but a short answer is that poverty declined because many poor workers managed to move from low-paying jobs to slightly higher-paying jobs, a process that was facilitated by the structural changes happening in the economy along with rapid economic growth. For a detailed analysis, see Osmani (2018).

[17] Another common countervailing force is to strengthen the hands of the workers in wage bargaining process through appropriate labor laws. Although some potentially useful laws exist in Bangladesh, their implementation is mostly ineffective.

Table 2.11 Proportion of households participating in SSNP by income quintiles: 2016 (percentage of households)

Income quintile	Urban	Rural	Bangladesh
1st quintile	20.9	30.8	28.7
2nd quintile	17.8	29.2	26.9
3rd quintile	14.8	30.0	26.6
4th quintile	10.0	28.8	22.4
5th quintile	8.3	24.6	17.0

Source Calculated by the authors from the raw data file of *Household Income and Expenditure Survey 2016*

There are two contrasting aspects of the safety net in Bangladesh. First, the orientation of the safety net programs as a whole is reasonably equitable, serving the poorer segments more compared with the rich. Second, the scope of the net and the size of the benefits delivered through it are exceedingly small, especially for the urban population. The combined result of these two features is that the safety net does very little to improve the equity of market outcomes in urban Bangladesh and in the country as a whole.

The equity orientation can be seen from Tables 2.11 and 2.12 based on the *Household Income and Expenditure Survey* of 2016. Table 2.11 shows that households in the poorer quintiles participate in safety net programs relatively more than the richer ones. Thus, in urban Bangladesh, the proportion of participating households goes down systematically from about 21% among the bottom quintile to just over 8% among the top. In this regard, the urban segment of the society does distinctly better than the rural segment where the variation among poorer and richer quintiles is much less pronounced.

Table 2.12 reveals similar progressivity in terms of distribution of safety net benefits across the income quintiles. The share of benefits accruing to them generally falls as we move from the poorer to the richer quintiles, except for the top quintile.

The real problem lies in the scope and size of the safety net programs, as can be seen from Tables 2.13 and 2.14. The limited nature of the scope of safety net in urban Bangladesh is demonstrated by a couple of features of Table 2.13. In the first

Table 2.12 Distribution of SSNP benefits across income quintiles: 2016 (Percentage share of total SSNP benefits)

Income quintile	Urban	Rural	Bangladesh
1st quintile	23.1	26.0	26.1
2nd quintile	20.3	18.5	19.4
3rd quintile	17.3	19.0	20.5
4th quintile	14.7	18.7	16.1
5th quintile	24.6	17.9	17.8
All households	100.0	100.0	100.0

Source Calculated by the authors from the raw data file of *Household Income and Expenditure Survey 2016*

Table 2.13 Participation in social safety net programs (SSNP) (% of households)

	2005	2016
Urban	5.5	10.1
Rural	15.6	34.5
Bangladesh	13.1	27.8

Source BBS (2011, 2019b)

Table 2.14 Contribution of SSNP to household income by income quintiles: 2016 (percentage of household income)

Income quintile	Urban	Rural	Bangladesh
1st quintile	1.30	3.26	2.66
2nd quintile	0.50	0.94	0.81
3rd quintile	0.29	0.66	0.58
4th quintile	0.17	0.45	0.31
5th quintile	0.10	0.17	0.13
All households	0.22	0.49	0.39

Source Calculated by the authors from the raw data file of HIES 2016

place, in 2016, only about one in ten urban households participated in some safety net program. This is an improvement from one in twenty in 2005, but it is still a very small number in absolute terms. Secondly, in relative terms the scope is more limited in the urban sector compared with the rural sector. Thus, in 2016, more than one in three households participated in rural areas as against only one in ten among the urban households. Thus, the safety net in urban Bangladesh is not just limited in scope in absolute terms, it is also discriminated against in favor of rural areas.

The same pattern holds for the size of benefits delivered by the safety net (Table 2.14). The contribution it makes to the average household income in the urban sector is quite measly—only about 0.22% of household income. And despite an evident progressivity in favor of the poorer quintiles, the benefits received by even the poorest urban quintile is also paltry, amounting to just 1.3% of their household income. For the rest of the quintiles, the contribution of safety net is even more negligible.[18]

[18] Once again, there is marked discrimination against the urban sector vis-à-vis the rural sector, but in absolute terms even the rural households do not receive any substantial benefit from the social safety net.

Table 2.15 Contribution of SSNP to distributional equity: 2016 (Gini coefficient)

	Without SSNP	With SSNP
Urban	0.498	0.498
Rural	0.456	0.454
Bangladesh	0.485	0.484

Source Calculated by the authors from the raw data file of HIES 2016

The paucity of safety net benefits relative to household income entails that, in spite of its equitable orientation, the safety net program cannot be expected to have any substantial impact on distributional equity. This is confirmed by Table 2.15, where we compare the Gini coefficient of actual income distribution with the Gini of the counterfactual income distribution after leaving out safety net benefits. The two Gini coefficients are seen to be identical for the urban economy, leading to the conclusion that the safety net program as it currently operates in Bangladesh has failed completely to provide any kind of countervailing force to the distributional outcome of market forces in urban Bangladesh.[19]

The failure of the safety net to improve the distributional outcome in urban Bangladesh stems from three major factors. First, the resources allocated to social protection is thoroughly inadequate in relation to the need, even allowing for the general limitation of government resources. Over the decade of the 2010s, budgetary resources earmarked for social protection (for the nation as a whole) has averaged around 1.7% of GDP and just under 10% of the annual budget.[20] Furthermore, there was no increase in the budgetary share for social protection during this period. There was a temporary upsurge during the COVID pandemic of 2020, but even the government efforts were geared more toward reviving economic growth than offering social protection to those who had lost, or were in danger of losing, their livelihoods.[21]

The second factor is gross mistargeting of resources. As can be seen from Table 2.12, less than half the benefits accrues to the bottom two quintiles of urban households. Thus, the larger part of the resources meant for social protection goes to those who are the least deserving of protection.[22] The third and final factor is the anti-urban bias of the safety net programs in Bangladesh, which we have already discussed. It is thus the unholy combination of paltry budgetary allocation, gross

[19] As can be seen from Table 2.5, the same conclusion holds for the country as a whole, including rural Bangladesh which has traditionally been favored by the safety net programs.

[20] Official statistics treats social protection as a subset of a more encompassing concept of safety net many of whose programs do not have a direct focus on helping the poor and vulnerable. Using this broader concept of safety net, the average allocation over the 2010s was about 2.3% of GDP and 13.5% of the budget. Source: https://mof.gov.bd/site/page/32220b73-846f-4a33-b4c0-a6650c918e25/nolink/SafetyNet.

[21] The anti-protection and pro-growth bias of the Covid-interventions of the Government of Bangladesh has been analysed in details by Osmani and Siddiquee (2021).

[22] Evidence for gross mistargeting of the social protection measures is also found by Sen et al. (2022) in their analysis of data from the *Multiple Indicators Clusters Survey 2018* carried out by the Bangladesh Bureau of Statistics.

mistargeting of resources and a palpable anti-urban bias that has rendered social safety net virtually impotent to make any impact on distributional equity in urban Bangladesh.

2.5 Non-income Dimensions of Urban Inequality

Urban inequality has so far been discussed in the space of income. This is of course standard practice in the traditional discourse on inequality, especially in the context of examining the relationship between growth and distribution, which was also the context of our preceding discussion. But any meaningful analysis of inequality cannot remain confined to the space of income alone. It is after all the palpable disparity in human well-being—in its various manifestations—that ultimately moves us when we despair of rising inequality. People's command over income plays an important role in determining their level of well-being. But income does not in itself constitute well-being; it is only a means toward well-being, along with various other means including actions taken by the state and non-state actors. As such, we ought to evaluate equality or the lack of it directly in the space of well-being. This is the idea propounded by Amartya Sen in his famous paper "Equality of What", in which he made a powerful case for focusing on basic capabilities as indicators of well-being when evaluating the state of inequality in any society (Sen 1980).[23] Some of these basic capabilities include achievements and the freedom to achieve satisfactory nutritional status, basic education, respectable clothing and shelter, etc. In this section, we look at some of these indicators of well-being inequality in urban Bangladesh and their evolution over time.

Unlike data on income or consumption, systematic data on the distribution of basic capabilities among the people are relatively hard to come by. Nonetheless, we are able to put together a tentative picture by combining data from multiple sources. In Bangladesh, these sources include mainly various rounds of the *Bangladesh Demographic and Health Survey* (BDHS), *Household Income and Expenditure Survey* (HIES), *Multiple Indicators Cluster Survey* (MICS) and *Bangladesh Urban Health Survey* (BUHS). We define inequality in capability as the ratio of the indicators of capability between the richest and the poorest quintiles of the relevant population, the quintiles being defined in terms of either wealth or income or expenditure depending on the nature of available data. The indicators are themselves defined in such a way that higher values of the ratio between the richest and the poorest quintiles indicate higher inequality.

We start by considering urban inequality in respect of nutritional status of children and mothers (Table 2.16). Data derived from various rounds of the *Bangladesh Demographic and Health Survey* (BDHS) show that, over the two decades from 1996/97 to 2017/18, inequality in nutritional status widened in urban Bangladesh according to most indicators. The inequality index of child stunting (indicator of long-term

[23] For further elaboration of this idea, see especially Sen (1992).

child undernutrition) increased from 1.87 to 2.13 during this period. Although there was a decline in inequality in child wasting (indicator of short-term child undernutrition, which is susceptible to various short-term stress factors), the indicator of overall child undernutrition, as measured by the proportion of underweight children, exhibits rising inequality. At the same time, inequality in maternal undernutrition, as measured by the proportion of underweight mothers, has also increased.

A couple of other features of these data may also be highlighted. First, most of the increase in inequality occurred in the first half of the period (1996/97–2007), the rising trend being significantly moderated though not completely eliminated in the second half. Second, urban inequality was generally higher than rural inequality at the beginning of the period, but owing to the moderation in its rising trend in the second half of the period, the level of urban inequality fell slightly below the rural level at the end of the period (Table 2.17).

While health inequality shows a rising trend, inequality of educational opportunities demonstrates a declining trend, according to the data culled from the *Household Income Expenditure Surveys* (HIES). The enrollment inequality has dropped for all levels of education during the decade between 2005 and 2016, and the schooling of household heads also shows sharply declining inequality (Table 2.18).

Several additional features of educational inequality are worthy of note. First, notwithstanding the overall decline of inequality, the level of inequality remains significantly higher at tertiary level compared with inequality at secondary and primary levels. Second, female enrollment has lower inequality compared with male

Table 2.16 Health inequality in urban Bangladesh: 1996/97–2017/18 (ratio between bottom and top quintiles in wealth)

Indicators	1996/97	2007	2017/18
Child stunting	1.87	2.15	2.13
Child wasting	1.84	1.37	1.47
Child underweight	1.76	2.15	2.22
Underweight mothers	2.12	2.80	2.97

Source Computed by the authors from raw data of various rounds of *Bangladesh Health and Demographic Surveys* (BDHS)

Table 2.17 Health inequality in rural Bangladesh: 1996/97–2017/18 (ratio between bottom and top quintiles in wealth)

Indicators	1996/97	2007	2017/18
Child stunting	1.75	2.26	2.52
Child wasting	1.33	1.49	1.18
Child underweight	1.66	1.93	2.51
Underweight mothers	1.73	3.41	5.09

Source Computed by the authors from raw data of various rounds of *Bangladesh Health and Demographic Surveys* (BDHS)

Table 2.18 Educational inequality in urban Bangladesh: 2005–2016 (ratio between top and bottom quintiles in expenditure)

Indicators	2005	2016
A. Children currently enrolled (%)		
Primary (age 6–10)		
Male	1.41	1.10
Female	1.27	1.09
Secondary (age 11–16)		
Male	1.95	1.34
Female	1.36	1.15
Tertiary (age 17–24)		
Male	11.99	2.77
Female	34.66	3.72
B. Years of schooling of household head	5.31	1.63

Source Computed by the authors from raw data of various rounds of *Household Income and Expenditure Surveys* (HIES)

enrollment at both primary and secondary levels but higher inequality at the tertiary level. Third, unlike in the case of health inequality, the level of educational inequality is higher in urban areas compared with rural areas, as judged by the most recent data (Table 2.19).

To understand more clearly the significance of these trends, we need to ask the same question that we asked in the context of income inequality (in Sect. 2.2): what is happing to the absolute level of achievement of the bottom rung of the population? When inequality widens, it can happen either because the bottom end of the distribution is getting absolutely worse off while the upper end is surging ahead,

Table 2.19 Educational inequality in rural Bangladesh: 2005–2016 (ratio between top and bottom quintiles in expenditure)

Indicators	2005	2016
A. Children currently enrolled (%)		
Primary (age 6–10)		
Male	1.31	1.08
Female	1.20	1.05
Secondary (age 11–16)		
Male	1.97	1.27
Female	1.48	1.13
Tertiary (age 17–24)		
Male	4.36	2.22
Female	7.44	3.52
B. Years of schooling of household head	4.26	1.36

Source Computed by the authors from raw data of various rounds of *Household Income and Expenditure Surveys* (HIES)

Table 2.20 Health-status of the bottom quintile (in wealth) in urban Bangladesh

Indicators (%)	1996/97	2007	2017/18
Child stunting	53.35	60.30	37.97
Child wasting	18.02	16.96	9.88
Child underweight	56.84	55.47	28.60
Underweight mothers	55.81	49.03	15.33

Source Computed by the authors from raw data of various rounds of *Bangladesh Health and Demographic Surveys* (BDHS)

or because the bottom end is also getting absolutely better off but at a slower pace compared with the upper end. The two scenarios are very different in terms of their implications for well-being as well as their underlying dynamics.

We have seen earlier that inequality has been increasing in child and maternal nutrition. When we examine, however, the absolute level of nutritional deprivation of the bottom quintile, we find a dramatic improvement in all the indicators (Table 2.20). The proportion of underweight children among the bottom quintile has declined impressively in two decades—from 57% in 1996/97 to 29% in 2017/18. The same story applies to maternal nutrition as well. While inequality in maternal undernutrition rose during the period, the share of underweight mothers among the bottom quintile has dramatically gone down from 56% to just 15%. Rising nutritional inequality in urban Bangladesh has thus been accompanied by absolute improvement in the nutritional status of the bottom rung of the population.

The educational scenario is slightly different in that we witness both relative and absolute improvement in the condition of the bottom end of the population. We have already seen that inequality has declined sharply in the sphere of education; we can now see from Table 2.21 that absolute levels of opportunities enjoyed by the bottom quintile have also improved.

Thus, while health and educational inequality show contrasting trends—with health inequality rising and educational inequality falling—in both cases, the bottom quintile has experienced tangible improvement in absolute terms. As in the case of income distribution, so in the case of non-income indicators, there is no evidence that the poorest segment of the urban society is getting absolutely worse off, although inequality is increasing in certain dimensions.

To understand the forces underlying these trends, it is useful to begin by recognizing the causal connection between income and non-income deprivations. We have seen that even though income inequality is rising over time, people in the bottom quintile have nevertheless experienced rising absolute levels of income in the recent decades. Higher incomes, other things remaining the same, would have helped achieve better non-income outcomes. In addition, there is evidence to suggest that access to various services have improved, even for the poor.

Access to school education is a well-known example. Initially spurred on by interventions by non-governmental organizations (NGOs), the Government of

Table 2.21 Educational status of the bottom quintile in urban Bangladesh

Indicators	2005	2016
A. Children currently enrolled (%)		
Primary (age 6–10)		
Male	70.14	90.25
Female	75.04	92.34
Secondary (age 11–16)		
Male	42.41	69.92
Female	56.95	81.72
Tertiary (age 17–24)		
Male	8.24	22.54
Female	3.43	10.38
B. Years of schooling of household head	1.66	6.10

Source Computed by the authors from raw data of various rounds of *Household Income and Expenditure Surveys* (HIES)

Bangladesh began to put serious emphasis on universalizing access to basic education since the 1990s, principally through stipend schemes. This has led to unprecedented rise in school enrolments, although serious questions remain about the quality of education offered. The urban poor have also benefitted from these initiatives, which, as we have seen, is reflected in better educational opportunities for the urban poor—both in absolute terms and relative to the urban rich.

The health sector has been neglected to some extent, especially relative to education, but evidence suggests that the urban poor have still experienced improving access to health-care services over time—in both absolute and relative terms. Table 2.22 shows that inequality between the richest and bottom quintiles in access to various types of health-care services has fallen in the recent years, and Table 2.23 shows that falling inequality has been accompanied by rising absolute levels of access to health-care for the bottom quintile.

All this helps explain how the urban poor are enjoying better health outcomes than in the past. But there still remains the question of why inequality in health outcomes is still rising in urban areas. Evidently, whatever improvement the urban poor are enjoying is occurring at a much slower pace compared with the rich. The question is as follows: why is this so? Part of the answer must lie in rising income inequality since inequality in the space of income would invariably have some effect on inequality in the space of health and nutritional outcomes. But this cannot be whole answer. After all, educational inequality declined despite rising income inequality. Why did health inequality fail to do so?

The clue to the answer lies, at least in part, in the role that urban slums play in the dynamics of urban inequality. The bulk of the poorest segment of the urban population live in slum areas. According to one survey, some 75% of all slum households belonged to the bottom two wealth quintiles of the overall urban population (NIPORT/ICDDRB 2013). This means that what happens to urban inequality is

Table 2.22 Inequality in access to health-care services: urban (ratio between top and bottom quintiles in wealth)

Indicators	2007	2017/18
% of children with fever who received proper medical treatment (BDHS)	2.010	1.581
% of children with diarrhea who were treated with ORT (BDHS)	1.166	1.156
% of children ever had vaccination (BDHS)	1.342	1.259
% of children had health card (BDHS)	1.259	1.106
% of women receiving 3+ ante-natal care (ANC) visits (BDHS)	7.26	2.66
% of women giving delivery at a medical facility (BDHS)	11.91	4.44
Multiple indicators cluster survey (MICS)	*2006*	*2019*
% of children having difficulty breathing with cough receiving treatment	1.235	1.077
% of children weighed at birth	5.096	2.835

Source Computed by the authors from raw data of various rounds of *Bangladesh Health and Demographic Surveys* (BDHS) and *Multiple Indicators Cluster Survey* (MICS)

Table 2.23 Access to health-care services for the bottom quintile: urban

Indicators		
Bangladesh demographic and health survey (BDHS)	*2007*	*2017/18*
% of children with fever who received proper medical treatment	26.23	29.05
% of children with diarrhea who were treated with ORT	73.68	78.21
% of children ever had vaccination	54.01	70.42
% of children had health card	53.48	79.17
% of women receiving 3+ ante-natal care (ANC) visits	7.49	28.92
% of Women giving delivery at a medical facility	3.95	18.39
Multiple indicators cluster survey (MICS)	*2006*	*2019*
% of children having difficulty breathing with cough receiving treatment	31.67	72.61
% of children weighed at birth	10.45	27.81

Source Computed by the authors from raw data of various rounds of *Bangladesh Health and Demographic Surveys* (BDHS) and *Multiple Indicators Cluster Survey* (MICS)

largely a reflection of the disparity between slum and non-slum parts of the urban society.

Slum-dwellers have traditionally been among the most deprived segments of the country—not just in comparison with non-slum urban dwellers but also in comparison with the rural people. Table 2.24, based on the nationally representative *Multiple Indicators Cluster Survey* (MICS) carried out by the Bangladesh Bureau of Statistic in 2009, presents the stark reality. By almost all criteria of non-income deprivation, slum-dwellers are found to have lagged behind even the rural people by a wide margin. Another study carried out in 2006 found that slum-dwellers were more likely to fall below the 'food poverty line' than both urban households in general and rural households (IFPRI/BBS/WFP 2008).

Table 2.24 Comparative living standards in urban, rural and slum areas: 2009

Indicators	Urban	Slum	Rural
U5MR (per 1000 live birth)	53	95	66
Skilled attendant at birth (%)	45	15	19
Population using improved sanitation (%)	54	9	54
Net attendance ratio in pre-school education (%)	26	13	22
Net attendance ratio in primary education (%)	84	65	81
Pupils starting grade 1 who reach grad 5 (%)	80	48	80
Drop out in primary education (%)	1	8	1
Net attendance ratio in secondary education (%)	53	18	48
Female youth literacy (%)	77	51	70
Birth registration (%)	53	28	54

Source UNICEF (2010), p. 11

There are a number of structural impediments that tend to create a hostile environment which makes it hard for the slum-dwellers to improve their living conditions.[24] To begin with, slums tend to emerge from the congregation of mainly rural migrants, most of whom used to belong to the poorest part of the rural society, and when they migrate they do so in a state of near-destitution, with very little education, and often suffering from life-long nutritional deprivation. Thus, their life begins in slums with a serious disadvantage. Secondly, their initial disadvantage is compounded by the apathy of government officials toward them. In the context of Bangladesh, it has been noted, for example, that officials deliberately refrain from extending support to the slums in the misguided belief that (a) the urban poor are better off than the rural people anyway, and (b) any support extended to the slum-dwellers will only serve to attract ever more migrants, thus further aggravating the existing problems of slum living (World Bank 2007).

But by far the most serious impediment they face is lack of security—security of both tenure and life. They can seldom formally rent, let alone own, the land on which they live—it is only at the mercy of others that they can live in whatever ramshackle structures they can manage to put up. In consequence, they are constantly faced with the threat of eviction. This in turn has a couple of pernicious consequences. First, lack of security and fear of eviction make the slum-dwellers easy prey to local goons, popularly known as *maastans* in Bangladesh, who are often linked to local politicians and work to their mutual advantage. The leverage the *maastans* enjoy because of their political backing on the one hand and the slum-dwellers' vulnerability on the other enables them to create a structure of extortion that is ruthlessly used to exploit the hapless slum-dwellers. Second, the lack of security of tenure and the possibility of eviction at any time discourage service providers from investing in the infrastructure

[24] Marx et al. (2013) offers a perceptive analysis of the pathways through which the multiple impediments faced universally by the slum dwellers can create a poverty trap from which escape can prove exceedingly difficult.

Table 2.25 Change in living standards in slum areas: 2006–2013

Indicators	2006	2013
Prevalence of stunting (%)	57	50
Prevalence of underweight (%)	46	43
Prevalence of wasting (%)	19	19
Neonatal mortality rate (%)	44	31
Child mortality rate (%)	63	49
Under-5 mortality rate (%)	81	57
Females with primary education (%)	13	15
Males with primary education (%)	11	14
Females with secondary education (%)	5	6
Males with secondary education (%)	13	11

Source NIPORT/ICCDRB (2013)

needed to provide necessary services. This is one of the reasons why the NGOs, who have been much praised for their sterling work in rural Bangladesh, have traditionally showed relative neglect toward the urban poor.[25]

There are encouraging signs, however, that many NGOs are shifting their focus toward urban squalor in recent years and finding innovative ways to serve the slum-dwellers (NIPORT/ICDDRB 2013: Their efforts, together with the effect of rising incomes, have been helping to push forward the quality of life in urban slums, as can be seen from Table 2.25.

This improvement is also reflected in rising absolute levels of well-being enjoyed by the bottom quintile of the urban population, as documented in Table 2.23. As we have noted, however, the improvement has not been fast enough to reverse the rising trend of health inequality, in particular. The various agencies of the government whose collective work has an effect on people's health must abandon their traditional attitude and engage seriously with the slum population. For this to happen, though, the first pre-requisite is to recognize and ensure the rights of slum-dwellers to live with security and dignity.

2.6 Conclusion

Bangladesh is currently going through a phase of rapid urbanization combined with rapid economic growth. Along with others, the urban poor have also benefitted from growing prosperity. They are earning higher levels of income; their educational opportunities are expanding, and their health and nutrition have also improved on the whole. But they are not always sharing in the general prosperity in an equitable

[25] For wide-ranging discussion of, and evidence on, multiple impediments faced by the people living in the slums of urban Bangladesh, see, among others, NIPORT/ICDDRB (2006, 2013), World Bank (2007), Habib (2009), Rashid (2009), UNICEF (2010), BBS (2015b), and WFP (2015).

manner. There is clear evidence of rising inequality in the distribution of income and health outcomes. Education is an exception, where inequality has narrowed considerably despite rising income inequality.

Widening inequality in the distribution of income is an outcome of a growth process which is founded on the use of cheap labor, made even cheaper by rapid influx of poor rural migrants. The gap that thus has been created between the real wage and productivity of labor is the primary economic force underlying rising income inequality. To that extent, the widening of income inequality can be attributed to the market forces unleashed by the growth process.

Yet, as the example of education shows, there is nothing inevitable about rising inequality. What matters is whether public action is used in a decisive manner as a countervailing force against the pure market outcome. In the sphere of education, the government of Bangladesh has taken determined action since the 1990s to universalize access to basic education across the country. Narrowing inequality in educational opportunity, despite rising inequality of income, is a consequence of this determined public action.

By the same token, the failure to take similar countervailing action in the case income and health is primarily responsible for widening inequality in these spheres. In the case of income, at least two types of actions were needed—providing adequate social protection to those who are being left behind by the market and to strengthen the hands of the poor in the wage bargaining process. The government has failed miserably on both counts. In the case of health, it was essential to provide basic healthcare services to the urban poor, especially to those living in urban slums. The slum-dwellers have traditionally been deprived of all kinds of public services for various reasons, the chief among which is the lack of tenurial security and the resulting threat of eviction which makes any investment in the infrastructure of service provision a risky proposition for the private sector. In recent years, some NGOs have been trying to fill the void through some innovative models. But determined public action remains conspicuously missing. Without public action, it would remain difficult to reverse the process of health inequality even if the urban poor enjoy a modest improvement in the absolute level of health outcomes.

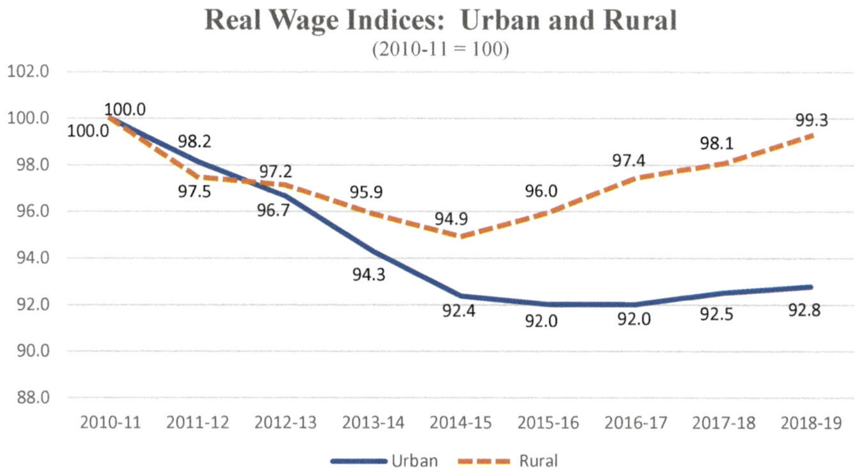

Fig. 2.1 Trends in real wages: urban and rural. *Source* The raw data on nominal wage rate indices and consumer price indices used for constructing real wage indices are from BBS (2020). *Note* BBS provides nominal wage rate indices only at the national level, but it also provide data separately for broad economic sectors, which was used for the present purpose. In this figure, rural wage is proxied by the index of agricultural wages and urban wage is proxied by a weighted average of wage rate indices for industry and services, the weights being the same as those used by BBS for constructing overall wage index from sectoral wage indices

Acknowledgments They are grateful to Zakir Hossain, Muntasir Murshed, and Jinnat Ara of BIDS for their excellent research support.

References

Ahmed S, Alamgir M, Mujeri MK (2017) Urbanization and land issues in Bangladesh. Bangladesh Economists' Forum

BBS (2008) Preliminary report of the agricultural census 2008. Bangladesh Bureau of Statistics

BBS (2011) Report of the household income and expenditure survey 2010. Bangladesh Bureau of Statistics

BBS (2015a) Revision and rebasing of wage rate index (WRI) from 1969–70 to 2010–2011. Bangladesh Bureau of Statistics

BBS (2015b) Census of slum areas and floating population 2014. Bangladesh Bureau of Statistics

BBS (2017) Bangladesh labor force survey 2016/17. Bangladesh Bureau of Statistics

BBS (2019a) Preliminary report of the agricultural census 2019. Bangladesh Bureau of Statistics

BBS (2019b) Final report on household income and expenditure survey 2016. Bangladesh Bureau of Statistics

BBS (2020) Consumer price index (CPI), inflation rate and wage rate index (WRI) in Bangladesh, July 2020. Bangladesh Bureau of Statistics

Ellis P, Roberts M (2016) Leveraging urbanization in South Asia. World Bank

Habib E (2009) The role of government and NGOs in slum development: the case of Dhaka city. Dev Pract 19(2):259–265

IFPRI/BBS/WFP (2008) Study of household food security in urban slum areas of Bangladesh, 2006. International Food Policy Research Institute, Bangladesh Bureau of Statistics and World Food Program
Kuznets S (1966) Modern economic growth: rate, structure and spread. Yale University Press
Marshall A (1890) Principles of economics. Macmillan
Marx B, Stoker T, Suri T (2013) The economics of slums in the developing world. J Econ Perspect 27(4):187–210
NIPORT/ICDDRB (2006) Bangladesh urban health survey 2006. National Institute of Population Research and Training (NIPORT) and International Centre for Diarrhoeal Disease Research, Bangladesh (ICDDR, B)
NIPORT/ICDDRB (2013) Bangladesh urban health survey 2013. National Institute of Population Research and Training (NIPORT) and International Centre for Diarrhoeal Disease Research, Bangladesh (ICDDR, B)
Osmani SR (2015) The growth-equity nexus in Bangladesh: an analysis of recent experience. Bangladesh Dev Stud XXXVIII(2):1–59
Osmani SR (2018) Aspects of the poverty scenario in Bangladesh during 2010–2016. Bangladesh Dev Stud XLI(3):1–34
Osmani SR (2021) Estimation of rural and urban GDP of Bangladesh: a proposed methodology with illustrative implementation (unpublished manuscript). National Accounts Division, Bangladesh Bureau of Statistics
Osmani SR, Siddiquee S (2021) Economic support in response to COVID-19 and the quest for political legitimacy. In: BIGD, State of governance in Bangladesh 2020–21. Governing COVID-19 in Bangladesh: realities and reflections to build forward better, 132–164. Brac Institute of Governance and Development
Palma JG (2011) Homogeneous middles vs. heterogeneous tails, and the end of the 'inverted-U': it's all about the share of the rich. Dev Change 42(1):87–153
Rashid SF (2009) Strategies to reduce exclusion among populations living in urban slum settlements in Bangladesh. J Health Popul Nutr 27(4):574–586
Sen A (1980) Equality of what? In: McMurrin S (ed) Tanner lectures on human values, vol 1. Cambridge University Press
Sen A (1992) Inequality reexamined. Clarendon Press
Sen B, Ali Z, Murshed M (2022) Poverty in the time of corona: trends, drivers, vulnerability and policy responses in Bangladesh. In: Alam S (ed) Background papers of the 8th five year plan, vol 4. General Economics Division, Bangladesh Planning Commission
UNICEF (2010) Understanding urban inequalities in Bangladesh. United Nations Children Fund
WFP (2015) Food insecurity and undernutrition in the urban slums of Bangladesh. World Food Program
World Bank (2007) Dhaka: improving living conditions for the urban poor. World Bank

Chapter 3
Evaluating Urban Multidimensional Inequalities in Peru from the Capability Approach

Jhonatan Clausen and Nicolas Barrantes

Abstract In this chapter, we evaluate multidimensional well-being disparities in urban areas in Peru, a middle-income country in one of the most unequal regions in the world: Latin America and the Caribbean. We rely on data from a specialized survey collected in this country in late 2018, which includes information on a broad set of non-monetary dimensions of human well-being. We use this data to specify, calibrate, and calculate an urban-specific measure of multidimensional well-being conceptually based on Amartya Sen's capability approach. This measure comprises a set of 12 dimensions we operationalize using 30 indicators of well-being achievement. Our main results reveal that the proportion of women in a situation of well-being sufficiency is lower than that of men and that the capital city of Lima shows higher values of well-being sufficiency than other urban regions in the Coast, the Highlands, and the Amazonia. This pattern of disparity holds valid even if we introduce changes to the multidimensional well-being sufficiency threshold, which corresponds to the minimum number of achievements that a person should accomplish to live with an adequate level of well-being.

Keywords Multidimensional well-being · Urban disparities · Amartya Sen · Latin America · Peru

3.1 Introduction

Latin America and the Caribbean is a world region where evaluating urban inequalities is particularly relevant. According to the estimates of the Population Division of the Department of Economic and Social Affairs of the United Nations (2019), the

J. Clausen (✉) · N. Barrantes
Department of Economics, Institute of Human Development of Latin America, Pontifical Catholic University of Peru, Lima, Peru
e-mail: jhonatan.clausen@pucp.edu.pe

N. Barrantes
e-mail: n.barrantes@pucp.pe

G. H. Tonon (ed.), *Urban Inequalities*, The Urban Book Series,
https://doi.org/10.1007/978-3-031-59746-6_3

proportion of people living in urban areas in this region will have increased from 49% in 1960 to 81% in 2021. This proportion is higher than in other significantly more affluent world regions such as, for instance, the European Union (75%) and only marginally lower than in North America (83%). At the same time, Latin America and the Caribbean are among the most unequal regions globally: 10% of its population concentrates 55% of the national income (Chancel et al. 2022). This is, therefore, both a highly urbanized and an unequal world region (Angotti and Irazábal 2017; Deneulin and Sánchez-Ancochea 2018; Leite et al. 2020).

The combination of high urbanization and income inequality in Latin America and the Caribbean has triggered an increase in the literature about urban inequalities focused on different countries within this region. One salient feature of this literature is that it seems to recognize that urban inequality manifests in different "planes" (Dammert Guardia et al. 2019b) beyond income inequality. These different planes or "dimensions" of well-being (Dammert Guardia et al. 2019a) might include healthcare (Lucero et al. 2016), education (Born et al. 2019), life expectancy (Bilal et al. 2019), access to parks and green areas (Moran et al. 2020; Rigolon et al. 2018), and housing conditions (Linares et al. 2016; Martinez 2016). Overall, this literature also shows that urban areas in Latin America and the Caribbean are significantly unequal, even if said inequality is evaluated from a non-income multidimensional perspective.

The studies above represent significant contributions to our current understanding of urban inequalities in Latin America and the Caribbean. In particular, studies such as Linares et al. (2016) in Argentina, and Deloitte et al. (2022) in Colombia, have relied on multidimensional indices of well-being/quality of life to explore urban inequalities. Yet, due to a lack of specialized data, some of these studies face limitations in including information on a broad range of dimensions, such as, for instance, the dimensions of the comprehensive proposal of Tonon (2016) to evaluate the quality of life in Latin America. This proposal includes dimensions not conventionally utilized in analyzing this region's urban quality of life: discrimination, community relations, participation in cultural activities, social support, personal security, politics, and institutions, among others (Tonon 2016).

In this chapter, we contribute to the abovementioned literature by conducting a comprehensive multidimensional evaluation of urban well-being disparities using information from a specialized dataset collected in urban areas in Peru, a Latin American middle-income country. Our evaluation relies on an urban-specific version of the Multidimensional Well-being Index for Peru (MWI-P) initially developed by Clausen and Barrantes (2022) based on Amartya Sen's capability approach, one of the most influential normative frameworks for the assessment of well-being and inequality. In this chapter, we consistently use the term urban "disparities" instead of urban "inequalities" because the MWI-P allows us to estimate well-being gaps between urban areas or urban populations, yet it is not a distribution-sensitive measure like other conventional measures of inequality. To the best of our knowledge, this is the first study focused on a Latin American middle-income country that relies on an individual-level comprehensive metric of well-being to explore both multidimensional gender disparities within urban areas as well as multidimensional disparities between urban geographical domains.

This chapter is structured as follows. Section 3.2 briefly introduces Amartya Sen's capability approach, which corresponds to the conceptual framework of our study. Section 3.3 describes our case study and justifies its significance for this research. Section 3.4 presents the specialized dataset we use in this chapter and the method we follow to specify and calibrate the MWI-P to urban contexts. Section 3.5 contains the main results of our evaluation. Finally, Sect. 3.6 discusses our main findings and concludes the chapter.

3.2 The Capability Approach and Well-Being in Urban Contexts

Amartya Sen's capability approach constitutes one of the most influential conceptual frameworks in the literature on well-being evaluation (White 2016). Providing a comprehensive description of this approach is beyond the scope of this chapter. However, one relatively succinct yet informative definition provided by Sen himself is that it corresponds to an "intellectual discipline that gives a central role to the evaluation of a person's achievements and freedoms in terms of his or her actual ability to do the different things a person has reason to value doing or being" (Sen 2009: 16). Sen labels these different achievable 'doings' and 'beings' as "functionings," whereas he uses the term "capabilities" to refer to the alternative combinations of functionings of a person (Sen 1993).

Scholars such as Frediani (2007) have relied on Sen's works to investigate how the capability approach could improve the urban quality of life in Salvador de Bahía (Brazil). Similarly, the research of Frediani (2015) used a capability lens to explore the relationship between the physical space and local abilities in urban areas in Salvador de Bahía, physical space and human opportunities in Kisumu (Kenya), as well as between physical space and people's aspirations in the peri-urban site of Los Pinos (Ecuador). Anand (2018) has also drawn in the capability approach to enrich the discussion on urban development and to illustrate how some of the potential benefits of living in urban areas, such as access to health and education, can be diminished by negative factors including lack of access to basic facilities, environmental pollution, poor social interaction, or physical insecurity.

In this chapter, we partially rely on the capability-based framework for urban integration developed by Deneulin et al. (2017). Based on a case study in Buenos Aires (Argentina), these authors highlight three features of the capability approach that, they argue, are particularly relevant to inform and shape policies oriented to reduce urban inequalities, which is the aim of the framework they developed. First, the capability approach proposes that inequalities and disparities should be evaluated in the multidimensional "space" of functionings and capabilities, not only using information from the unidimensional "space" of income. Second, the approach not only focuses on the situation of the individuals in terms of the functionings and capabilities they can enjoy but also on the institutions and social arrangements that can improve or deteriorate their situation. Third, the capability approach gives a central role to people's agency to trigger changes that expand their sets of valued functionings and capabilities. Specifically, in this chapter, we draw on the first and third features highlighted by Deneulin et al. (2017) inasmuch our evaluative analysis adopts a multidimensional perspective that incorporates the aspect of agency both to calibrate the measurement tool we use, as well as one of the dimensions ("agency") we include in our measure of multidimensional well-being.

3.3 Case Study: Urban Areas in Peru

According to the Population Division of the Department of Economic and Social Affairs of the United Nations (2019), Latin America and the Caribbean is the second region in the world with the most significant number of megacities—urban agglomerations with more than 10 million inhabitants—and Lima, the Peruvian capital city, has recently become one of the six megacities in the region. In addition to being highly urbanized, the distribution of the urban population in Latin America and the Caribbean is particularly uneven. Espinoza et al. (2022) estimate that this region has the most significant number of "primate cities," which concentrate significantly high proportions of the urban population of the countries in which they are located. Furthermore, these authors identify that Peru constitutes an extreme case in the region insofar as the population size of Lima is currently ten times larger than the population living in the second-largest city in this country.

The rapid urbanization process in Peru began in the twentieth century. It was triggered by a nascent dynamic of national industrialization, increased public investment in road infrastructure, and the development of international trade circuits (Espinoza et al. 2022). Urbanization in Peru accelerated even more during the second half of the twentieth century. The estimates of Sánchez (2015) based on data from National Censuses show that between 1940 and 1993, the share of the population living in urban areas increased from 35 to 70%. In addition, although urbanization took place across different parts of the Peruvian territory and some intermediate cities emerged in coastal areas and the Amazonia, Lima has played a prominent leading role in this process inasmuch its population multiplied by ten over this period. Currently, the 2017 Peruvian National Census data reveal that Lima concentrates 41% of the urban population, which, in turn, represents 79% of the total national population (INEI 2018).

The continued expansion of the urban population in Peru has been accompanied by a considerable decline in consumption-based monetary poverty from 2007 to 2019 (see Fig. 3.1). In contrast, the degree of inequality in the distribution of household expenditures in urban areas—measured using the Gini coefficient—has remained relatively unchanged since 2011 (see Fig. 3.2). Yet, more recently, the multiple crises caused by the COVID-19 pandemic generated an abrupt rise in urban monetary poverty in 2020 that has not yet been fully reversed. This sudden increase might be explained by the high vulnerability to monetary poverty—the proportion of people living at risk of falling into monetary poverty—which, before the pandemic, was above 30% (INEI 2022).

Recent well-being reductions among the urban population in Peru are not confined to the monetary sphere. One of the most salient examples of non-monetary deprivations that have affected urban people in this country during the COVID-19 pandemic corresponds to the dimension of health. Peru has been one of the most severely affected countries by COVID-19 mortality worldwide (Clausen 2020), and urban populations have accounted for the largest number of COVID-related deaths in this country. The research of Clausen and Barrantes (2021) shows that more than 50% of the total excess deaths from non-violent causes in 2020 occurred in the metropolitan area of Lima. These authors also show an unequal pattern of mortality in Lima:

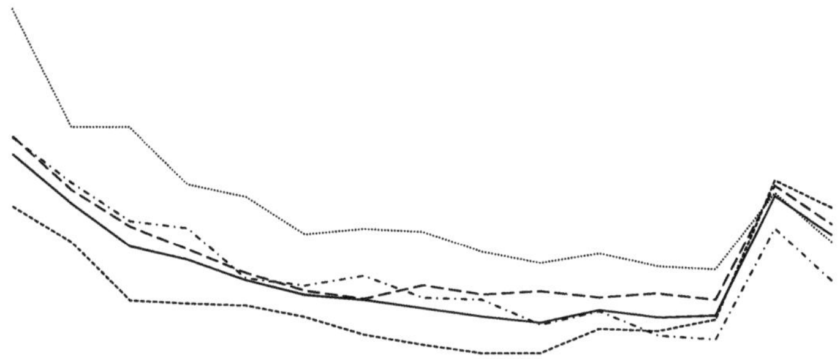

	2007	2008	2009	2010	2011	2012	2013	2014	2015	2016	2017	2018	2019	2020	2021
Urban	30.1	25.4	21.3	20.0	18.0	16.6	16.1	15.3	14.5	13.9	15.1	14.4	14.6	26.0	22.3
Lima (capital city)	25.1	21.7	16.1	15.8	15.6	14.5	12.8	11.8	11.0	11.0	13.3	13.1	14.2	27.5	24.9
Urban Coast	31.7	27.4	23.7	23.0	18.2	17.5	18.4	16.3	16.1	13.7	15.0	12.7	12.3	22.9	17.9
Urban Highlands	31.8	26.7	23.2	21.0	18.7	17.0	16.2	17.5	16.6	16.9	16.3	16.7	16.1	27.0	23.3
Urban Amazon	44.0	32.7	32.7	27.2	26.0	22.4	22.9	22.6	20.7	19.6	20.5	19.3	19.0	26.3	21.6

Fig. 3.1 Incidence of monetary poverty (%), Peruvian urban domains, 2007–2021. *Source* INEI (2019, 2022)

	2007	2008	2009	2010	2011	2012	2013	2014	2015	2016
Urban	0.36	0.34	0.34	0.33	0.32	0.32	0.32	0.32	0.32	0.31
Lima (capital city)	0.36	0.33	0.35	0.33	0.33	0.33	0.33	0.32	0.32	0.32
Urban Coast	0.34	0.31	0.31	0.31	0.30	0.30	0.30	0.30	0.30	0.29
Urban Highlands	0.36	0.36	0.34	0.34	0.32	0.32	0.33	0.33	0.33	0.32
Urban Amazon	0.36	0.34	0.34	0.34	0.33	0.33	0.32	0.32	0.31	0.31

Fig. 3.2 Inequality in the distribution of household expenditure measured with the Gini coefficient, Peruvian urban domains, 2007–2016. *Source*: INEI (2017)

districts with lower values of the Human Development Index and with a higher incidence of multidimensional non-income deprivations—such as overcrowding, lack of access to piped water, lack of a refrigerator, and use of polluting cooking fuels—presented relatively higher excess mortality rates compared to less deprived districts. Moreover, the severity of this COVID-induced "human development crisis" in Peru seems to trigger a process of forced migration that led people who had moved to Lima to look for better opportunities to return to their original hometowns (Boyd et al. 2021).

While the literature on urban inequalities in Peru is not particularly abundant, the currently available studies show that this is far from being a minor issue in this country. Even though some of the abovementioned studies rely on data beyond income inequality, they include information on relatively conventional non-income indicators of well-being/quality of life. There is, therefore, room for exploring urban disparities in Peru, relying on a more comprehensive tool for well-being evaluation. In the next section, we introduce this tool and justify its relevance for this research.

3.4 Data and Method

We use data from a dedicated survey on multidimensional well-being collected by the Institute of Public Opinion (Instituto de Opinión Pública, IOP) and the Institute of Human Development of Latina America (Instituto de Desarrollo Humano de América Latina, IDHAL) at the Pontifical Catholic University of Peru (Pontificia Universidad Católica del Perú, PUCP) in late 2018. The sample contains information from 1600 individuals aged 18 or older who were randomly selected, relying on a strategy that used urban blocks and dwellings as the first and second sampling units, respectively. This sample is representative of Peruvian adults living in a group of 21 out of the 25 country administrative regions in Peru. This group of 21 regions concentrates 80% of the total urban population. All the estimates we present in this chapter use sampling weights to reflect the structure of this population.

We utilize the Alkire and Foster (2011) method (AF method) to estimate an index that reflects aggregate levels of well-being in urban areas. Even though the AF method has been usually applied to measure multidimensional poverty, previous studies have also used it to calculate multidimensional well-being indexes (Alkire and Kovesdi 2020; Centre for Bhutan Studies and GNH Research 2016). Specifically, we follow the proposal of Clausen and Barrantes (2022), who designed a Multidimensional Well-being Index for Peru (MWI-P), which applied the AF method to well-being measurement at the national level. In this chapter, we specify and calibrate said index to reflect urban Peru's reality.

Like the MWI-P, the measure we estimate in this chapter includes 12 dimensions which we operationalize through 30 indicators of well-being achievement (see Table 3.1), consistent with the proposal of Tonon (2016). In this chapter, we use the sufficiency criteria for these 30 indicators of achievement proposed by Clausen and Barrantes (2022) (see the third column in Table 3.1). If the individual i satisfies the sufficiency criterion corresponding to the indicator j, then $I_{ij} = 1$ which mean that they show an adequate level of achievement in the indicator j, where $i = 1, \ldots, n$ and $j = 1, \ldots, 30$. Conversely, $I_{ij} = 1$ when the individual i does not satisfy the sufficiency criterion for the indicator j which corresponds to a situation in which they do not meet an adequate level of achievement in the indicator j. Then, we ask urban respondents to use 12-step ladders (see Fig. 3.3) to reveal the importance they assign to each one of the 12 dimensions of our measure. We use this information to calculate the normalized relative importance of each dimension for each individual in our urban sample. The weight we assign to each dimension of our measure corresponds to the average value of the normalized relative importance of said dimension. We uniformly distribute the weight of each dimension across all the indicators we use to operationalize it. Figure 3.4 shows the values of these dimensional weights. The sum of the 12 weights of our index is equal to 100%. We argue that calibrating dimensional weights based on the values of the urban population are consistent with the proposal of Deneulin et al. (2017), which emphasizes the importance of considering an agentic perspective to approach human well-being in urban contexts.

We assign each individual i with a "sufficiency score" that corresponds to the weighted sum of the 30 indicators of well-being achievements of our measure, $c_i = w_1 I_{i1} + w_2 I_{i2} + \cdots + w_{30} I_{i30}$, where w_j is the weight of indicator j we explained above. This sufficiency score takes values between 0 and 100%. We use this score to identify which individual is living in a situation of well-being sufficiency or if they are living in well-being insufficiency. To conduct this identification process, we set a multidimensional well-being sufficiency threshold, k, which we define as the number of weighted dimensions in which an individual should exhibit adequate achievements to be considered as living in a multidimensional well-being sufficiency.

We choose an urban-specific value of k based on information on self-reported freedom from our urban sample. We first ask survey respondents to answer the following question in Samman's (2007: p. 465) review of psychological and subjective well-being indicators: "How true is the following statement for you? I feel free to decide for myself how to lead my life." Response options were "not at all true," "somewhat true," "fairly true," and "completely true." This question is oriented to capture one aspect of human agency/autonomy. It was initially included by Ryan and Deci (2000) in a set of short-form scales that are intended to measure the "basic psychological needs" comprised in their "self-determination theory." Then, we define the value of the threshold, k, as the average sufficiency score (c_i) for respondents who answered "fairly true" or "completely true" to the self-reported freedom question we mentioned above. Following this procedure, we estimate that $k = 59.2\%$, which means that an individual is identified as living in well-being sufficiency if they show adequate levels of achievement in more than seven out of the 12 weighted dimensions of our measure (or in a combination of well-being achievement indicators with an equivalent sum of weights). We argue that choosing a value of the parameter k using the information on self-reported freedom is also consistent with the capability-based agentic perspective proposed by Deneulin et al. (2017) to approach urban integration.

Once every individual in our sample is identified as living in well-being sufficiency or insufficiency, we use an AF method-based family of measures to estimate aggregate

Table 3.1 Dimensions, indicators of well-being achievement, and sufficiency criteria

Dimensions	Indicators	Sufficiency criteria (adequate level of achievement)
Health	Medical attention	The individual suffered an illness but got medical attention or, if they were not sick, they claim they would be able to get medical attention
	Health insurance	The individual has some type of medical insurance, excluding the non-contributory basic public health insurance
Education	Educational attainment	The individual has completed at least 11 years of basic education
	School attendance	The individual lives in a household in which all members aged 6–17 years old have completed school or are attending school
Employment and social security	Employment	The individual is not unemployed nor underemployed, they receive a payment and work in safe conditions
	Social security	The individual is aged 18–64, they are affiliated to a pension system, and regularly contribute to said system. If the individual is aged 65 or older, they receive a pension or claim that they will receive a pension if they decide to retire
Housing	Housing materials	The individual lives in a dwelling with adequate floor, walls, and roof materials
	Housing tenure	The individual has a house title which is legally recognized by the Government or rents a house whose owner has a house title legally recognized by the Government
	Overcrowding	The individual lives in a house in which the number on inhabitants per room (excluding the restroom, the kitchen, the hallway, and the garage) is lower than three
Basic facilities	Water and sanitation	The individual lives in a house that is connected both to the public water network as well as to the public sewerage network
	Energy	The individual lives in a house that has 24-h electricity supply every day. Also, the fuel they use for cooking is not polluting
	Communication technology	The individual has both access to at least some type of telephone, as well as access to the Internet
Social connectedness	Financial support	The individual manifests that they could get financial assistance from family or friends in case of need

(continued)

Table 3.1 (continued)

Dimensions	Indicators	Sufficiency criteria (adequate level of achievement)
	Care support	The individual manifests that, in case of need, they could rely on other trustworthy people who could take care of their children, older parents or family members suffering illnesses
	Emotional support	The individual manifests that they can count on other people to talk about personal issues
Living without shame and humiliation	Freedom from discrimination	The individual manifests that they have not feel discriminated by other people over the last three months prior to the survey
	Freedom from shame	The individual manifest that never or only in rare occasions they have felt shamed, ridiculed, repressed/afraid to speak out or act, humiliated, stupid, treated like a child, or that people have made fun of them
	Fair and respectful treatment	The individual manifests that they always feel that other people threat them with respect and fairly
Environment	Pollution-free environment	The individual manifests that, over the last 12 months, they have not frequently experienced air pollution or bad smells, nor they have witnessed pollution on rivers, canals, lakes, accumulation of garbage on the streets, roads, pavements, or other public spaces
	Protection from environmental shocks	The individual manifests that, over the last 12 months, neither they nor any member of their household have been significantly affected by droughts, torrential rains, pests, floods, mudslides, frosts, or earthquakes
Empowerment	Power to improve personal situation	The individual manifests that it would be very easy or easy for them to make important changes to improve their personal situation if they would like to do it
	Power to improve family's situation	The individual manifests that it would be very easy or easy for them to make important changes to improve their family situation if they would like to do it
	Power to improve community's situation	The individual manifests that it would be very easy or easy for them to make important changes to improve their community situation if they would like to do it
Sleep and recreation	Sleep	The individual manifests that, over the last week, they have slept at least seven hours every day

(continued)

Table 3.1 (continued)

Dimensions	Indicators	Sufficiency criteria (adequate level of achievement)
	Free time and recreational activity	The individual manifests that, over the last week, they have had at least one hour of leisure time every day
Security	Freedom from victimization	The individual manifests that, over the last 12 months, they have not been victim of robberies or violent assaults
	Community safety	The individual manifests that, over the last 12 months, the neighborhood in which they live their neighbors have not been robbed nor assaulted, and there have not been shootings or street fights
	Freedom from the fear of crime	The individual manifests that, over the last 12 months, their fear of being a victim of crime have not prevented them from conducting activities such as going out alone at night, leaving their house empty at night, sending their children to school, wearing certain clothes or items, among other daily activities
Citizenship and governance	Trust in institutions	The individual trusts in the local municipality, the provincial municipality, the regional government, the police, the army, the parliament, the law courts, and the presidency of the republic
	Perception of civil liberties	The individual considers that there is a sufficient degree of freedom of speech, press, political expression, and respect for human rights

Source Clausen and Barrantes (2022)

levels of well-being in urban areas. The first is the "incidence of multidimensional well-being sufficiency," also called "headcount ratio of well-being sufficiency,"H, which reflects the percentage of people living in well-being sufficiency. The second is the "average level of adequate achievements for people in well-being insufficiency," A, which is the average value of the individual sufficiency score (c_i) of people whose sufficiency score is below the multidimensional well-being threshold (k). The third is the "Multidimensional Well-being Index" (MWI) or "adjusted multidimensional well-being headcount ratio," which, following Alkire and Kovesdi (2020) and Clausen and Barrantes (2022), can be defined as $\text{MWI} = H + (1 - H) * A$, ranging between 0 and 1. Higher values of this adjusted multidimensional well-being headcount ratio reflect higher aggregate well-being levels, which, in turn, could indicate a higher proportion of people living in well-being sufficiency or a higher average value of the sufficiency score of people living in well-being insufficiency. In the Results section, we present our estimates for H, A and the MWI for the Peruvian urban

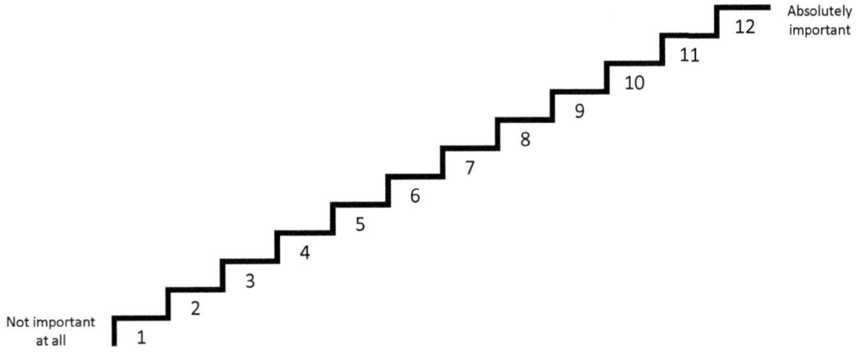

Fig. 3.3 Survey question on the importance each respondent assigns to the dimension of "security." On a scale of 1 ("not important at all") to 12 ("absolutely important"), how important are the following aspects to you? [Living in an environment without street and domestic violence]. *Source* 2018 IOP and IDHAL's survey

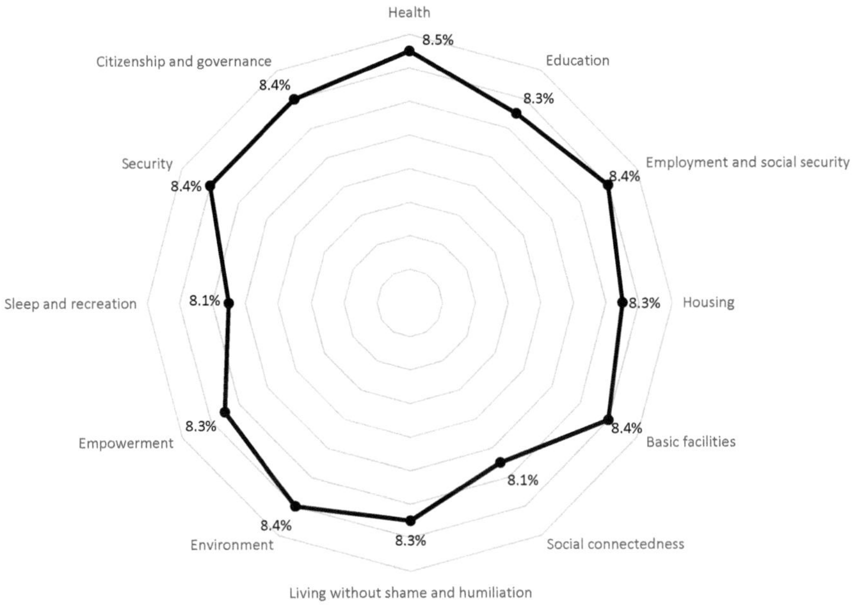

Fig. 3.4 Weights of each dimension based on urban population values. *Source* Authors' calculations based on 2018 IOP and IDHAL's survey

population. Since the AF method allows us to estimate the contribution of these subgroups to the adjusted multidimensional well-being headcount ratio, we also disaggregate our results within urban populations according to gender (men/women) and between four geographical domains (Lima [Peruvian capital city]/urban Coast/urban Highlands/urban Amazon).

The adjusted multidimensional well-being headcount (or MWI) can be decomposed to calculate an "adjusted multidimensional ill-being headcount," also called M_0. Following Alkire (2016), this latter measure of ill-being reflects levels of well-being insufficiency and can be expressed as $M_0 = 1 - \text{MWI}$. The M_0 measure also corresponds to the proportion of people who do not meet the multidimensional well-being sufficiency threshold (multidimensional ill-being headcount ratio) multiplied by the weighted average number of indicators in which this group could not achieve sufficiency. Inasmuch M_0 is also based on the AF method, it can be broken down to identify the dimensions that contribute the most to multidimensional ill-being or multidimensional well-being insufficiency. The dimensional breakdown property is particularly relevant from a policymaking perspective since it provides useful information to prioritize interventions aimed at promoting achievements in the dimensions that contribute the most to ill-being, which, in turn, can increase the proportion of people living in a situation of well-being sufficiency. Below we present each dimension's percentage contribution to the "adjusted multidimensional ill-being headcount."

3.5 Results

In Table 3.2, we present a dashboard showing the incidence of achievements in each one of the 30 indicators of well-being we include in our multidimensional analysis. In particular, Table 3.2 focuses on achievement disparities between men and women in urban areas and disparities between Lima and urban areas in the coastal, highland, and Amazonian domains. Table 3.2 shows that the total urban population in Peru has an incidence of achievement lower than 50% in 10 out of the 30 indicators of the dashboard: "trust in institutions" (1.6%), "pollution-free environment" (12.5%), "perception of civil liberties" (19.9%) "freedom from the fear of crime" (31.3%), "employment" (25.3%), "social security" (28.4%), "fair and respectful treatment" (38.1%), "health insurance" (42%), and "free time and recreational activities" (45.3%). Additionally, we identify gender disparities within urban areas that situate women in a status of disadvantage in three out of these ten indicators: "social security," "freedom from the fear of crime," and "health insurance." Nevertheless, we also find gender disparities in other dimensions that are not among the ten indicators with lower levels of achievement. These correspond to "educational attainment" and "freedom from shame."

Results in Table 3.2 also reveal disparities between urban geographical domains. Lima is in a situation of relative advantage in seven indicators: "perception of civil liberties," "protection from environmental shocks," "communication technology," "water and sanitation," "housing materials," "social security," and "health insurance." However, Lima is clearly disadvantaged in other indicators such as "freedom from victimization" and "community safety."

Figure 3.5 presents information on the incidence of multidimensional well-being sufficiency, which corresponds to the proportion of individuals who have adequate

Table 3.2 Incidence of achievements by indicator; urban population and subgroups according to gender and geographic domains

Dimensions	Indicators	Urban (%)	Gender		Geographic domains			
			Men (%)	Women (%)	Lima (capital city) (%)	Urban coast (%)	Urban highlands (%)	Urban Amazon (%)
Health	Medical attention	74.1	72.2	75.8	76.0	70.2	73.9	72.3
	Health insurance	42.0	44.7	39.5	50.3	31.9	35.3	30.4
Education	Educational attainment	78.0	82.3	74.1	82.3	70.7	80.9	64.0
	School attendance	96.0	95.8	96.1	95.0	97.7	96.4	96.5
Employment and social security	Employment	25.3	28.3	22.6	27.3	21.0	25.3	25.0
	Social security	28.4	36.4	21.1	34.3	23.4	22.4	17.3
Housing	Housing materials	84.8	84.4	85.1	92.0	75.2	79.9	74.5
	Housing tenure	64.1	64.4	63.7	65.0	72.0	57.2	51.3
	Overcrowding	88.6	91.6	85.8	88.3	90.2	89.8	83.5
Basic facilities	Water and sanitation	61.4	60.9	61.9	78.3	45.0	44.8	32.7
	Energy	95.5	94.8	96.2	98.0	94.2	95.6	82.1
	Communication technology	45.5	46.9	44.2	56.0	36.0	35.6	24.7
Social connectedness	Financial support	76.9	78.2	75.6	77.8	73.2	79.2	75.8
	Care support	77.4	79.2	75.7	79.3	76.0	74.0	76.6
	Emotional support	86.0	87.4	84.8	86.8	85.8	84.4	85.3
Living without shame and humiliation	Freedom from discrimination	89.4	89.6	89.3	89.0	89.5	88.8	93.1
	Freedom from shame	58.3	61.7	55.1	56.5	58.6	57.4	71.4
	Fair and respectful treatment	38.1	37.3	38.9	38.3	42.4	30.1	43.6
Environment	Pollution-free environment	12.5	11.7	13.3	14.5	5.2	12.2	20.8
	Protection from environmental shocks	64.9	64.5	65.3	84.5	45.2	39.8	46.5

(continued)

Table 3.2 (continued)

Dimensions	Indicators	Urban (%)	Gender		Geographic domains			
			Men (%)	Women (%)	Lima (capital city) (%)	Urban coast (%)	Urban highlands (%)	Urban Amazon (%)
Empowerment	Power to improve personal situation	91.5	93.1	90.0	94.0	85.1	91.2	93.4
	Power to improve family's situation	90.1	90.9	89.4	94.5	82.1	87.4	89.8
	Power to improve community's situation	73.0	73.2	72.8	75.3	70.4	67.1	78.4
Sleep and recreation	Sleep	56.9	53.4	60.2	57.8	58.6	48.7	65.6
	Free time and recreational activity	45.3	45.8	44.9	42.5	44.4	46.8	63.6
Security	Freedom from victimization	65.0	64.5	65.5	59.5	67.7	74.1	74.1
	Community safety	53.4	54.4	52.6	50.3	49.3	64.4	61.1
	Freedom from the fear of crime	31.3	37.0	26.0	36.0	22.2	20.7	49.4
Citizenship and governance	Trust in institutions	1.6	1.9	1.3	1.5	2.0	0.8	2.7
	Perception of civil liberties	19.9	21.9	18.1	24.3	14.6	14.4	18.6

Source Authors' calculations based on 2018 IOP and IDHAL's survey

levels of achievement in at least seven out of the 12 dimensions of our measure. Overall, 46.7% of people living in urban areas live in multidimensional well-being sufficiency. These results also reveal gender disparities insofar as the proportion of men in a situation of well-being sufficiency (49.3%) is higher than that of women (44.2%). Additionally, our findings reveal that Lima, the capital city, presents a significantly higher incidence of multidimensional well-being (55.0%) compared to the urban Amazon (41.1%), the urban Highlands (38.4%), and the urban Coast (35.5%). Both the average level of achievements for people in well-being insufficiency (see Fig. 3.6) and the adjusted multidimensional well-being headcount (see Fig. 3.7) also show the abovementioned patterns of gender and geographical disparities.

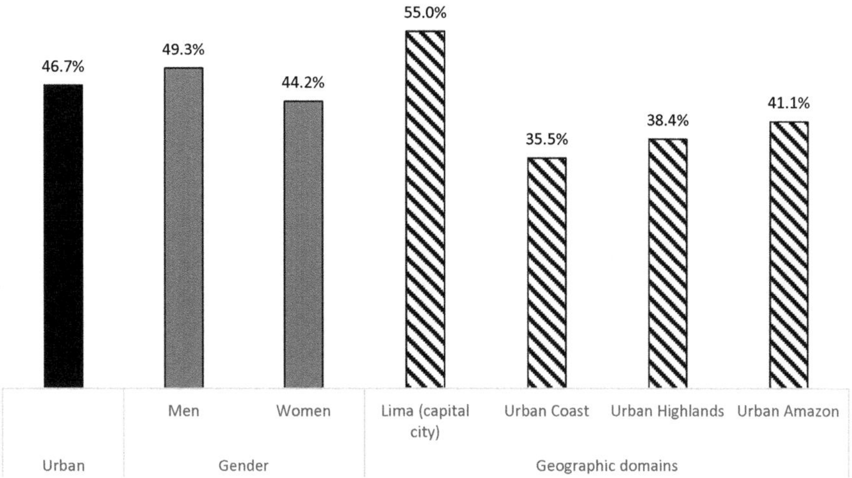

Fig. 3.5 Incidence of multidimensional well-being sufficiency; urban population and subgroups according to gender and geographic domains. *Source* Authors' calculations based on 2018 IOP and IDHAL's survey

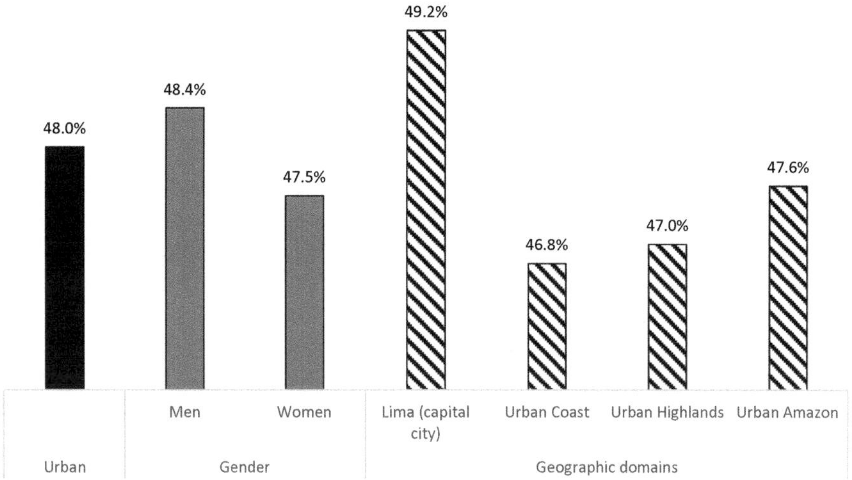

Fig. 3.6 Average level of achievements for people in well-being insufficiency; urban population and subgroups according to gender and geographic domains. *Source* Authors' calculations based on 2018 IOP and IDHAL's survey

In Fig. 3.8, we decompose the adjusted multidimensional well-being headcount into different subgroups according to gender and geographical domains. We then compare the contribution of each subgroup to the adjusted multidimensional well-being headcount ratio with their share of the population. We find a pattern of slight overrepresentation in terms of gender since the contribution of men to the adjusted

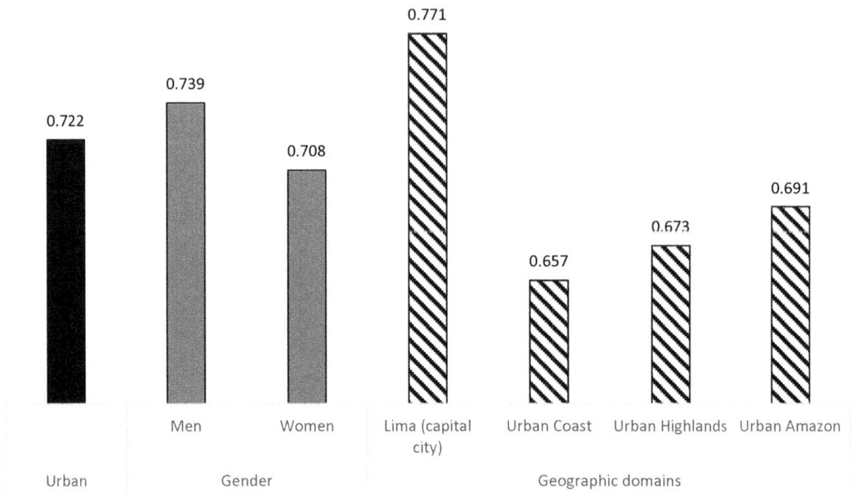

Fig. 3.7 Adjusted multidimensional well-being headcount; urban population and subgroups according to gender and geographic domains. *Source* Authors' calculations based on 2018 IOP and IDHAL's survey

multidimensional well-being headcount ratio is marginally higher than the share of the population they represent. Similarly, find evidence of a moderate pattern of overrepresentation in terms of geographical domains since the contribution of the population living in Lima to the adjusted well-being headcount ratio is higher than their share of the population.

One potential criticism of our results is that the abovementioned disparities might be contingent on the multidimensional well-being sufficiency threshold, k, that we selected based on information about the degree of self-perceived freedom of the individuals in our sample. In Figs. 3.9 and 3.10, we show information on the adjusted multidimensional well-being headcount values for different subgroups using a range of alternative values for the sufficiency threshold. We find that gender and geographical disparities are robust to different parameter values, meaning that disparities remain even if we change the parameter we use to determine if a person lives in a situation of well-being sufficiency.

In Fig. 3.11, we take advantage of the property of dimensional breakdown to measure the contribution of lack of achievement in each dimension to the overall situation of ill-being. The dimensions that contribute the most to the situation of well-being insufficiency at the urban national level are: "citizenship and governance" (15.2%), "employment and social security" (14.2%), "environment" (11.6%), "security" (9.8%), and "health" (9.6%). These results seem to reflect a situation that Deneulin et al. (2017), based on Wolff and de-Shalit (2007), would refer to as clustered disadvantages. According to these findings, urban populations living in well-being insufficiency likely have inadequate jobs, lack access to health insurance, and poor health conditions. In addition, they live in polluted and insecure areas and

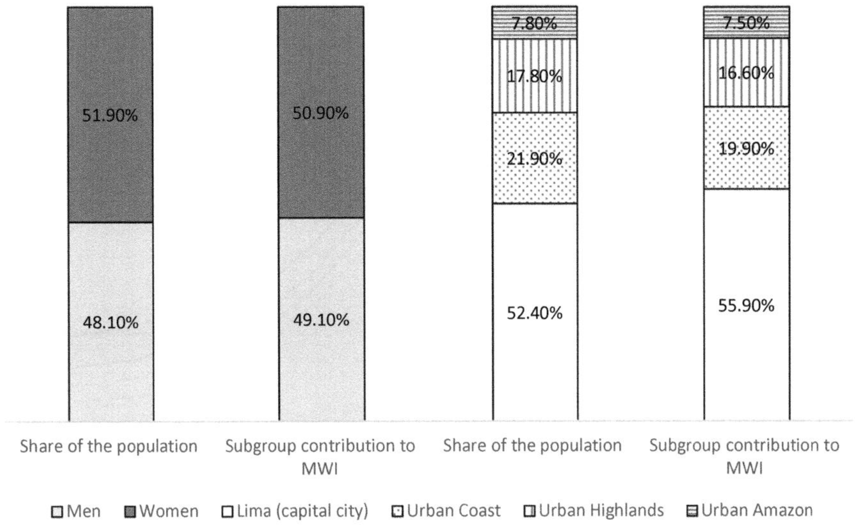

Fig. 3.8 Contribution of urban subgroups according to gender and geographic domains to the adjusted multidimensional well-being headcount. *Source* Authors' calculations based on 2018 IOP and IDHAL's survey

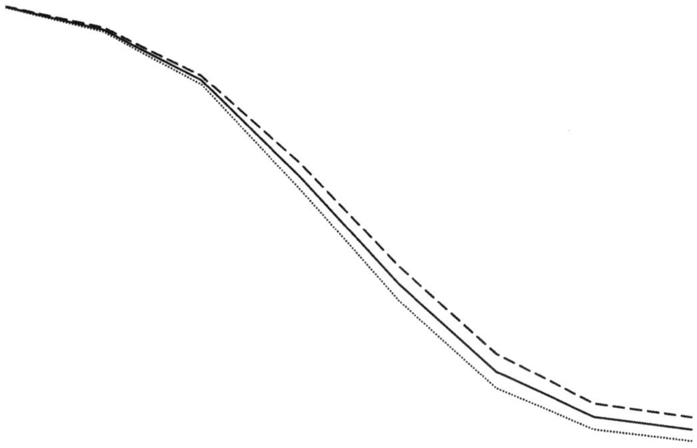

	3/12	4/12	5/12	6/12	7/12	8/12	9/12	10/12
——— Urban	0.996	0.974	0.924	0.830	0.725	0.639	0.595	0.582
— — - Men	0.996	0.976	0.929	0.843	0.742	0.656	0.608	0.594
··········· Women	0.996	0.972	0.920	0.818	0.709	0.623	0.583	0.571

Fig. 3.9 Adjusted multidimensional well-being headcount; *k* from 3/12 to 10/12; urban population and subgroups according to gender. *Source* Authors' calculations based on 2018 IOP and IDHAL's survey

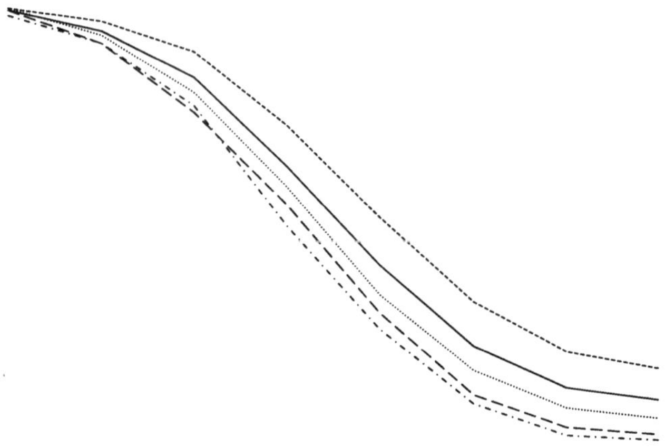

	3/12	4/12	5/12	6/12	7/12	8/12	9/12	10/12
Urban	0.996	0.974	0.924	0.830	0.725	0.639	0.595	0.582
Lima	0.998	0.984	0.952	0.874	0.776	0.686	0.633	0.616
Urban Coast	0.990	0.961	0.895	0.767	0.657	0.578	0.544	0.539
Urban Highlands	0.996	0.961	0.887	0.789	0.674	0.587	0.553	0.545
Urban Amazon	0.998	0.969	0.909	0.808	0.693	0.613	0.574	0.563

Fig. 3.10 Adjusted multidimensional well-being headcount; *k* from 3/12 to 10/12; urban population and subgroups according to geographic domains. *Source* Authors' calculations based on 2018 IOP and IDHAL's survey

do not trust public institutions. These results are not particularly different for men and women, yet comparing geographical domains shows some relevant disparities. The contribution of the dimensions of "environment" and "basic facilities" to the adjusted headcount of multidimensional ill-being in Lima is lower than in the Coast, the Highlands, and the Amazonia urban areas. Conversely, the contribution of insufficiency in the dimensions of "security," "living without shame and humiliation," and "sleep and recreation" is more prominent in Lima than in other urban domains. This heterogeneity indicates that how urban population experience ill-being (or insufficiency of well-being) varies depending on the geographical domain in which they live. Policies aimed at reducing well-being disparities between geographical domains should incorporate this information to prioritize achievements in the dimensions that contribute the most to the adjusted multidimensional ill-being headcount ratio in each domain, respectively.

3.6 Analysis

In this chapter, we relied on Amartya Sen's capability approach to evaluate multidimensional well-being disparities in urban areas in Peru, a Latin American middle-income country. Specifically, we partially relied on the capability-based framework

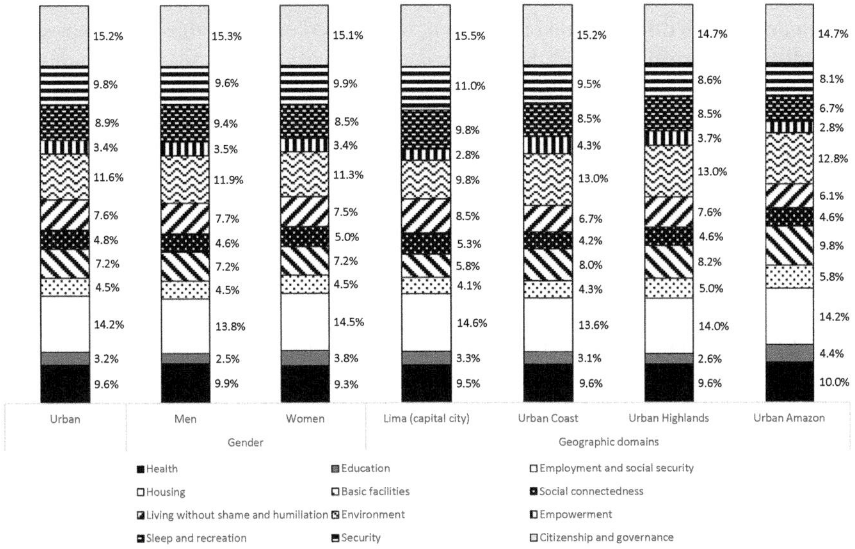

Fig. 3.11 Contribution of dimensions to the adjusted multidimensional ill-being headcount; urban population subgroups according to gender and geographic domains. *Source* Authors' calculations based on 2018 IOP and IDHAL's survey

for urban integration developed by Deneulin et al. (2017), which the authors originally applied to the case of Buenos Aires (Argentina). We conducted our empirical analysis using an urban-specific version of the MWI-P (Clausen and Barrantes 2022) which we calibrated and estimated using data from a specialized survey collected in urban areas in Peru in late 2018. To the best of our knowledge, this is the first study focusing on a middle-income country that used an individual-level comprehensive metric of well-being to explore both multidimensional gender disparities within urban areas as well as multidimensional disparities between urban geographical domains.

Our main results revealed disparities in a range of dimensions of well-being that place women living in urban areas in Peru in a situation of relative disadvantage compared to men. Particularly, we found that some of these disparities correspond to indicators not conventionally included in the literature on urban development but consistent with pioneering proposals such as the set of dimensions selected by Tonon (2016) to explore the urban quality of life in Latin America. Such is the case of "freedom from shame," linked to community relations, and "freedom from the fear of crime," related to personal security. In addition, we found that the proportion of women in a situation of well-being sufficiency is lower than that of men and that the capital city of Lima shows higher values of well-being sufficiency than other urban regions in the Coast, the Highlands, and the Amazonia. This pattern of disparity held valid even if we introduced changes to the multidimensional well-being sufficiency threshold, which corresponds to the minimum number of achievements a person should accomplish to live with an adequate level of well-being. Additionally, based on our analysis of dimensional breakdown, we showed evidence that the composition of well-being insufficiency in these four geographical domains was heterogeneous, which calls for adopting flexible and context-dependent well-being-enhancing policy strategies. Our findings are also consistent with the analysis of Anand (2018), inasmuch poor achievements in "security" and "environment" are part of the group of dimensions that contribute the most to the adjusted multidimensional ill-being headcount ratio of Lima, the only megacity in Peru.

Like all research, this study is subject to some limitations. First, although we relied on a well-being-specialized dataset including a broad set of indicators, our data's sample size, and characteristics did not allow us to conduct a georeferenced spatial analysis of urban disparities within each geographical domain. Future research could benefit from collecting georeferenced information focusing on specific cities to further explore multidimensional spatial disparities within urban areas. Second, throughout this chapter, we have consistently used the term "multidimensional disparities" instead of "multidimensional inequalities" because our empirical strategy mainly relied on identifying gender multidimensional well-being gaps within urban domains or multidimensional well-being gaps between urban geographical domains. Evaluating multidimensional inequalities would involve relying on distribution-sensitive measures. However, as Campo Seth and Santos (2020) acknowledge, applying these kinds of measures to the multidimensional space of functionings and capabilities involves a series of challenges that further methodological research should address. Third, due to data constraints, our analysis was restricted to people aged 18 or older. This means our results did not provide information on well-being disparities among children and adolescents. This limitation could be overcome by designing and conducting well-being-specialized surveys focused on this group of the urban population, as well as protocols to collect said information in an age-relevant and ethical way.

3.7 Conclusion

We began this chapter by recognizing that high-income inequality and disparities have accompanied the rapid urbanization that most Latin American countries have experienced.

Our research showed that said disparities also manifest in other non-income dimensions of well-being that are not conventionally measured due to data limitations but are important for urban populations.

A comprehensive approach to urban development policies in this region could significantly benefit from adopting a broad understanding of well-being that, instead of relying only on information about what people who live in urban areas have, focuses on what they are actually able to do and to be in a broad set of life domains.

References

Alkire S (2016) The capability approach and well-being measurement for public policy. In: Adler M, Fleurbaey M (eds) The Oxford handbook of well-being and public policy. Oxford University Press

Alkire S, Foster J (2011) Counting and multidimensional poverty measurement. J Public Econ 95(7–8):476–487. https://doi.org/10.1016/j.jpubeco.2010.11.006

Alkire S, Kovesdi F (2020) A birdseye view of well-being: exploring a multidimensional measure for the United Kingdom. OPHI Research in Progress 60a, vol 3. Oxford Poverty and Human Development Initiative (OPHI), University of Oxford

Anand PB (2018) Cities and the capability approach. In Comim F, Anand PB, Fennell S (eds) New frontiers of the capability approach. Cambridge University Press, pp 519–546. https://doi.org/10.1017/9781108559881.022

Angotti T, Irazábal C (2017) Planning Latin American cities: dependencies and "Best Practices", vol 44. SAGE Publications Sage CA, Los Angeles, CA, pp 4–17

Bilal U, Alazraqui M, Caiaffa WT, Lopez-Olmedo N, Martinez-Folgar K, Miranda JJ, Rodriguez DA, Vives A, Diez-Roux AV (2019) Inequalities in life expectancy in six large Latin American cities from the SALURBAL study: an ecological analysis. Lancet Planetary Health 3(12):e503–e510. https://doi.org/10.1016/S2542-5196(19)30235-9

Born D, Colamarco V, Delamónica E, Minujín A (2019) South American children's quality of life: intra-urban disparities along life-cycle indicators. Appl Res Qual Life 14(3):799–817. https://doi.org/10.1007/s11482-018-9607-2

Boyd C, Vásquez T, Yancari J (2021) Migración interna y desarrollo humano en el Perú pos-COVID-19. In: Iguíñiz J, Clausen J (eds) COVID-19 & Crisis de Desarrollo Humano en América Latina. IDHAL. Universidad Católica del Perú, pp 447–461

Centre for Bhutan Studies and GNH Research (2016) A compass towards a just and harmonious society: 2015 GNH survey report. Centre for Bhutan Studies and GNH Research

Chancel L, Piketty T, Saez E (2022) World inequality report 2022. Harvard University Press

Clausen J (2020) Waking up from the middle-income country dream: COVID-19 and the human development crisis in Peru. J Ciencias Sociales 8:205–210. https://doi.org/10.18682/jcs.vi15.4314

Clausen J, Barrantes N (2021) ¿Cómo se asocian el riesgo multidimensional y los efectos de la COVID-19? Evidencia a nivel distrital para las provincias de Lima y el Callao en Perú. In: Iguíñiz J, Clausen J (eds) COVID-19 & Crisis de Desarrollo Humano en América Latina. Instituto de Desarrollo Humano de América Latina de la Pontificia Universidad Católica del Perú, pp 111–132

Clausen J, Barrantes N (2022) Developing a comprehensive multidimensional wellbeing index based on what people value: an application to a middle-income country. Appl Res Qual Life. https://doi.org/10.1007/s11482-022-10064-w

Dammert Guardia M, Delgadillo V, Erazo J (2019a) Bibliografía sobre América Latina: nuevas desigualdades urbanas. Andamios 16(39):255–262

Dammert Guardia M, Delgadillo V, Erazo J (2019b) La ciudad, espacio de reproducción de las desigualdades. Andamios 16(39):7–13

Deloitte, Fundación Corona, Red Colombiana de Ciudades Cómo Vamos, & Social Progress Imperative (2022) Reporte Metodológico y de Resultados del Índice de Progreso Social (IPS) para 15 Ciudades de Colombia 2016–2020

Deneulin S, Lépore E, Mitchell A, Suárez AL (2017) A framework for urban integration: the case of Buenos Aires. In: Capability-promoting policies. Policy Press, pp 111–130

Deneulin S, Sánchez-Ancochea D (2018) Urban inequality, youth and social policy in Latin America: introduction to special section. Oxf Dev Stud 46(1):3–9. https://doi.org/10.1080/13600818.2017.1383375

Espinoza A, Fort R, Espinoza M (2022) Reorganizar el Perú: ciudades intermedias y desarrollo. In: Balarin M, Cueto S, Fort R (eds) El Perú Independiente: Ensayos para un desarrollo con bienestar. Grupo de Análisis para el Desarrollo Económico, pp 287–310
Frediani AA (2007) Amartya Sen, the World Bank, and the redress of urban poverty: a Brazilian case study. J Hum Dev 8(1):133–152. https://doi.org/10.1080/14649880601101473
Frediani AA (2015) Space and capabilities: approaching informal settlement upgrading through a capability perspective. In: The city in urban poverty. Springer, pp 64–84
Instituto Nacional de Estadística e Informática (INEI) (2017) Evolución de la pobreza monetaria 2007–2016. Informe Técnico. INEI
Instituto Nacional de Estadística e Informática (INEI) (2018) Perú: Perfil Sociodemográfico. Informe Nacional. Censos Nacionales 2017: XII de Población, VII de vivienda y III de Comunidades Indígenas. INEI
Instituto Nacional de Estadística e Informática (INEI) (2019) Evolución de la pobreza monetaria 2007–2018. Informe Técnico. INEI
Instituto Nacional de Estadística e Informática (INEI) (2022) Evolución de la pobreza monetaria 2010–2021. Informe Técnico. INEI
Leite C, Acosta C, Militelli F, Jajamovich G, Wilderom M, Bonduki N, Herling T (2020) Social urbanism in Latin America. In: Acosta LC, Militelli F, Jajamovich G, Wilderom M, Bonduki N, Somekh N, Herling T (eds) Social urbanism in Latin America: cases and instruments of planning, land policy and financing the city transformation with social inclusion, pp 3–21. https://doi.org/10.1007/978-3-030-16012-8_1
Linares S, Mikkelsen CA, Velázquez GA, Celemín JP (2016) Spatial Segregation and quality of life: empirical analysis of medium-sized cities of Buenos Aires Province. In: Tonon G (ed) Indicators of quality of life in Latin America. Springer International Publishing, pp 201–218. https://doi.org/10.1007/978-3-319-28842-0_8
Lucero PI, Mikkelsen CA, Ares SE (2016) How are we today? Quality of urban life in Argentina first contributions from the households permanent survey (2003–2012). In: Tonon G (ed) Indicators of quality of life in Latin America. Springer International Publishing, pp 81–108. https://doi.org/10.1007/978-3-319-28842-0_4
Martinez J (2016) Mind the gap: monitoring spatial inequalities in quality of life conditions (case study of Rosario). In: Tonon G (ed) Indicators of quality of life in Latin America. Springer International Publishing, pp 151–172. https://doi.org/10.1007/978-3-319-28842-0_6
Moran MR, Rodríguez DA, Cotinez-O'Ryan A, Miranda JJ (2020) Park use, perceived park proximity, and neighborhood characteristics: Evidence from 11 cities in Latin America. Cities 105:102817. https://doi.org/10.1016/j.cities.2020.102817
Population Division of the Department of Economic and Social Affairs of the United Nations (2019) World urbanization prospects 2018. Highlights. United Nations
Rigolon A, Browning MHEM, Lee K, Shin S (2018) Access to urban green space in cities of the global south: a systematic literature review. Urban Sci 2(3):67
Ryan RM, Deci EL (2000) Self-determination theory and the facilitation of intrinsic motivation, social development, and well-being. Am Psychol 55(1):68–78. https://doi.org/10.1037//0003-066x.55.1.68
Samman E (2007) Psychological and subjective well-being: a proposal for internationally comparable indicators. Oxf Dev Stud 35(4):459–486. https://doi.org/10.1080/13600810701701939
Sánchez A (2015) Migraciones Internas en el Perú. Organización Internacional para las Migraciones
Sen A (1993) Capability and well-being. In: Nussbaum M, Sen A (eds) The quality of life. Oxford University Press, pp 30–53. https://doi.org/10.1093/0198287976.003.0003
Sen A (2009) Capability: reach and limit. In: Chiappero-Martinetti E (ed) Debating global society: reach and limits of the capability approach. Fondazione Giangiacomo Feltrinelli, pp 15–28
Seth S, Santos ME (2020) Multidimensional inequality and human development. In: Chiappero-Martinetti E, Qizilbash M, Osmani S (eds) The Cambridge handbook of the capability approach. Cambridge University Press, pp 392–416

Tonon G (2016) Indicators of quality of life in Latin America. Springer. https://doi.org/10.1007/978-3-319-28842-0
White SC (2016) Introduction: the many faces of wellbeing. In: Cultures of wellbeing. Springer, pp 1–44. https://doi.org/10.1057/9781137536457_1
Wolff J, de-Shalit A (2007) Disadvantage. Oxford University Press. https://doi.org/10.1093/acprof:oso/9780199278268.001.0001

Chapter 4
Housing Inequalities in an International Financial Center: The Case of Hong Kong

Daniel T. L. Shek and Lindan Tan

Abstract While Hong Kong is an urban international financial center with high GDP per capita, it has been criticized as a city with inequalities in many areas, especially economic and housing inequalities. In this chapter, based on an integration of concepts on social inequalities, we briefly review political, cultural, and opportunity inequalities. The review suggests that although inequalities in these areas are not very serious, there is room for improvement. We then conduct an in-depth examination of economic and housing inequalities in Hong Kong. We found that income inequality has been worsening since the 1980s, resulting in roughly one-fifth of the Hong Kong people living in poverty. In particular, it is progressively harder for young people to have an upward social mobility. Besides, wealth inequalities are serious in Hong Kong. Regarding housing inequality, there are disturbing phenomena, including small living spaces, such as sub-divided flats and "cage homes". Low housing affordability and long waiting queue for public housing are also perpetual housing issues faced by the people of Hong Kong since 1997. The contributing factors and consequences of economic and housing inequalities on human development and the possible remedies are discussed.

Keywords Social inequality · Economic inequality · Housing inequality · Hong Kong · International financial center

---

D. T. L. Shek (✉) · L. Tan
Department of Applied Social Sciences, The Hong Kong Polytechnic University, Hung Hom, Hong Kong, People's Republic of China
e-mail: daniel.shek@polyu.edu.hk

L. Tan
e-mail: lindan.tan@polyu.edu.hk

G. H. Tonon (ed.), *Urban Inequalities*, The Urban Book Series,
https://doi.org/10.1007/978-3-031-59746-6_4

4.1 Introduction

According to Dworkin (2002), equality is regarded as "the endangered species of political ideals" (p. 1), which highlights the great difficulty in achieving equality in reality. With particular reference to Hong Kong which is a capitalistic society and an international financial center, urban inequalities are not remote. Although Hong Kong ranked fourth in the Human Development Index in 2021 (UNDP 2022) and has a huge fiscal reserve (HK$1,025.3 billion, roughly = US$130.6 billion) in January 2022, (HKSAR Government 2022), around 226,340 people lived in sub-divided flats in 2020 (Transport and Housing Bureau 2021b), with a disturbing number of underprivileged people renting "cage homes".

There are several sections in this chapter. First, based on a review of the concepts and models in inequality, we build up a conceptual framework on different types of inequality which guides our discussion. Second, we briefly review inequalities in the political, cultural, and opportunity domains. Third, we conduct an in-depth analysis of economic inequality and housing inequity in Hong Kong. Under economic and housing inequalities, we also discuss the contributing factors and consequences, as well as the possible remedies.

4.2 Definitions and Frameworks

Different academics and organizations have constructed different frameworks on the conception of social inequality: OECD (2022b) developed a model of inequality involving eight key elements; UNDESA (2020) introduced an inequality framework including "economic inequality" and "inequality of opportunity" (p. 21); Fraser (2009) used "socioeconomic distribution", "legal-cultural recognition", and "political representation" (p. 30) in defining inequality; Baker et al. (2006) framework focused on inequality in the "economic", "cultural", "political", and "affective" social systems (p. 418). A summary of different frameworks on inequality is presented in Table 4.1.

Conceptually speaking, different definitions of inequality almost overlap when a society is understood as a meso-concept that incorporates political, economic, cultural, and civic domains where several areas of inequality are identified. These include political (e.g., legal rights), cultural (e.g., access to education), opportunity (e.g., gender equality), and economic (e.g., income distribution) inequalities (see Fig. 4.1). In this chapter, our review of urban inequalities in Hong Kong focuses on these four areas.

Table 4.1 Different conceptual frameworks on social inequality

Framework	Domains	Sub-domains
Fraser's framework of social justice	Maldistribution	Socioeconomic structures
	Misrecognition	Cultural relations
	Misrepresentation	Political structures
Baker et al.'s equality system framework	Economic system	Goods and services
	Cultural system	Cultural practices and products
	Political system	Collectively binding decisions
	Affective system	Relationships of love, care, and solidarity
UNDESA's inequality framework	Economic inequality	Income inequality; labor; and capital inequality
	Inequality of opportunity	Group-based disparities in poverty; health; education; employment
OECD's inequality framework	Income inequalities	Income
	Gender inequalities	Education; employment; entrepreneurship
	Inequality and tax	Transparency; tax compliance
	Health inequalities	Living and working conditions; behavioral factors; access to and quality of health care
	Education inequalities	Education investment
	Inequalities and innovation	Social inclusiveness; industrial inclusiveness; territorial inclusiveness
	Regional disparities	Urban advantage
	Well-being inequalities	Housing, work-life balance, civic governance, environmental quality; subjective well-being, etc

Sources Authors' own based on information from Baker et al. (2006), Fraser (2009), OECD (2022b), UNDESA (2020)

4.3 A Brief Overview of Social Inequalities in Hong Kong

4.3.1 Political Inequality in Hong Kong

Since 1997, Hong Kong SAR (HKSAR) has had "a high degree of autonomy and enjoy executive, legislative and independent judicial power" ("Basic Law" 1990 Article 2) under "One Country, Two Systems" (OCTS). The "Basic Law" (BL) was enacted to safeguard "the previous capitalist system and way of life" (BL Article 5), and the political equality reflected in the participation right and legal just. In more detail, as equality of participation in politics is reflected in the equal right of each

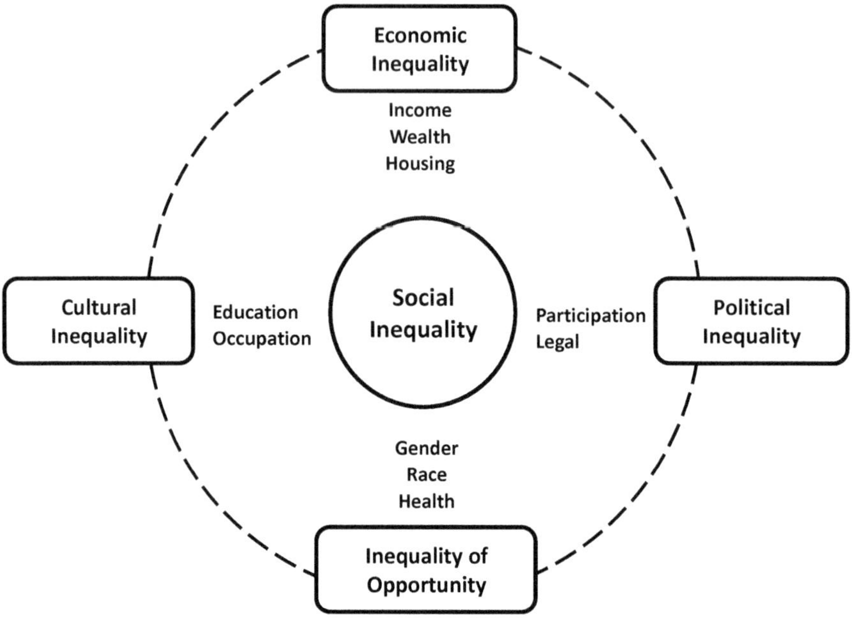

Fig. 4.1 Inequality framework and domains. *Source* Authors' own

individual to influence policymaking, the right of "universal suffrage" (BL Article 45) as the ultimate aim is an excellent manifestation of it. Politically speaking, people in Hong Kong have enjoyed more political rights compared to the pre-1997 period. Under British rule, the Hong Kong governor was nominated by the British government without consulting the people of Hong Kong. Under the Hong Kong SAR, the chief executive is elected by the Hong Kong people via the Election Committee and appointed by the Central Government. Besides, since the promulgation of the "District Councils Ordinance" (Cap. 547) in 1999, District Council elections have been operated in accordance with Western democratic practices, where all District Council members (except for a few ex-officio members) were elected directly by the electorate on universal suffrage (District Councils Ordinance 1999). Unfortunately, with the social unrest taking place from 2019 to 2020, there has been a change in the election system in Hong Kong (e.g., the "Improving Electoral System" initiative) in 2020. While this has been criticized as a "regression" of the democratic elements in the electoral system in Hong Kong, some counter-argue that this is a necessity in view of the political chaos in Hong Kong in 2020, which underscores the need to have "patriots" in the legislative system. Besides, some also counter-argue that Western democracy is not the only democratic system in the world, and democracy has to take the cultural and social contexts into account. Furthermore, the rise of post-Western world order is accompanied by calls for non-Western forms of democracy (Youngs 2015).

Regarding legal equality, the legal system in Hong Kong follows the Common Law system, which enjoys a high reputation internationally. According to the World Justice Project Rule of Law Index (WJP 2021), Hong Kong ranked the 19th out of 139 economies for its satisfactory performance in eight domains, including "constraints on government powers", "absence of corruption", "order and security", and "civil justice", among others (p. 9). Besides, although legal charges are very high in Hong Kong, the government provides legal aid to those who have financial difficulty in lawsuits. In short, Hong Kong has a very sound Common Law legal system in which Hong Kong people enjoy a high level of legal rights.

4.3.2 Cultural Inequality in Hong Kong

Hong Kong is prestigious for its outstanding educational accessibility and quality. Education is compulsory and free for students at government and government-aided primary and secondary schools, and a minimal fee charged in higher education. The World Economic Forum ranked Hong Kong 14th out of 137 economies in the pillar of "higher education and training" (P. 141) in 2017 with a net primary education enrollment rate of 94.6% and the gross tertiary education enrollment rate of 68.5% (Schwab and World Economic Forum 2017). However, there are achievement gaps between kids based on their family's income and Chinese proficiency among ethnic minorities (Oxfam 2018). Besides, Hong Kong is faced with a highly unbalanced career distribution since the bulk of the manufacturing industry was relocated north and replaced by the financial economy, which has grown to be one of the main economic pillars, accounting for 21.2% of GDP, while agriculture, fishing, mining, and quarrying accounted for only 0.1% in 2019 (C&SD 2021c).

4.3.3 Inequality of Opportunity in Hong Kong

Hong Kong was categorized as Group 2 of five groups, meaning medium–high equality in Gender Development Index (GDI) achievements while 63 economies, including Norway, Singapore, United States are classified as Group 1, indicating high equality in GDI in 2019 (UNDP 2020). Although Hong Kong has moved up to Group 1 in 2021 in the recently issued report (UNDP 2022), it is still inadequate. For example, the gap between the median monthly employment earnings of employed men and women in Hong Kong has increased significantly from HK$3,500 in 2008 to HK$5,000 in 2020 (C&SD 2021f). Nevertheless, females in Hong Kong have a comparatively high and improving status in comparison with other Asian societies. In Hong Kong, the general consensus tends to be that gender is inconsequential (Schnurr and Mak 2011), and females are even more "favored" in careers and family. Although there is a gender-based wage gap in Hong Kong, the disparity between men and women in the corporate context stems from "extra-firm mobility" rather than "intra-firm gender discrimination" (Lam and Dreher 2004, p. 791). Besides, gender discrimination is prohibited in the "Sex Discrimination Ordinance" (Cap. 480) and the Equal Opportunities Commission (EOC) was established in 1996.

Ethnic minorities make up about 8% (or 0.62 million persons) of the population, mainly Filipino and Indonesian (C&SD 2022a) in Hong Kong. The "Race Discrimination Ordinance" (Cap. 602) was gazette in 2008, and the Steering Committee on Ethnic Minorities Affairs was set up in 2018 with a faithful commitment to the elimination of racial discrimination. However, minority groups remain in disadvantaged socioeconomic positions and are unintentionally but virtually marginalized. One out of every three South Asian youngsters lives in poverty (Oxfam 2018), exacerbated by limited school choices and career prospects.

Hong Kong prides itself on public sector non-refusal policy and universal health coverage, supported by high subsidies. However, in comparison with South Korea, healthcare access in Hong Kong varied throughout the socioeconomic ladder and it is more pro-rich (Lu et al. 2007), particularly in regard to delivering non-acute outpatient care for the patient with long-term chronic illnesses under the dual-track health system, while the public–private partnerships (PPP) initiated in recent years do not seem to work well. In short, the general observation derived from the above-mentioned three pillars is that inequality is not really serious in Hong Kong, although there are areas deserving improvement.

4.4 Economic Inequality in Hong Kong

4.4.1 Unequal Sharing of the Fruits of Economic Development

Hong Kong is the third global financial center ranked after New York and London (Wardle and Mainelli 2021) and ranked 15th in the world in terms of nominal GDP per capita (Statistics Times 2020). Hong Kong's per capita GDP in 2004 was HK$194,140 (roughly = US$24,735), and in 2021 it was HK$387,110 (roughly = US$49,321) (C&SD 2022c), representing an increase of 99.4%. However, the Real Index of Payroll per person after deducting inflation has only increased from 105.5 in 2004 to 124 in 2021 (C&SD 2022b), constituting an increase of only 17.5% in 17 years. These figures indicate that employees have been unable to equally share the fruit of economic boom. This is particularly the case for young workers with university degrees. New Forum and New Century Forum (2015) found that from 1993 to 2013, the overall median income of university-educated workers rose from HK$23,684 (roughly = US$3,018) to HK$24,240 (roughly = US$3,088), merely with a 2.3% increase. As for the top 10% of university students, their income fell from HK$65,789 (roughly = US$8,382) to HK$60,817 (roughly = US$7,749), with a drop of 7.6%. In short, there is limited upward social mobility that provokes passive and even furious sentiments in young generations (Shek 2020).

Hong Kong has also been criticized for its low minimum wage. On May 1, 2011, the government implemented the statutory minimum wage which was originally established at HK$28 (roughly = US$3.57) per hour and gradually raised to HK$37.5 (roughly = US$4.79) per hour in 2019, with the latest adjustment to HK$40 (roughly = US$5.11) per hour from 1 May 2023 onwards (Labor Department 2023). The minimum wage can be considered to be low when compared to other developed economies, including South Korea, New Zealand, and France, the minimum wage has been raised to 9,160 won (roughly = US$7.98) per hour (South Korea's Ministry of Employment and Labor 2022), US$21.20 per hour (Employment New Zealand 2022), and €11.07 (roughly = US$11.11) per hour (Business France 2022), respectively.

4.4.2 *Inequality in Income Distribution*

Since the 1990s, the Gini coefficient in Hong Kong has been maintained at a level of 0.5 or above. It was 0.430 in 1971 and then rose continuously to 0.518 in 1996 and 0.539 in 2016 (HKSAR Government 2017), far exceeding the social income distribution inequality threshold of 0.4. In 2016, Hong Kong was included among the regions with unequal income distribution as its Gini coefficient exceeded 0.5. When comparing Hong Kong's Gini Index internationally based on data collected by the World Bank (World Bank 2022), it is also at a high level among many high-income economies (see Fig. 4.2).

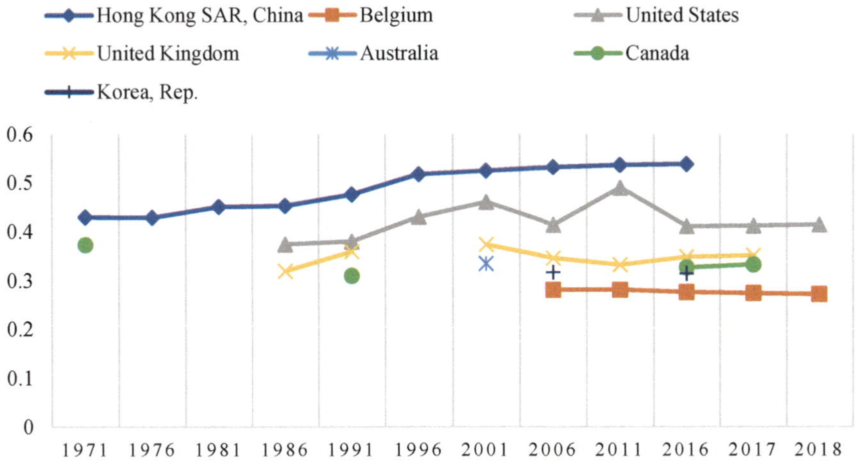

Fig. 4.2 Gini index in developed economies. *Source* Authors' own based on data from HKSAR Government (2017); World Bank (2022)

Decile dispersion ratio statistics also point to the disturbing income disparity. In 2006, the top decile's median monthly original household income (HK$76,250, roughly = US$9,715), excluding foreign domestic helpers, was 33.9 times more than the bottom decile's (HK$2,250, roughly = US$287) in Hong Kong. This gap widened to 43.9 times in 2016. In addition, the lowest decile's median monthly income climbed by only 13.8% from 2006 to 2016, while those in the highest decile increased by 47.5% (C&SD 2017). Furthermore, the P90/P10 differences rose from 13.1 in 2006 to 16.1 in 2011 and then fell slightly to 14.0 in 2016 (C&SD 2017), compared to 2.9 in Denmark, 3.4 in France, 4.2 in the United Kingdom, 5.7 in Korea, and 6.3 in the United States in 2016 (OECD 2022a).

Income inequality is also characterized by a clear geographical and spatial distribution in Hong Kong. First, median monthly household income varies by district. Second, in-work poverty which is a severe issue in most developed economies, also exists in Hong Kong, with an increasing proportion of individuals in Hong Kong in poverty, in particular those who are low-skilled employees although they have actively participated in the labor force (Cheung and Chou 2016). The 2020 Report on Annual Earnings and Hours Survey (C&SD 2021a) shows that in May–June 2020, the highest monthly median wage for employees by industry category was HK$30,000 (roughly = US$3,822) for education and public administration (excluding government), and the lowest was HK$13,000 (roughly = US$1,656) for estate management, security, and cleaning services.

4.4.3 Wealth Inequality Phenomena

Despite its material affluence, wealth is very unevenly distributed in Hong Kong, and the disparity between the rich and the poor has been widening day by day (Shek and Siu 2019). In 2020, the density of millionaires in Hong Kong stood at 8.3%, and Hong Kong ranked third in the world with its mean wealth per adult (US$503,340), behind Switzerland (US$673,960) and the United States (US$505,420) (Shorrocks et al. 2021). Besides, Hong Kong has the highest number of Rolls-Royce Cars per kilometers/capita (Torne 2021). According to Credit Suisse (Stierli et al. 2014), Hong Kong was regarded as a developed economy with "very high inequality" closely followed by the United States and Switzerland. The top decile in these three developed economies shared over 70% of the wealth in 2014 (77.5%, 74.6%, and 71.9%, respectively).

According to Forbes Rich List 2021, the 50 richest billionaires in Hong Kong, with a combined net worth of US$331 billion in 2020, up 7.5% from US$308 billion in the previous year (Forbes 2021). Meanwhile, Hong Kong's fiscal reserves in 2019–20 were HK$1,160.3 billion (US$147.8 billion) (C&SD 2021c), which can be interpreted as the top 50 tycoons' wealth was approximately 2.2 times Hong Kong's reserves. Besides, in 2020, Hong Kong saw a record of 515,000 millionaires, approximately 8.7% of the population, with a net worth of HK$10 million (roughly = US$1.27 million) or above (Citibank 2021). However, the poverty population in Hong Kong, unfortunately, increased to 1.653 million in 2020, with a poverty rate (before policy intervention) of 23.6%, which is higher than many other developed economies (C&SD 2021e). In 2020, women's and men's poverty rates were 8.2% and 7.6%, respectively, while 27% of children suffered from poverty. Furthermore, 583,600 elders were categorized as impoverished, accounting for 45% of the total aged population, compared to 366,000 elders who lived poor in 2009 (C&SD 2021e). "Poverty in the Midst of Affluence" (Goodstadt 2014) is one of Hong Kong's "unfortunate" landscape.

Contrary to common belief, Hong Kong's social mobility is extremely limited, preventing people from getting out of poverty (Forrest et al. 2004). Actually, the basis of Hong Kong's economy and employment structure implies a fall in job mobility in the labor market, and opportunities to improve one's fortunes by hard effort are fast dwindling. Hong Kong tops the Economist's Crony Capitalism Index issued in 2014 which assesses the ratio of a billionaire's wealth to the country's GDP, with a higher ratio indicating the greater likelihood of suffering from crony capitalism (The Economist 2014), with the super-rich in Hong Kong utilizing high degree of economic rent at their fingertips to reap profits. Given that capital returns are significantly bigger than labor returns (Piketty 2014), economic inequality is reasonably predicted to continue to increase in the future.

Obviously, economic inequality has been linked to intergenerational mobility (Jerrim and Macmillan 2015). The situation in Hong Kong follows the Great Gatsby curve, which states that intergenerational mobility is lower in societies with rising levels of economic inequality. Peng et al. (2019) discovered that intergenerational economic transmission is consistently robust among high-income individuals. In other words, the socioeconomic status of this group is largely determined by the wealth he/she inherits from his/her family, and such fortunes tend to be passed on from generation to generation. As inherited wealth would dominate and far outnumber wealth gained through a lifetime's labor effort, capital concentration will reach unprecedented heights (Piketty 2014).

4.4.4 Factors Contributing to Income and Wealth Inequality

As a laissez-faire economy, the Hong Kong government has deliberately limited the breadth and depth of its involvement in the social policy arena (Wu and Chou 2017). It has adopted a residual welfare system with limited redistribution (Chan and Chou 2018). Historically, the Hong Kong Government upholds neoliberal rhetoric such as "full respect for free market" and "take responsibility for your own personal well-being", refusing to establish an adequate social security system that would alleviate economic inequalities caused by financialization (Lee et al. 2018). Although expenditure on social welfare has grown from HK$40,519 million (roughly = US$5,162 million) in 2010–11 to HK$100,066 million (roughly = US$12,749) in 2020–21 (revised estimates), representing a rise from 2.3% to 3.7% of GDP, its share in total public expenditure has fallen from 12.6% to 11.6% (C&SD 2021b).

Besides, the strong ties between the government and the commercial and financial elites in Hong Kong have translated into a low tax regime, concentrated mainly on revenues related to land development, which produced great wealth but also resulted in a very lopsided distribution of wealth (Wissink et al. 2017). The tax system is particularly favorable to the wealthy without capital gains or inheritance taxes, as well as no dividend income taxes or progressive taxation, with the bulk of the tax burden falling on the shoulders of middle- and lower-income groups. Meanwhile, the region, which was established as a free port, is exempt from import and export customs tariffs.

Globalization in the past decades has also contributed to economic inequality in Hong Kong. While globalization indeed has contributed greatly to economic growth, it has also exacerbated economic disparity within nations (Potrafke 2015). The damage done to Hong Kong's equality by globalization is concentrated in its restructuring of the economy. Since the 1980s, Hong Kong has steadily reoriented its economy from a re-export manufacturing hub to a regional and global service center for specialized business services, along with the relocation to mainland China for much of the manufacturing that had contributed to its economic boom in the 1970s. The disintegration and hollowing of manufacturing industries have forcibly cast the surplus labor force into competition for unskilled and semi-skilled employment with other low-skilled employees (Cheung and Chou 2016). Marginal workers, who are primarily women, elderly individuals, new immigrants, and ethnic minorities who are excluded from the dominant labor market (Lee and Wong 2004), have emerged. This economic restructuring process exacerbates social division and expands the wage gap between skilled and unskilled employees (Chiu and Lui 2004).

Furthermore, Lui (2019) argued that one of the key causes for the expanding household income dispersion is the aging population along with the rise in single-person households. In 2020, the elderly population was estimated to be 1.3 million, or roughly one-fifth of the entire population, with approximately nine in ten of the elderly being economically inactive (C&SD 2021e). At the same time, since most residents do not have a retirement plan and the "Mandatory Provident Fund" (MPF) is insufficient to ensure a decent retirement, this might result in not just increased wealth disparity but also major societal problems.

In addition, the ongoing rise of the international migrant population has intensified downward wage pressure in the low-skilled labor market (Cheung and Chou 2016), lowering average earnings and increasing income inequality. Additionally, immigrants from mainland China may also intensify the rich-poor divide. As Hong Kong's economy began to transform in the 1980s, new immigrants began to be widely perceived by the public and media as being one of the main causes of exacerbating income disparity in Hong Kong due to their mostly low-paying jobs (Xiaobin et al. 2004). In 2016, Mainlanders who had resided in Hong Kong for less than seven years accounted for 2.4% of the overall population, and the median monthly income (HK$10,250, roughly = US$1,306) was only about 66.1% of the overall working population (HK$15,500, roughly = US$1,975) (C&SD 2016). At the same time, many Hong Kong husbands in cross-border marriages come from lower socioeconomic strata and are generally manual workers (Hung and Fung 2016), while their

Mainland spouses, on average, have lower education levels and lower incomes than the Hong Kong average. However, some researchers argued that new immigrants, who represent a relatively minor portion of the total population and labor force, are not the main factor contributing to income inequality in Hong Kong. Instead, it is the overall economic restructuring that has occurred without the essential backing from government and the requisite support stem from social policies (Xiaobin et al. 2004).

Finally, the "Hong Kong dream" also contributes to economic inequalities. Hong Kong as a place where over half of the billionaires are self-made (Wissink et al. 2017) was once reputed to be a place where anyone could have a better life if they worked hard, performed well in school, and had an entrepreneurial spirit, regardless of their parents' background (Chew 2014). The "Hong Kong Dream", or the prospect of upward mobility, may be a crucial influence on Hong Kong inhabitants' high regard for hard labor, while the Chinese cultural focus on social harmony and the significance of self-reliance and family care above government assistance is strongly engrained in Hong Kong. However, today's youth, on the other hand, regularly lament the "death of the Hong Kong dream", claiming that those from wealthy backgrounds have a greater chance of succeeding (Chew 2014). The tycoon class, which had previously been seen as living icons of Hong Kong's social mobility (Forrest and Xian 2018), was increasingly identified as the manipulators of economic inequality. Besides intergenerational transmission of wealth/poverty, social conditions are also unfavorable for poor people to stand up on their own feet, such as suspension of issuance of hawker licenses. For example, the total number of licensed hawkers decreased from 50,000 in 1974 to 9,200 in 2000 (Legislative Council Secretariat 2017) and to only 5,598 at the end of December 2021 (FEHD 2022).

4.4.5 Consequences of Economic Inequality and Related Remedies

High economic inequality implies both absolute and relative deprivation. Evidence demonstrates that belonging to a deprived group makes it more difficult to get out of a deprived environment (Lundberg and Startz 1998). Lewis's (1966) "culture of poverty" paradigm posits that economic stress first has a negative impact on marital quality, which in turn affects parenting and family processes, and subsequently generates further child abuse and neglect that damages children and adolescents' physical and mental health, and eventually resulting in a poverty-perpetuating belief. This view suggests a vicious chain of damage within the family triggered by intractable poverty caused by economic inequality. There is a Chinese proverb that says, "poor couples have a lot to mourn," and the stress of survival due to economic disadvantage can cause emotional distress and conflict between spouses (Shek and Ng 2014). Moreover, their psychological well-being could be worse if they perceive that the triggers for economic inequality come more from external factors beyond their control

(Shek 2004). Parents with poor mental health who are preoccupied with dealing with financial stress and spousal conflict often have less energy to care for and accompany their children (Shek 2008). At the same time, poverty can reduce parenting patience and enthusiasm, and rather than negotiating and reasoning; financially disadvantaged parents are more prone to utilize harsh discipline (McLoyd 1990), thereby reducing warmth in the parent–child relationship and increasing conflict (Shek 2005a).

Also, economically disadvantaged adolescents exhibited less perceived family interactions and positive parenting (Shek and Tsui 2013). Besides, economic disadvantage brings about damage not only to mental health but also to physical health. Oxfam (2011) found that one in six poor families with children (16%) are highly food insecure and suffer from hunger on a regular basis. In the long run, experiences of economic disadvantage undermine the healthy mental and physical development of children and adolescents. In conjunction with unfavorable parenting practices or poor neighborhoods, their chance of engaging in risky behaviors such as substance abuse would increase (Shek and Lin 2014).

Growing economic inequality can lead to a loss of in-group trust, less engagement in the community, impairment of community cohesion, and intensification of undesirable behaviors such as violence and drug use. First, economic stratification negatively affects community cohesion. When people realize they are in a heterogeneous group with diverse background constituents, they tend to self-segregate and avoid inter-community relations (Alesina and La Ferrara 2000). Besides, people with economic troubles may be limited in their skills to participate in community life and tend to show lower social cohesion than those without economic difficulties (Vergolini 2011). Secondly, higher inequality is associated with many types of social pathologies (Goya-Tocchetto and Payne 2022), including mental health issues (e.g., schizophrenia) and adolescent developmental problems (Burns et al. 2014).

As economic inequality rises, social instability of different types increases because the social fabric that is essential to maintaining societal internal stability collapses. First, research consistently shows a positive association between the degree of inequality and the extent of crime (Pablo Fajnzylber et al. 2002). Second, Hong Kong society is at risk of suicide triggered by economic stress (Shek 2005b). There is a clear socioeconomic gradient in suicide observed in Hong Kong, with the poorest quintile having over twice the suicide rate of the least poor (Hsu et al. 2015). Third, economic inequality can fuel social unrest and diminish social harmony. Analyzing the protests in Hong Kong, Shek (2020) argued that many social unrests are triggered by quality of life issues.

In alleviating economic inequity and relieving poverty in Hong Kong, the Commission on Poverty (CoP) and the Social Welfare Department (SWD) play a vital role. Since its re-establishment in 2012, CoP has been dedicated to fighting poverty and strengthening support for the disadvantaged. Its universal initiatives include the establishment of the official poverty line, formulation and implementation of the "Low-Income Working Families Allowance (LIFA) Scheme", targeting young people with the aim of improving the upward mobility of those with grassroot background, and holding of a territory-wide retirement security counseling campaign for the elderly, in addition to furthering the work of the "Community Care Fund" (CCF) and social innovation (CoP 2021).

Hong Kong has developed a relatively well-rounded social security system designed to build a "financial safety net" for those who are unable to sustain themselves, including the "Comprehensive Social Security Assistance (CSSA) Scheme", the "Social Security Allowance (SSA) Scheme", and the "Criminal and Law Enforcement Injuries Compensation Scheme". Among them, the CSSA recurrent expenditure increased by about 21.4% from HK$17.4 billion (roughly = US$2.22 billion) in 2010–11 to HK$21.2 billion (roughly = US$2.70 billion) in 2020–21 (C&SD 2021d). However, there are several criticisms of such welfare measures. First, CSSA may play a limited role because there has been a decline in the share of CSSA expenditure in total recurrent government expenditure from 7.8% in 2010–11 to 4.5% in 2020–21 (C&SD 2021d). Second, people may abuse CSSA because it can provide minimal but adequate support for poor people. Actually, abuse of CSSA was thought to be a severe concern by 59.6% of the 1,026 respondents surveyed by Oxfam and PolyU (2007). Third, the welfare benefits system is blamed for encouraging reliance and fraud (Wong and Lou 2010). Chung (2010) found that 74.3% of respondents opined that CSSA would reduce the motivation of the recipients in the workplace. In a society that prides itself on "self-reliance" and "hard work", while the provision of public financial assistance to the elderly and the disabled is generally supported, the provision of assistance to the unemployed and low-income earners is considered less justified. Some researchers even argued that to avoid abuse of CSSA, the government and the media may deliberately portrait a negative image of those who are welfare-dependent, stigmatizing CSSA recipients (Oxfam and PolyU 2007; Wong and Lou 2010).

4.5 Housing Inequality in Hong Kong

4.5.1 Phenomena on Housing Inequality

According to Piketty (2014), capital gains, such as real estate and the stock market, rather than labor income, are the primary causes of wealth inequality in the twenty-first century. This is particularly the case in Hong Kong, where taxes are low and government redistribution effort is weak. Housing has indeed been a historically thorny problem in Hong Kong and has even become more acute after the restoration of Hong Kong to China (Shek 2020). Our Hong Kong Foundation (2022) also cautioned that the struggle of being forced into a costly but confined and densely crowded landscape has long afflicted Hong Kong residents, who have almost reached the ultimate tolerance limit.

There are several burning issues surrounding housing inequalities in Hong Kong. The first phenomenon is "cramped" living space. In 2021, about 224,608 domestic households have a floor area of accommodation of only less than 20 m^2, accounting

Fig. 4.3 Monster Building in Quarry Bay known as "pigeonhole". *Source* "Monster Building Quarry Bay Hong Kong" by blieusong is marked with CC BY-SA 2.0. https://www.flickr.com/photos/blieusong/48128094843

for 8.4%, while 98,461 households have a floor area of more than 100 m^2, accounting for 3.7% (C&SD 2022a). Such cramped living spaces are evident in one modular building after another, known as "pigeonholes" (see Fig. 4.3).

In the past two decades, there has been a trend for owners to divide their flats into multiple units, which are referred to as sub-divided units (SDUs). Around 226,340 people lived in SDUs in 2020, representing an increase of 7.9% since 2016. All people living in SDUs had only a median area of 6.6 m^2 per capita, with a median monthly rent of HK$4,800 (roughly = US$612). The rent makes up about one-third of the monthly revenue for SDU households (Transport and Housing Bureau 2021b). For those who even cannot afford to rent sub-divided flats, they are forced to rent smaller "cage homes" (bed space apartments), both of which have poor environmental and hygiene conditions. Based on data from the SoCo (2009), nearly 100,000 individuals lived in cage homes, cubicles, or small partitioned flats in 2009. Back in the early 1990s, the UN Commission on Human Rights remarked that Hong Kong's cage dwellings is an "insult to human dignity" (Shing 2021).

The second phenomenon is the housing affordability issue. Hong Kong has ranked at the top of the list of the world's most unaffordable cities for twelve consecutive years, and the median housing price to median family income ratio has risen from 11.4 in 2010 to 23.2 in 2021, according to the Demographia International Housing Affordability Report (Demographia 2011; URI and FCPP 2022), revealing that Hong Kong's household income growth has heavily lagged behind the city's surge in house prices. It could be inferred that to afford an average-priced accommodation in 2021, an average-earning Hong Kong family must save approximately 23.2 years of income (without eating or drinking). Meanwhile, Hong Kong's "median multiple" (i.e., housing price-to-income ratio) far exceeds that of other developed economies such as Australia (8), Canada (6), Singapore (5.8), the United Kingdom (5.1), and the United States (5) in 2021 (URI and FCPP 2022). From 1989 to 2019, the overall property price index increased from 40.2 to 383.0 (i.e., an increase of 852.7%), while the overall rental index increased from 69.4 to 194.4, representing an increase of 180.1% (New Forum and New Century Forum 2020), resulting in heavy costs of home ownership and rentals.

The third phenomenon is that unlike developed economies where the home ownership rate usually exceeds 60%, the home ownership rate in Hong Kong in 2020 is only 51%, which is lower than that in the United States (63%), EU (69%), South Korea (59%) and Singapore (90%) (Legislative Council Secretariat 2021). Meanwhile, a larger proportion of completed private residential units remain vacant and unoccupied. In 2021, the vacancy rate for private housing was 7.5% for large units and 3.8% for small/medium units (Rating and Valuation Department 2021).

In particular, home ownership rates among younger age groups have fallen, affected by deteriorating property affordability and the government's prudent mortgage restrictions. In 1997, young people under the age of 35 used to make up 22.1% (198,100 persons) of the total number of heads of owner-occupied households, but this dropped by two-thirds in 2019 to just 7.6% (98,200 persons) (Legislative Council Secretariat 2021). Besides, only 18% of college students were confident and willing to afford a private home in the future, but also expected parental support (Castro Campos et al. 2016). This supports the conjecture of "hereditary capitalism" that has indeed been occurring in Hong Kong for many years in terms of home ownership, where hard effort and meritocracy are less significant than parents' socioeconomic status (Lee et al. 2018).

The fourth phenomenon is that there is a huge demand for public housing. Public rental housing is hailed as a hallmark of Hong Kong's social welfare and indeed 2,175,360 people, representing 29.3% of the population, live in public rental housing in Hong Kong in 2021 (C&SD 2022a) and enjoy such benefits. At the end of June 2021, the average monthly public rental housing rent was about HK$2,276 (roughly = US$290), remaining at an affordable level for most residents (Transport and Housing Bureau 2021a). Unfortunately, although public housing is claimed to be a "safety net" for low-income families, the waiting time for this "safety net" is incredibly long. The average waiting period for general applicants to be assigned a flat was six years as of June 2022, which is much longer than the pledged duration of three years. Besides, the waiting list is already piled with a whopping 254,600 applications, comprising 144,200 general applications and 98,400 Quota and Points System (QPS) applications (Hong Kong Housing Authority 2022). The regression in housing development is becoming an obstacle to Hong Kong social development as well as a structural factor of poverty.

4.5.2 Factors Contributing to Housing Inequality

There are several factors contributing to housing inequality in Hong Kong. The first factor is limited land that can be used for public housing programs. By the end of 2020, with a size of 1,114 km^2, only roughly 345 km^2, or 30.97% of Hong Kong's land had been developed, including residential, commercial, industrial, institutional, transport, agricultural land, and other built-up lands. The remaining 69.03% is mostly woodland, shrubland, grassland, and wetland, which are mostly protected areas under statutory protection. Among the developed land, only 79 km^2 are residential, accounting for only 7.09% of the land area (C&SD 2021b). Hence, people begin to question whether such a large area of country parks or protected areas is necessary and proposed that a portion of the country parks be converted for housing construction in the face of land supply shortage. Furthermore, there are criticisms that the government has deliberately or intentionally underdeveloped the land outside Hong Kong Island and Kowloon, and the region nearby the border between Hong Kong and mainland China (Wissink et al. 2017).

The second factor is related to non-Hong Kong buyers who may be regarded as scapegoats. Foreign nationals and mainland Chinese citizens are vulnerable easily to becoming scapegoats for the intensified competition in Hong Kong's housing market and the rising house prices. Since 2008, with the rapid economic development of mainland China, many cash-rich investors have started to buy property in Hong Kong (Wissink et al. 2017). As such, mainland Chinese purchasers accounted for a staggering 49.3% of new property transactions worth more than HK$12 million (roughly = US$1.53 million) in the first quarter of 2011 (Li 2014). As a result, the government imposed a 15% stamp duty on non-Hong Kong permanent residents in 2012, which has reduced the number of Mainland Chinese buyers. Nevertheless, as suggested by Wissink et al. (2017), because of inadequate supply, high property prices may not necessarily be the result of an influx of mainland Chinese buyers.

Another factor leading to housing inequalities in Hong Kong is its high dependence on land finance and property taxation for economic and social development, as well as the "real estate hegemony" (Wong 2015, p. 1). Described as a real estate economy (Haila 2000), Hong Kong differs from other developed economies in that it has progressively developed a real estate-led growth model since the 1980s. Real estate is even more important than the finance and insurance sectors (Smart and Lee 2003), with real estate as a highly leveraged financial asset to complement other financial markets and ultimately stimulate economic growth as a whole. Real estate, along with ownership of premises, and professional and business services, accounted for 21.4% of Hong Kong's GDP in 2019 (C&SD 2021b). In addition, profits and salaries tax from the construction industry and stamp duty also contribute to government revenue. It is therefore hardly surprising that the Hong Kong government has a substantial and perpetually fresh interest in the maintenance of high land-related prices, given the dependence of its revenue on land-related revenues (Wissink et al. 2017). In addition, hurt by the Asian financial turmoil and the SARS outbreak, the government took measures to revitalize the private property market, including the cessation of the "Home Ownership Scheme" and the termination of the "Tenants Purchase Scheme" (Lee et al. 2018). Obviously, how to continuously pinpoint the optimal balance between the government-intervened public housing regime and the market-led private housing market in a timely and dynamic manner is a worthwhile consideration for Hong Kong's housing economy, as it may be the key to breaking the housing dilemma. Once private housing prices fall severely and the price gap with public housing narrows significantly, the government tends to interrupt the public housing policy by "abandoning the cart and protecting the marshal", so that the public housing policy is effectively "kidnapped" by the private property market.

In reality, most of Hong Kong's billionaires have made their fortunes via real estate investment. By controlling real estate, they also control over much of the key sectors of the city's economy. "Property" accounts for 71% of multimillionaires' total assets, followed by "cash and deposits" (15%), "stocks" (7%), "funds and bonds" (5%), and "other investment products" (3%) (Citibank 2021, p. 1). The housing problem in Hong Kong has become so severe under the public policy of real estate abduction and the "real estate hegemony" (Wong 2015, p. 1) that the general public, especially young people, are very frustrated with their housing prospects.

4.5.3 *Consequences of Housing Inequality and Remedial Measures*

Growing housing inequality and chronic unmet housing needs can seriously damage individuals' physical health and psychological health (Shek 2020). A historical cohort study conducted by Dedman et al. (2001) showed that childhood housing conditions characterized by poor ventilation were related to adult mortality. Besides, Solari and Mare (2012) showed that crowded housing impaired the physical health of children. Worse still, housing deprivation tends to create a sense of psychological insecurity for individuals who would then feel trapped, lonely and adrift (Bower et al. 2021). With specific reference to the Chinese culture, housing serves not only as a consumer and investing commodity, but also as a status symbol, and the lack of a house is perceived as a major disadvantage in the competition for marriage partners. Furthermore, renting, both public and private, is perceived as unstable, insecure, and even stigmatized (Colic-Peisker and Johnson, 2012).

Besides, housing issues damage family health by creating tension, clashes, and conflict under one roof (Shek 2020). For young people, they don't seem to have much choice and are probably forced to settle in split-level apartments with safety and sanitation issues or forced to live with their parents (Shek 2020). According to Forrest and Xian's (2018) survey, three-quarters of the Hong Kong younger aspire to relocate out and stop living with their parents when they got married, and 40% expected to move out once they own enough money to be self-sufficient. However, Li (2014) noticed that 74% of individuals living in privately held residences were owned by their parents, while just 11% were owned by the young person themselves.

Housing inequality poses serious damage to political health (Shek 2020). Housing inequality would aggravate disparities between communities by creating residential segregation and intensifying the homogeneity of economic traits of residents within the community, thus triggering group grievances about the living environment in contrast to upscale communities. Besides, Saunders (1978) found that homeownership could have a significant influence on people's political attitudes. High housing prices have led young people to resent ineffective government housing policies (Shek 2020) and to a greater distrust of the government. Historically, many protests against the so-called "real estate hegemony" have erupted in Hong Kong (Wong 2015), with the most recent 2019 protest being an example of the anger and hopelessness of young people ignited by the Extradition Bill (Shek 2020).

The Hong Kong government has realized that housing issues lie at the crux of numerous political, economic, and social problems: "housing is not only a matter of supply and demand, but also an issue of social justice and allocation of resources ... unresolved housing problems may jeopardize the long-term prosperity and stability of the society" (HKSAR Government 2020, p. 43). In fact, as early as 1997, the first Chief Executive, Tung Chee-haw, was concerned about the housing problem. He proposed to provide at least 85,000 residential units each year, with the goal of enabling 70% of all families to purchase their own houses within 10 years, a plan known as the "85,000 policy" that was shelved after the outbreak of the Asian financial crisis (Chan 2018). Subsequent chief executives have also tried to tackle the problem of high property prices, but with little success. In her 2020 policy address, Chief Executive Carrie Lam stated that the government would "increase housing supply with a proportion of 70% for public housing" (HKSAR Government 2020, p. 51). To increase land supply and address the core problem of insufficient land for housing construction, the Hong Kong government has targeted 330 hectares of land, including brownfield sites, for the provision of 316,000 public housing units, and other measures such as the redevelopment of old districts, rezoning of industrial areas and reclamation have also been mentioned (HKSAR Government 2020). However, to be followed up by the sixth administration.

4.6 Conclusion

After a review of social inequalities in Hong Kong with reference to the political, cultural, and opportunity pillars, we conclude that inequalities in these areas are not very serious but there is much room for improvement. Politically speaking, people in Hong Kong SAR have enjoyed more political rights compared to the pre-1997 period. Although the change in the electoral system following the social unrest has been criticized as a "regression" in democracy, some argue that this highlights the need for "patriots" in the legislative system, which is a reasonable expectation in different places in the world. Nevertheless, the change in the political system has to be tested through time. In terms of legal equality, it is safeguarded by the Common Law system in Hong Kong. Regarding cultural inequality, Hong Kong is famous for its outstanding educational accessibility and quality. However, Hong Kong has faced a highly unbalanced career distribution during its transformation to a service economy. In terms of inequality of opportunity in Hong Kong, women in Hong Kong enjoy gender equality as compared to other Asian societies, although female leadership in politics is yet to be strengthened. Besides, minorities in Hong Kong are still disadvantaged socioeconomically mostly due to language barriers. Moreover, Hong Kong prides itself on universal health coverage, but healthcare access still varies throughout the socioeconomic ladder.

Economic inequality is obvious in Hong Kong. Regarding income inequality, Hong Kong has been criticized not only for its low minimum wage compared to other developed economies but also for its severe wage depreciation. Moreover, the city's Gini index is among the highest worldwide. In terms of wealth inequality, it is widening day by day, with the poverty rate increased to 23.6% in 2020 (C&SD 2021e). Besides, social mobility is extremely limited, and the "Hong Kong dream" of starting a business from scratch has been increasingly shattered, while the element of "economic rent" has become increasingly prominent and hereditary capitalism has risen, making it extremely difficult for people to escape from poverty. Regarding housing inequality, cramped living space (such as sub-divided flats and caged homes), rocket high property price, low housing affordability, and long waiting queue for public housing are disturbing issues for Hong Kong for a long time. Economic inequalities are particularly acute under COVID-19 (Shek et al. 2023).

In this chapter, we also outline the possible macro factors leading to economic and housing inequalities in Hong Kong, which would eventually impair human development, particularly the development of children and adolescents. Remedies for inequalities in these two areas are also briefly discussed in this chapter.

Acknowledgements The preparation for this work is financially supported by the Li and Fung Endowed Professorship and the Research Matching Fund of the Research Grants Council (ZH4C and ZECL).

References

Alesina A, La Ferrara E (2000) Participation in heterogeneous communities. Q J Econ 115(3):847–904. https://doi.org/10.1162/003355300554935

Baker J, Lynch K, Cantillon S, Walsh J (2006) Equality: putting the theory into action. Res Publ 12(4):411–433. https://doi.org/10.1007/s11158-006-9020-9

Basic Law (1990). https://www.basiclaw.gov.hk/en/basiclaw/facts.html

Bower M, Buckle C, Rugel E, Donohoe-Bales A, McGrath L, Gournay K, Barrett E, Phibbs P, Teesson M (2021). 'Trapped', 'anxious' and 'traumatised': COVID-19 intensified the impact of housing inequality on Australians' mental health. Int J Hous Policy, 1–32. https://doi.org/10.1080/19491247.2021.1940686

Burns JK, Tomita A, Kapadia AS (2014) Income inequality and schizophrenia: increased schizophrenia incidence in countries with high levels of income inequality. Int J Soc Psychiatry 60(2):185–196. https://doi.org/10.1177/0020764013481426

Business France (2022, August 1) Increase of the minimum wage in France. Retrieved from: https://www.welcometofrance.com/en/increase-of-the-minimum-wage-in-france
C&SD (2016) Thematic report: persons from the Mainland having resided in Hong Kong for less than 7 years. Retrieved from: https://www.censtatd.gov.hk/en/EIndexbySubject.html?scode=459&pcode=B1120101
C&SD (2017) Thematic report: household income distribution in Hong Kong. Retrieved from: https://www.censtatd.gov.hk/en/EIndexbySubject.html?scode=459&pcode=B1120096
C&SD (2021a) 2020 report on annual earnings and hours survey. Retrieved from: https://www.censtatd.gov.hk/en/EIndexbySubject.html?scode=210&pcode=B1050014
C&SD (2021b) Hong Kong annual digest of statistics 2021 edition. Retrieved from: https://www.censtatd.gov.hk/en/EIndexbySubject.html?pcode=B1010003&scode=460
C&SD (2021c) Hong Kong in figures 2021 edition. Retrieved from: https://www.censtatd.gov.hk/en/EIndexbySubject.html?scode=460&pcode=B1010006
C&SD (2021d) Hong Kong monthly digest of statistics: statistics on comprehensive social security assistance scheme, 2010 to 2020. Retrieved from: https://www.censtatd.gov.hk/tc/EIndexbySubject.html?pcode=FA100078&scode=390
C&SD (2021e) Hong Kong poverty situation report 2020. Retrieved from: https://www.censtatd.gov.hk/en/EIndexbySubject.html?pcode=B9XX0005&scode=461
C&SD (2021f) Women and men in Hong Kong—key statistics 2021 edition. Retrieved from: https://www.censtatd.gov.hk/tc/EIndexbySubject.html?pcode=B1130303&scode=180
C&SD (2022a) 2021 population census—summary results. Retrieved from: https://www.census2021.gov.hk/en/publications.html
C&SD. (2022b). Table 23: Real indices of payroll per person engaged by industry section (1st Qtr 1999 = 100). Retrieved from: https://www.censtatd.gov.hk/en/web_table.html?id=23
C&SD. (2022c). Table 31: Gross Domestic Product (GDP), implicit price deflator of GDP and per capita GDP. Retrieved from: https://www.censtatd.gov.hk/en/web_table.html?id=31
Castro Campos B, Yiu CY, Shen J, Liao KH, Maing M (2016) The anticipated housing pathways to homeownership of young people in Hong Kong. Int J Hous Policy 16(2):223–242. https://doi.org/10.1080/14616718.2015.1130605
Chan H (2018, June 13) Ex-Hong Kong leader CY Leung defends controversial 1990s housing policy, says crisis could have been averted. Hong Kong Fress Press. Retrieved from: https://hongkongfp.com/2018/06/13/ex-hong-kong-leader-cy-leung-defends-controversial-1990s-housing-policy-says-crisis-averted/
Chan L-S, Chou K-L (2018) A survey of asset poverty among older adults of Hong Kong. Soc Indic Res 138(2):605–622. https://doi.org/10.1007/s11205-017-1674-5
Cheung KC-K, Chou K-L (2016) Working poor in Hong Kong. Soc Indic Res 129(1):317–335. https://doi.org/10.1007/s11205-015-1104-5
Chew A (2014, October 7) The end of the Hong Kong 'dream'. ALJAZEERA. Retrieved from: https://www.aljazeera.com/features/2014/10/7/the-end-of-the-hong-kong-dream
Chiu SWK, Lui T-L (2004) Testing the global city-social polarisation thesis: Hong Kong since the 1990s. Urban Stud 41(10):1863–1888. https://doi.org/10.1080/0042098042000256297
Chung KW (2010) Negative public perception on welfare recipients and its implications for social security in Hong Kong. J Asian Public Policy 3(2):200–206. https://doi.org/10.1080/17516234.2010.504328
Citibank (2021) Citibank announces results of Hong Kong affluent study 2020. Retrieved from: https://www.citibank.com.hk/english/info/news-room.htm
Colic-Peisker V, Johnson G (2012) Liquid life, solid homes: young people, class and homeownership in Australia. Sociology 46(4):728–743. https://doi.org/10.1177/0038038511428754
CoP (2021, June 25) Welcome message. Retrieved from: https://www.povertyrelief.gov.hk/eng/welcome.html
Dedman DJ, Gunnell D, Smith GD, Frankel S (2001) Childhood housing conditions and later mortality in the Boyd Orr cohort. J Epidemiol Community Health 55(1):10–15. https://doi.org/10.1136/jech.55.1.10

Demographia (2011) 7th annual demographia international housing affordability survey: 2011. Retrieved from: http://www.demographia.com/db-dhi-index.htm
District Councils Ordinance, Cap. 547 (1999) 19 March 1999. https://www.elegislation.gov.hk/hk/cap547?xpid=ID_1438403413041_002
Dworkin R (2002) Sovereign virtue: the theory and practice of equality. Harvard University Press
Employment New Zealand (2022, April 1) Current minimum wage rates. Retrieved from: https://www.employment.govt.nz/hours-and-wages/pay/minimum-wage/minimum-wage-rates/
FEHD (2022, July 8) Hawker control. Retrieved from: https://www.fehd.gov.hk/english/pleasant_environment/hawker/overview.html
Forbes (2021, February 24) Total wealth of tycoons on 2021 Forbes Hong Kong rich list rises amid pandemic. Forbes Press Releases. Retrieved from: https://www.forbes.com/sites/forbespr/2021/02/24/total-wealth-of-tycoons-on-2021-forbes-hong-kong-rich-list-rises-amid-pandemic/?sh=78dc5bc26c44
Forrest R, La Grange A, Yip N-M (2004) Hong Kong as a global city? Social distance and spatial differentiation. Urban Studies 41(1):207–227. https://doi.org/10.1080/0042098032000155759
Forrest R, Xian S (2018) Accommodating discontent: youth, conflict and the housing question in Hong Kong. Hous Stud 33(1):1–17. https://doi.org/10.1080/02673037.2017.1342775
Fraser N (2009) Scales of justice: reimagining political space in a globalizing world. Columbia University Press
Goodstadt LF (2014) Poverty in the midst of affluence: how Hong Kong mismanaged its prosperity. Hong Kong University Press
Goya-Tocchetto D, Payne BK (2022) How economic inequality shapes thought and action. J Consum Psychol 32(1):146–161. https://doi.org/10.1002/jcpy.1277
Haila A (2000) Real estate in global cities: Singapore and Hong Kong as property states. Urban Studies 37(12):2241–2256. https://doi.org/10.1080/00420980020002797
HKSAR Government (2017) Half-yearly economic report 2017. Retrieved from: https://www.legco.gov.hk/yr16-17/english/panels/fa/papers/facb1-1374-1-e.pdf
HKSAR Government (2020, November 25) The chief executive's 2020 policy address. Retrieved from: https://www.policyaddress.gov.hk/2020/eng/policy.html
HKSAR Government (2022, February 28) Financial results for the 10 months ended January 31, 2022. Retrieved from: https://www.info.gov.hk/gia/general/202202/28/P2022022800418.htm
Hong Kong Housing Authority (2022, 12 August) Number of applications and average waiting time for public rental housing. Retrieved from: https://www.housingauthority.gov.hk/en/about-us/publications-and-statistics/prh-applications-average-waiting-time/
Hsu C-Y, Chang S-S, Lee EST, Yip PSF (2015) Geography of suicide in Hong Kong: spatial patterning, and socioeconomic correlates and inequalities. Soc Sci Med 130:190–203. https://doi.org/10.1016/j.socscimed.2015.02.019
Hung SL, Fung KK (2016) Understanding social capital in migrant women from mainland China to Hong Kong. J Soc Serv Res 42(1):1–14. https://doi.org/10.1080/01488376.2015.1065783
Jerrim J, Macmillan L (2015) Income inequality, intergenerational mobility, and the Great Gatsby Curve: is education the key? Soc Forces 94(2):505–533. https://doi.org/10.1093/sf/sov075
Labor Department (2023) Statutory minimum wage. Retrieved from: https://www.labor.gov.hk/eng/public/content2_smw.htm
Lam SSK, Dreher GF (2004) Gender, extra-firm mobility, and compensation attainment in the United States and Hong Kong. J Organ Behav 25(7):791–805. https://doi.org/10.1002/job.264
Lee KM, To BHP, Yu KM (2018) Financialization and economic Inequality in Hong Kong: the cost of the finance-led growth regime. In: Fong BCH, Lui T-L (eds) Hong Kong 20 years after the handover: emerging social and institutional fractures after 1997. Springer, pp 127–151
Lee KM, Wong H (2004) Marginalized workers in postindustrial Hong Kong. J Comp Asian Dev 3(2):249–280. https://doi.org/10.1080/15339114.2004.9678402
Legislative Council Secretariat (2017) Hawker control in Hong Kong. Retrieved from: https://www.legco.gov.hk/research-publications/english/1718issh05-hawker-control-in-hong-kong-20171201-e.pdf

Legislative Council Secretariat (2021) Socioeconomic implications of home ownership for Hong Kong. Retrieved from: https://www.legco.gov.hk/research-publications/english/2021rb02-socioeconomic-implications-of-home-ownership-for-hong-kong-20210301-e.pdf
Lewis O (1966) The culture of poverty. Sci Am 215(4):19–25
Li J (2014) I am not leaving home: post-80s' housing attitudes and aspirations in Hong Kong. Urban Research Group—CityU on Cities Working Paper Series. City University of Hong Kong. Retrieved from: https://www.cityu.edu.hk/cityuoncities/upload/file/original/705520140620145010.pdf
Li S (2014, November 26) Local buyers push Hong Kong home prices to new highs as mainland demand slumps. South China Morning Post. Retrieved from: https://www.scmp.com/news/hong-kong/article/1648791/local-buyers-push-home-prices-higher
Lu J-FR, Leung GM, Kwon S, Tin KYK, Van Doorslaer E, O'Donnell O (2007) Horizontal equity in health care utilization evidence from three high-income Asian economies. Soc Sci Med 64(1):199–212. https://doi.org/10.1016/j.socscimed.2006.08.033
Lui HK (2019) Ageing population and rising income inequality in post-handover Hong Kong. Rev Integr Bus Econ Res 8(1):51–63
Lundberg S, Startz R (1998) On the persistence of racial inequality. J Law Econ 16(2):292–323. https://doi.org/10.1086/209890
McLoyd VC (1990) The impact of economic hardship on black families and children: psychological distress, parenting, and socioemotional development. Child Dev 61(2):311–346. https://doi.org/10.1111/j.1467-8624.1990.tb02781.x
New Forum, & New Century Forum (2015) 香港各世代大學生收入比較研究報告. Retrieved from: https://www.ncforum.org.hk/file/upload/file_828_4jy.pdf
New Forum, & New Century Forum (2020) 香港各世代大學學歷勞工 (1989–2019) 住屋承擔能力研究報告. Retrieved from: https://www.ncforum.org.hk/file/upload/file_1007_E6X.pdf
OECD (2022a) Income inequality (indicator). Retrieved from: https://data.oecd.org/inequality/income-inequality.htm
OECD (2022b) Inequality. Retrieved from: https://www.oecd.org/inequality.htm
Our Hong Kong Foundation (2022) The immediate actions to resolve Hong Kong's housing conundrum: cut the red tape now! Retrieved from: https://ourhkfoundation.org.hk/ebook/Land&Housing_Report_2022/Eng/mobile/index.html
Oxfam (2011) Oxfam Hong Kong survey reveals 1 in 6 poor families caught in a hunger trap. Retrieved from: https://www.oxfam.org.hk/tc/f/news_and_publication/1326/content_10644en.pdf
Oxfam (2018) Hong Kong inequality report. Retrieved from: https://www.oxfam.org.hk/en/news-and-publication/inequality-alarming-as-city-s-richest-earn-44-times-more-than-poorest
Oxfam & PolyU (2007) Perception and utilization of the CSSA: a study on the views of the public and the lower income people. Oxfam Hong Kong
Hong Kong Polytechnic University. Retrieved from: https://www.oxfam.org.hk/tc/f/news_and_publication/1323/content_18557en.pdf
Fajnzylber P, Lederman D, Loayza N (2002) Inequality and violent crime. J Law Econ 45(1):1–39. https://doi.org/10.1086/338347
Peng C, Yip PSF, Law YW (2019) Intergenerational earnings mobility and returns to education in Hong Kong: a developed society with high economic inequality. Soc Indic Res 143(1):133–156. https://doi.org/10.1007/s11205-018-1968-2
Piketty T (2014) Capital in the twenty-first century (A. Goldhammer, Trans.). Cambridge Massachusetts: Belknap Press of Harvard University Press
Potrafke N (2015) The evidence on globalization. World Econ 38(3):509–552. https://doi.org/10.1111/twec.12174
Rating and Valuation Department (2021) Hong Kong property review 2021. Retrieved from: https://www.rvd.gov.hk/en/publications/property_market_statistics.html
Saunders P (1978) Domestic property and social class. Int J Urban Reg Res 2(1–3):233–251. https://doi.org/10.1111/j.1468-2427.1978.tb00749.x

Schnurr S, Mak B (2011) Leadership in Hong Kong: is gender really not an issue? Gend Lang 5(2):343–371. https://doi.org/10.1558/genl.v5i2.343
Schwab K, World Economic Forum (2017) The global competitiveness report 2017–2018. Retrieved from: www.weforum.org/gcr
Shek DTL (2004) Beliefs about causes of poverty and psychological well-being of parents and adolescents experiencing economic disadvantage in Hong Kong. Am J Family Therapy 32(3):239–254. https://doi.org/10.1080/01926180490425658
Shek DTL (2005a) Perceived parental control processes, parent—child relational qualities, and psychological well-being in Chinese adolescents with and without economic disadvantage. J Genet Psychol 166(2):171–188. https://doi.org/10.3200/gntp.166.2.171-188
Shek DTL (2005b) Social stress in Hong Kong. In: Estes RJ (ed) Social development in Hong Kong: the unfinished agenda. Oxford University Press, pp 213–222
Shek DTL (2008) Economic disadvantage, perceived family life quality, and emotional well-being in Chinese adolescents: a longitudinal study. Soc Indic Res 85(2):169–189. https://doi.org/10.1007/s11205-007-9087-5
Shek DTL (2020) Protests in Hong Kong (2019–2020): a perspective based on quality of life and well-being. Appl Res Qual Life 15(3):619–635. https://doi.org/10.1007/s11482-020-09825-2
Shek DTL, Leung JTY, Tan L (2023) Social policies and theories on quality of life under COVID-19: in search of the missing links. Appl Res Qual Life 18(3):1149–1165. https://doi.org/10.1007/s11482-023-10147-2
Shek DTL, Lin L (2014) Personal well-being and family quality of life of early adolescents in Hong Kong: do economic disadvantage and time matter? Soc Indic Res 117(3):795–809. https://doi.org/10.1007/s11205-013-0399-3
Shek DTL, Ng CSM (2014) Family quality of life, personal well-being, and risk behavior in early adolescents in Hong Kong: Related phenomena and research gaps. In: Shek DTL, Sun RCF, Ma CMS (eds) Chinese adolescents in Hong Kong: family life, psychological well-being and risk behavior. Springer Singapore, pp 1–18
Shek DTL, Siu AMH (2019) "Unhappy" environment for adolescent development in Hong Kong. J Adolesc Health 64(6):S1–S4. https://doi.org/10.1016/j.jadohealth.2019.01.007
Shek DTL, Tsui P-F (2013) Developmental outcomes of economically disadvantaged adolescents in Hong Kong: a replication. Int J Adolesc Med Health 25(4):395–403. https://doi.org/10.1515/ijamh-2013-0037
Shing TH (2021, September 3) Can Hong Kong get rid of its atrocious subdivided flats anytime soon? Think China. Retrieved from: https://www.thinkchina.sg/can-hong-kong-get-rid-its-atrocious-subdivided-flats-anytime-soon
Shorrocks A, Davies J, Lluberas R (2021) Global wealth report 2021. Retrieved from: https://www.credit-suisse.com/about-us/en/reports-research/global-wealth-report.html
Smart A, Lee J (2003) Financialization and the role of real estate in Hong Kong's regime of accumulation. Econ Geogr 79(2):153–171. https://doi.org/10.1111/j.1944-8287.2003.tb00206.x
SoCo (2009, March 11) Cage home exhibition. Retrieved from: https://soco.org.hk/en/cage-home_exh/
Solari CD, Mare RD (2012) Housing crowding effects on children's wellbeing. Soc Sci Res 41(2):464–476. https://doi.org/10.1016/j.ssresearch.2011.09.012
South Korea's Ministry of Employment and Labor (2022) Minimum wage. Retrieved from: https://www.moel.go.kr/english/
Statistics Times (2020, October 23) List of countries by projected GDP per capita. Retrieved from: https://statisticstimes.com/economy/countries-by-projected-gdp-capita.php
Stierli M, Shorrocks A, Davies JB, Lluberas R, Koutsoukis A (2014) Global wealth report 2014. Retrieved from: https://www.credit-suisse.com/about-us/en/reports-research/global-wealth-report.html
The Economist (2014, March 15) Planet plutocrat. Retrieved from: https://www.economist.com/international/2014/03/15/planet-plutocrat

Torne R (2021, May 29) Here's why Hong Kong has the most rolls royces per capita. HotCars. Retrieved from: https://www.hotcars.com/heres-why-hong-kong-has-the-most-rolls-royces-per-capita/
Transport and Housing Bureau (2021a) Hong Kong: the facts (housing). Retrieved from: https://www.thb.gov.hk/eng/publications/housing/hongkongthefacts/Hong-Kong-Factsheet.pdf
Transport and Housing Bureau (2021b) Report of the task force for the study on tenancy control of subdivided units. Retrieved from: https://www.thb.gov.hk/eng/contact/housing/studyOnTenancyControl_Report.pdf
UNDESA (2020) World social report 2020: inequality in a rapidly changing world. Retrieved from: https://www.un.org/development/desa/dspd/world-social-report/2020-2.html
UNDP (2020) Human development report 2020: The next frontier, human development and the anthropocene (978-92-1-126442-5). Retrieved from: https://hdr.undp.org/content/human-development-report-2020
UNDP (2022) Human development report 2021–22: uncertain times, unsettled lives: shaping our future in a transforming world. Retrieved from: http://report.hdr.undp.org.s3-website-us-east-1.amazonaws.com/
URI & FCPP (2022) Demographia international housing affordability 2022 edition. Retrieved from: http://www.demographia.com/db-dhi-index.htm
Vergolini L (2011) Social cohesion in Europe: how do the different dimensions of inequality affect social cohesion? Int J Comp Sociol 52(3):197–214. https://doi.org/10.1177/0020715211405421
Wardle M, Mainelli M (2021) The global financial centres index 30. Retrieved from: https://www.longfinance.net/publications/long-finance-reports/global-financial-centres-index-30/
Wissink B, Koh SY, Forrest R (2017) Tycoon city: political economy, real estate and the super-rich in Hong Kong. In: Forrest R, Koh SY, Wissink B (eds) Cities and the super-rich. Palgrave Macmillan, New York, pp 229–252
WJP (2021) World justice project: rule of law index 2021. Retrieved from: https://worldjusticeproject.org/our-work/research-and-data/wjp-rule-law-index-2021
Wong C-K, Lou VW-Q (2010) I wish to be self-reliant: aspiration for self-reliance, need and life satisfaction, and exit dilemma of welfare recipients in Hong Kong. Soc Indic Res 95(3):519–534. https://doi.org/10.1007/s11205-009-9524-8
Wong SH-W (2015) Real estate elite, economic development, and political conflicts in postcolonial Hong Kong. China Rev 15(1):1–38. Retrieved from: http://www.jstor.org/stable/24291927
World Bank (2022) Gini index. Retrieved from: https://data.worldbank.org/indicator/SI.POV.GINI
Wu AM, Chou K-L (2017) Public attitudes towards income redistribution: evidence from Hong Kong. Soc Policy Adm 51(5):738–754. https://doi.org/10.1111/spol.12192
Xiaobin Z, Li Z, Kelvin STO (2004) Income inequalities under economic restructuring in Hong Kong. Asian Surv 44(3):442–473. https://doi.org/10.1525/as.2004.44.3.442
Youngs R (2015) Exploring non-western democracy. J Democr 26(4):140–154. https://doi.org/10.1353/jod.2015.0062

Chapter 5
Urban Inequalities and Diversities in Germany

Peter Krause

Abstract Germany has emerged over centuries as a central European country marked by political shifts that have resulted in deep regional fragmentation. The political burdens of two world wars led, in the late 1940s, to a separation of the country into the communist German Democratic Republic (GDR) and the western Federal Republic of Germany (FRG), a separation that ended with German (re)unification in 1990. The current federal constitution of Germany, which assigns considerable political power to the country's 16 independent states (*Bundesländer*), incorporates several elements of these historical heritages. Consequently, urban developments in Germany reflect the cultural and political backgrounds as historical diversified regions and cities. The first part of this chapter briefly describes these historical impacts on regional and urban developments in Germany. The second part describes the database—the German Socio-Economic Panel (SOEP)—the distinction between urban and rural areas, with variables and measurement designs for (multidimensional) empirical applications. The third part discusses measurement approaches for the genuine characteristics of urban well-being, and the resulting tensions between inequalities and urban diversities. It goes on to explore fixed-fuzzy approaches to capture (multidimensional) fuzzy-based measures of well-being and deprivation in urban living conditions and documents a new measurement framework using Gini-based inequalities and diversities, with a special focus on (multidimensional) between-group differentiations as inequality-based diversity applications. The fourth part provides empirical results on urban population and income developments. The fifth part discusses further multidimensional results on urban inequality, well-being, and deprivation. The chapter ends by summarizing the conceptual and empirical outcomes (part six) and pointing the way toward further discussions (part seven).

Keywords Urban developments · Deprivation and well-being · Inequality and diversity · Gini

---

P. Krause (✉)
Department Socio-Economic Panel (SOEP), German Institute for Economic Research (DIW Berlin), Berlin, Germany
e-mail: pkrause@diw.de

G. H. Tonon (ed.), *Urban Inequalities*, The Urban Book Series,
https://doi.org/10.1007/978-3-031-59746-6_5

5.1 Regional and Urban Development in Germany

In the remarkable 2014 exhibition in London and Berlin, *Germany. Memories of a Nation,* Neil MacGregor characterized the history of Germany, in contrast to other European countries, as "inherently fragmented" but containing "a large number of widely shared memories, awareness and experiences." He continued: "'Germany? Where is it? I do not know where to find such a country' wrote Goethe and Schiller in 1796. In Germany both geography and history have always been unstable. Borders move." (MacGregor 2014: XLIII). This is strikingly reflected, according to MacGregor, in the broad range of former regional currencies as well as the cultural importance of several (formerly German-speaking) cities that now lie outside German borders. German unification in 1990 was the latest of Germany's historical shifts, when the nation's capital was moved to Berlin in the former East.

Cultural identifications in Germany are still strongly linked to regional legacies as well as to common national and cultural backgrounds. These identifications are now consolidated in the country's 16 federal states, which enjoy a high degree of cultural and political sovereignty. We simplify these regional historical identifications by assigning German states to three geographical designations: the Southwest, Northwest, and East.

Southwest (SW) Germany (Bavaria, Baden-Württemberg, Rhineland Palatinate, Saarland, Hesse) is mainly below the Roman Limes and the average mountain East–West lines, which are located almost in the geographic middle of the country along historical within-German language isoglosses. SW Germany contains several historic trade routes and (small) manufacturing centers, a number of independent cities (*freie Reichsstädte*), higher rates of Catholicism, and at least in its western parts, areas with different rules for how inherited land was to be shared or divided among children (Bartels et al. 2020). The resulting fragmentation of land use is still visible in the landscape from a bird's-eye view.

Northwest (NW) Germany (North Rhine-Westphalia, Lower Saxony, Schleswig Holstein, Hamburg, Bremen) has former coal mining and steel industries in its western parts (Ruhr region) and several trading centers in the historic Hanseatic League cities on the North and Baltic Seas. In 1648, the Peace of Westphalia was negotiated and signed in two NW German cities, Münster and Osnabrück, by a European council after thirty years of war throughout German territories, laying the foundations for the BeNeLux countries (Belgium, Netherlands, and Luxembourg) and resolving political conflicts between Protestant and Catholic regions.

East Germany (Thuringia, Saxony, Saxony-Anhalt, Brandenburg, Mecklenburg Western Pomerania, and Berlin) was, under the Warsaw Pact, a separate German country (German Democratic Republic, GDR) that belonged almost entirely to the former communist regimes. From 1961 to the fall of the Berlin Wall in November 1989, East Germany was divided from West Germany by the Iron Curtain. The city of Berlin was also divided under the Four-Power Agreement, with the western part occupied and administered by France, the United Kingdom, and the United States, and the Eastern part occupied by the USSR and serving as the capital of the

GDR. Hawes (2018: 223) claims that the regional differentiation between East and West Germany has further historical roots in periods prior to the twentieth century: "West Germany (1949–1990) was extraordinarily similar to Germania as planned by Caesar Augustus c. 1AD, to East Francia at the Treaty of Verdun in 843AD, and to the confederation of the Rhine in 1808." East Germany also has several historic world cultural heritage sites bearing great importance for German culture—among them Wittenberg, famous for its close connection to Martin Luther; Weimar, where Goethe and Schiller lived and wrote; and Museum Island in Berlin.

Urban developments in Germany are embedded in regional developments. The population explosion following the industrial revolution and improvements in medical care came later to Germany than to other European countries and reshaped the country's urban structure (Heuer 1977; Blotevogel O.J.). Berlin's population grew from just 55,200 in 1709 to 1,888,850 in 1900. For comparison, Paris grew from 515,000 to 2,763,100 between 1700 and 1901 and London from 575,000 to 6,506,900 (leaving out surrounding metropolitan areas). The rapid rise in the German population since the nineteenth century was closely related to urban developments. As illustrated in Fig. 5.1 the West German population increased from 20 to 60 million in the 100 years between 1870 and 1970, a period spanning two world wars and significant waves of refugees resulting from territorial shifts.

In 1871, around 12 million people—more than 60% of Germany's population—were living in small municipalities of fewer than 2000 inhabitants (Krause 1984). The absolute number of individuals living in small villages remained remarkably stable up

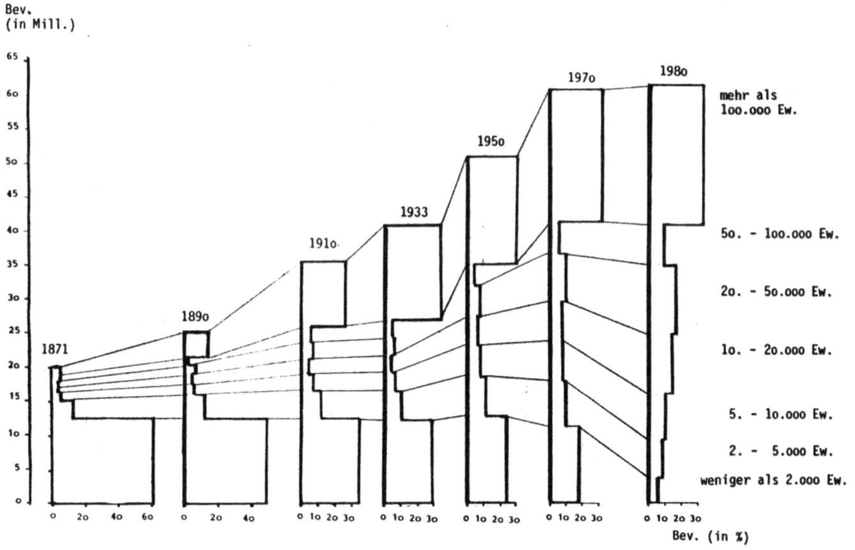

Fig. 5.1 Urban population developments in West Germany, 1871–1980. *Source* Krause (1984: 49); Data from national statistics (1972, 1980) adjusted to West German territory. (Bev.: population; Ew.: inhabitants; mehr als: greater than; weniger als: less than)

to the administrative redistricting of the 1970s. The percentage of population living in large cities of more than 100,000 increased disproportionately from around 5% to more than 30% in 1933, but remained relatively constant thereafter. Despite the absolute increase in the country's population of about 20 million inhabitants between 1933 and 1970/80—driven by post-WW-II refugees, immigration from communist East Germany (up to 1961), workforce immigration, and high birth rates—the percentage of West Germans living in large cities remained constant at below one third of the total population. After the boom in urbanization up to the 1930s, for almost five decades urban growth in Germany was no longer dominated by large cities. Several mid-sized cities instead became urban centers, with a disproportionate increase in their populations up to the 1980s.

The official population rates for cities and municipalities do not reflect the effects of urban complexes created by converging metropolitan areas. Urban world population rates have been steadily increasing—from about 33% in 1960 to 56% in 2021. According to the United Nations' World Urbanization Prospects (2018 Revision), Germany's urban population increased between 1990 and 2010 from about 72 to about 76% of the total population and remained relatively stable up to 2020. In 2021, 78% of German population lived in urban areas. This compares to 81% in France, 84% in the United Kingdom, 83% in US, 92% in Japan, and 75% in Russia (data from the World Bank website).

5.2 Database, Indicators, and Design

5.2.1 Database

For the empirical analyses of urban development, we use the German Socio-Economic Panel (SOEPv37),[1] with its 1,052,596 (continuous) individual observations over the period 1984–2020, which is approximately 15,000–50,000 annual observations of individuals in private households per year. The SOEP (2021) is a representative longitudinal scientific survey launched in West Germany in 1984 (Wagner et al. 2007). Every year since then, in each SOEP household, a personal interview has been conducted with each individual aged 16 and older. In addition, a household questionnaire is completed by the head of the household or the household member who knows the most about household matters. All households and adult household members are surveyed annually through personal interviews. The main topics covered in the questionnaire are demographics as well as indicators of living conditions, including economic welfare and satisfaction with life (happiness) and different life domains. Differentiated information is collected on personal and

[1] SOEPv35, https://doi.org/10.5684/soep.v37.

household incomes and further indicators of wealth, housing, labor market issues, health, and leisure (Goebel et al. 2018; Giesselmann et al. 2019). The SOEP was extended to cover the East German population in 1990, so that the transformation of living conditions following unification can be analyzed from the beginning of this process.

5.2.2 Indicators

Urban Population: We differentiate urban population development in the SOEP according to three dimensions:

- *Regional background*—grouping of 16 states
  - Northwest/Southwest/city-states/East (four categories)
  - Northwest/Southwest/East (three categories)
- *Municipal population*—large cities vs. small villages
  - > 500,000/100–500,000/20–100,000/2–20,2000/< 2000
  - large cities (> 500,000)
- *Urban versus rural areas*—from a (global) urban population perspective, we prioritize urban areas as our main distinction, with a special emphasis on large cities within urban areas as a
  - *joint urban indicator (large cities/urban areas/rural areas)*

Periods and Timing: We observe urban development over a period of 36 years from 1985 to 2020 (the urban area information is not available for 1984) annually and by decades and half-decades before and after unification. For East Germany, data are only available since 1990, annual income data since 1992.

- *Timing and periods* 1985–2020 (2017)
  - 1985-89/1990-94/1995-99/2000-04/2005-09/2010-14/2015-20
  - 1985-89/1990-99/2000-09/2010-20.

As not all variables have been surveyed each year, we concentrate on 2017 for our multidimensional analyses.

Subgroups: The consideration of subgroups is a key aspect in our measurement design. We use the *joint urban indicator (3)* and *regional background (3)* as our main concepts to characterize urban development.

- *Regional background (3) and joint urban indicator (3)*

Individual well-being: For unidimensional analyses, we concentrate on economic and subjective well-being—both measures are further used as key variables to derive dimensional weighting schemes.

- *Economic well-being*: Annual household-net-equivalent incomes (annual previous year incomes at 2020 price levels, adjusted at income-years ($t - 1$); revised OECD-scale)
- *Subjective well-being*: life satisfaction (0–10 scale, 0 = completely dissatisfied, 10 = completely satisfied).

Multidimensional well-being: For multidimensional applications, we refer to two types of designs. First, we concentrate on multidimensional economic well-being, examining two sub-dimensions of economic resources: income and assets. These (sub) dimensions are (as illustrated in Fig. 5.2) further differentiated by aggregation levels, ranging from subjective (intra-individual) evaluations to regional backgrounds. The indicator design defines the reference frame for the analyses; open cells (without indicators) are not considered for interpretation. Cells with several indicators are accordingly down-weighted (by design weights; more details on multidimensional weighting are presented below). We use two indicators for household incomes: annual household equivalent incomes (yinc) and monthly household-net-incomes after (eligible) housing costs (minc_ehc), and two indicators for (equivalized) assets—gross assets (ass_groq) and (positive) assets net of debts (ass_netq). Further indicators across other aggregation levels refer to satisfaction with household income (sat_hinc), personal income (pinc), and regional GDP per capita rates, indicating the regional economic background. Multidimensional analyses of economic well-being refer to the total population, including children. Subjective values for children who are too young to be SOEP respondents are derived from joint income evaluations by their parents; for personal incomes, child allowances are considered individually. All indicators are available for 2017.

The second type of multidimensional indicator design refers to more generalized well-being applications considering further dimensions (as illustrated in Fig. 5.3). The design of dimensions and sub-dimensions is inspired by several types of well-being-related concepts. First, we distinguish (in the tradition of Maslow 1943) two types of basic needs dimensions—(daily) life and home—as absolute necessities, each of them further differentiated by three sub-dimensions: the former by health, nutrition, and recreation, and the latter by housing, household production, and environment and nature.

The following three types of economic, social, and cultural well-being resources reflect Bourdieu's (1984) capital differentiation on the one hand, and Allardt's (1993) distinction between having, loving, and being on the other. Economic resources

Dimensions	***Sub-dimension***	**Aggregation Levels**				
			intra-personal	personal	social	regional
Econ.Resources	*income*		sat-hinc	pinc	yinc	reg_gdp_pc
					minc_ehc	
	assets				ass_netq	
					ass_groq	

Fig. 5.2 Indicator design for multidimensional economic well-being. *Source* Author

Dimensions	*Sub-dimension*		Aggregation Levels				
			intra-personal	personal	social	regional	gener. & global
Life	health	**D1**	sat-health	health-status			
	nutrition						
	recreation						
	pain						
Home	housing	**D2**	sat-home	hroom1_pc			
				hroom2_pc			
	production						
	environment						
	safety						
Econ.Resources	income	**D3**	sat-hinc	pinc	yinc	reg_gdp_pc	
					minc_ehc		
	assets				ass_netq		
					ass_groq		
	consumption						
	debts						
Soc.Resources	governance	**D4**					
	participation			pol_int	cul_int		
				engage	cul_act		
	qol		sat-life				
	exclusion						
Cult.Resources	skills	**D5**		educy			
	tools						
	doings			emp	wi	reg_emp	
	handicaps			uem			

Fig. 5.3 Indicator design for generalized multidimensional well-being. *Source* author (See also Table 5.17 in the Appendix for further details on indicators)

provide the economic foundation and are further differentiated by income, assets, and consumption. Social resources provide the institutional framing, and social capabilities—and are differentiated by governance, participation, and quality of life. Cultural resources encompass current work and achievements and are further differentiated by skills, tools, and doings. Following the principle that there is often an asymmetric relationship between positive aims and negative avoidances (Krause 2016, 2019) (in dynamic contexts referred to as loss aversion)—for instance, satisfaction versus worries, socializing versus loneliness, or emotions versus anxiety—we added another negatively connotated fourth sub-dimension to each of the five main dimensions, adding pain to life, safety to home, debts to economic resources, exclusion to social resources, and handicaps to cultural resources.

The differentiation of each dimension and sub-dimension by aggregation levels may be regarded as an application and extension of Wolfgang Glatzer's and Wolfgang Zapf's welfare conception for social reporting (Glatzer and Zapf 1984) as the joint application of objective and subjective well-being indicators along different domains, and the immanent differentiation in social indicators perspectives for micro, meso, and macro aggregation levels (Noll 2018). Aggregation levels indicate by a generalized socio-ecological perspective interrelated elements of individual, ecological, and social spheres (Bronfenbrenner 1979; Bohmann 2021: 16f.). We included a further conceptual category for global and generational impacts on current living conditions and current responsibilities to explicitly address sustainability and social resilience as an important aspect of current living conditions. Due to a lack of indicators, we are not yet able to consider the global and generational level for our empirical analyses of urban inequalities. The full lists of indicators are documented in the Appendix.

5.3 Measurement Frameworks

5.3.1 *Multidimensional Well-Being—Capabilities and Deprivations*

Inequalities have received a great deal of attention over recent years (Atkinson 2015, 2019; Piketty and Zucman 2015) and are associated above all with tensions between rich and poor. One-dimensional poverty approaches are now established as FGT measures (Foster et al. 1984, 2010) and further developed for multidimensional application (Maasoumi 1986; Aaberge and Brandolini 2015). Following the impressive work at the Oxford Poverty and Human Development Initiative (OPHI) on multidimensional poverty measurement with the Alkire and Foster measurement framework (Alkire and Foster 2011a; b), we describe urban inequalities first as (multidimensional) deprivation, operationalized according to the capability approach (CA) of Sen (1999) and Nussbaum (2000) as a lack of capabilities. We also use the same measurement technique for positive well-beings (Peichl and Pestel 2013)—which may be characterized as opportunities, and which are referred to here simply as aspects of well-being—to indicate, according to the CA, the underlying latent capabilities in the functional space of well-being indicators.

We extend both views—on multidimensional deprivation and multidimensional well-being—with a fixed-fuzzy approach to capture partial deprivation above the poverty line and partial well-being below the wealth line (Krause 2019) as well. The fixed-fuzzy approach operates—in contrast to common fuzzy measures (Betti and Verma 2007; Mueller 2019), which are often criticized as too volatile—with fixed lines, using the poverty line z as the lower limit of full deprivation and the wealth line was the upper limit of full well-being. We further define upper limits u indicating the end of deprivation and lower limits l indicating a complete lack of well-being. To derive upper bounds for deprivation u and lower bounds for well-being l, the same kind of conceptual assumptions about identification are applied as for the definition of common poverty and wealth thresholds (z, w).

To avoid volatility due to (marginal) shifts in lines, it is essential for fuzzy applications to explicitly define the zones of partial deprivation and zones of prosperity by upper and lower bounds as well (Fig. 5.4 indicates the underlaying design). Fixed-fuzzy approaches are more robust than common dummy-based poverty and wealth thresholds when it comes to capturing deprivation and well-being around the reference lines, as even in the case of conceptual difficulties addressing exact lines, at

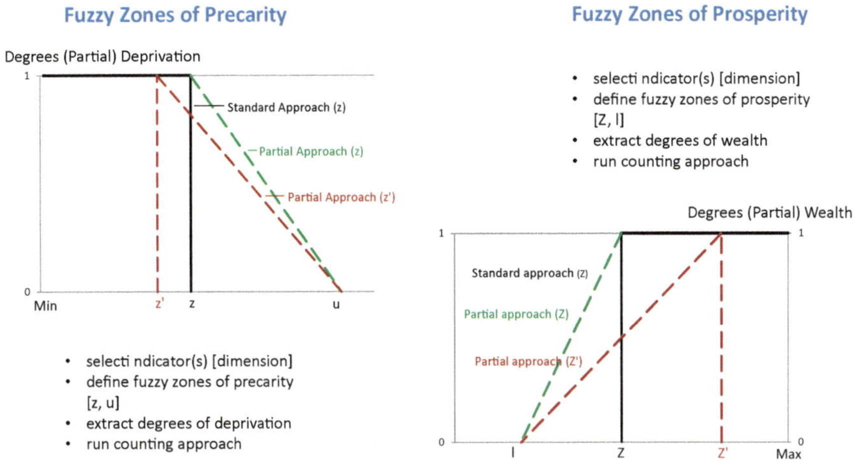

Fig. 5.4 Fixed-fuzzy approach to deprivation and well-being. *Source* Author

least a substantial partial degree of deprivation and well-being is considered. In the standard dummy-based view on poverty and wealth, in contrast, even a small distance above the poverty line and below the wealth line is treated as completely non-poor or non-rich.

Following Sen's distinction between indication and aggregation (1976, 1979), the calculation of deprivation and well-being includes four basic steps:

- Definition of concepts: define dimensions and select indicators
- Indication of concepts: define lines (for each indicator or dimension)
- Indication of measures: apply lines or fuzzy zones (to each indicator or dimension)
- Aggregation of measures: run aggregation (across indicators or dimensions).

The identification of poverty and wealth lines for the standard approach $(hz_i|hw_i)$ is accomplished with dummy variables (as summarized in Fig. 5.5, top). For identification of partial deprivation and partial wealth, we use a fully parameter-driven exponential membership function (as illustrated in Fig. 5.5, bottom), which includes an approximation to the standard dummy-view ($\tau \geq 100$) as a special case, with a default option ($\varepsilon = 1$) for an almost linear decrease in degrees of deprivation or prosperity between upper and lower limits. Values of 1 are assigned for deprivations at and below poverty line z as full poverty; values of 0 indicate no deprivation for indicators at and above the upper limit u. Correspondingly, we apply 1 for full well-being at and above the wealth line w and 0 as almost no well-being for indictor values at and below their lower bounds l. The choice of the membership function is again a matter of conceptual or empirical strategies as part of the identification process as it is the case for the standard dummy-based approach, in which a complete disappearance of any deprivation immediately above the poverty line and non-well-being below the wealth line are assumed a priori. We use the default option ($\varepsilon = 1$) indicating an almost linear decrease between upper and lower bounds.

Identification of Poverty

$$hz_i = 1 \text{ if } y_i < z$$

$$hz_i = 0 \text{ if } y_i \geq z$$

with

y_i individual indicator for poverty
z poverty threshold
hz_i individual poverty score [0, 1]

Identification of Wealth

$$hw_i = 1 \text{ if } x_i > w$$

$$hw_i = 0 \text{ if } x_i \leq w$$

with

x_i individual indicator for wealth
w wealth threshold
hw_i individual wealth score [0, 1]

Identification of Partial Deprivation

$$pd_i = exp\left\{\varepsilon\left[1-\left(\frac{y_i}{z}\right)^{\tau\left[\left(\frac{y_i}{z}\right)\right]}\right]\right\}$$

with

pd_i individual score of partial deprivation for indicator y_i
y_i individual value of deprivation in indicator y
z threshold of poverty/deprivation for indicator y
τ parameter for the type of the baseline identification function
ε parameter for the shape of the identification function

Identification of Partial Well-Beings

$$pw_i = exp\left\{\varepsilon\left[1-\left(\frac{x_i}{w}\right)^{-\tau\left[\left(\frac{x_i}{w}\right)^{-1}\right]}\right]\right\}$$

with

pw_i individual score of partial wealth for indicator x_i
x_i individual value of wealth in indicator x
w threshold of wealth for indicator x
τ parameter for the type of the baseline identification function
ε parameter for the shape of the identification function

Fig. 5.5 Fixed-fuzzy approach for identification of deprivation and well-being. *Source* Author

The definition of upper and lower bounds is sensitive to empirical outcomes, as the fuzzy zones usually catch major parts of the population. Each threshold for upper and lower bounds has to be specified for each variable separately. The parameters for the membership functions are generated from thresholds (from ratios $b = u/z$ for partial deprivation and $q = l/w$ for partial well-being) based on the conceptual settings, which may even vary between individuals if varying thresholds (e.g., due to different regional contexts) are considered (which is not done for the present empirical analysis). In general, we operate with asymmetric settings for fuzzy-based deprivation and well-being so deprivations usually end below the wealth line, whereas prosperities end at the poverty line, meaning that the poverty line corresponds to the lower bound of well-being (for a brief empirical discussion on incomes, see also Krause 2019).

Multidimensional aggregation is processed using the Alkire and Foster measurement framework (Alkire et al. 2015) as a dual-cutoff-censored counting approach[2] (by default we use censoring cut-off lines at cc = 33.3%), accordingly applied to the standard dummy-based approaches for deprivation and wealth as well as to the fixed-fuzzy applications including partial deprivation and partial well-being. The

[2] See also Rippin (2014, 2015) for an alternative multidimensional measurement approach.

aggregations may use ordinal or cardinal specifications; we restrict our analyses here to ordinal settings.

5.3.2 *Inequality and Diversity*

Living conditions in urban areas differ from those in rural areas in the much higher density of the population on the one hand and the greater (multi-cultural) diversity on the other. Urban inequalities are therefore closely related to urban diversity. Inequality reflects differences in living conditions as tensions between having and not-having along the (vertical) rich-poor dimension, whereas diversity reflects (horizontal) between-group variations in ways of living.

To capture these specific aspects of urban inequalities, we specify simplified variations of the Gini inequality index as subindices and derive two Gini-based diversity indices. Since its introduction by Gini (1936), the Gini inequality index has been further developed and used intensively (Mukhopadhyay and Sengupta 2021), analyzed critically with regard to sophisticated methods of decomposition (Yitzhaki 1994; Frick et al. 2006), and applied in multidimensional settings (Decancq and Lugo 2012). We refer here to very basic distinctions for between-group decompositions.

The Gini index can be formulated either based on a Lorenz-type specification as the ratio of two (graphical) semi-areas or as the ratio between empirical distribution in x relative to the hypothetical equal distribution of x. The ratio runs from 0 (total equality: each unit in the population has the same value in x) to 1 (total inequality: just one unit in the population takes all of x). Another identical formulation defines the Gini coefficient as the relative sums of absolute differences in variable x between any units in the population.

5.3.3 *Calculation of the Gini Coefficient*

We refer to the second formulation of the Gini coefficient (5.1) by sum equations as a starting point using individuals i as units of population N. This specification of the Gini index has a double sum: the first sum summarizes, for each individual i in population N, any absolute difference in x to any other individual j in N $|x_i - x_j|$. The second sum aggregates these cumulative differences for each i in total population N. As in the first specification, the Gini index represents the ratio of the empirical sums of absolute differences to the corresponding hypothetical sums of equal differences.

A standard formulation of the Gini index G for variable (x) is

$$G = \frac{\sum_{i=1}^{n} \sum_{j=1}^{n} \left| x_i - x_j \right|}{2n \sum_{i=1}^{n} x_i}, \text{ with } \sum_{i=1}^{n} = N \text{ and } \sum_{j=1}^{n} = N. \tag{5.1}$$

However, this simplified standard Gini formulation of absolute differences does not work precisely, especially for small populations. Consider the following distribution of variable x for three individuals i in population $N = 3$ and a total sum of $x = 10$, such that for individuals I, $x_1 = 0$, $x_2 = 0$, and $x_3 = 10$, taking the total sum of x. The Gini index should be 1, if one unit takes all values of x. According to the simplified standard formula, the Gini index would instead be $G = 40/2 * 3 * 10 = 0.6666$.

That is the case, as by definition for $i = j$ the absolute difference of $|x_i - x_j|$ is always 0, so that for the real sum of differences only $N - 1$ of total population is considered. If we modify the Gini accordingly using $(N - 1)$ instead of N, we get the corrected Gini index $G = 40/(2 * (3–1) * 10) = 1.0$. The $(N - 1)$ modification of the standard Gini index is very common and sensitive to small population sizes. We therefore use the $(N - 1)$ formulation for any subgroup-related differentiations and population adjustments.

For empirical survey data we usually re-weight each population unit by its relative representation in population w_i. The population-weighted $N - 1$ formulated Gini index is based on weighted differences $w_i * w_j * (|x_i - x_j|)$. We use the generalized $N - 1$ formulation of Gini as

$$\mathrm{GG} = \frac{\sum_{i=1}^{n}\sum_{j=1}^{n}\left(\left|x_i - x_j\right|\right) * w_i * w_j}{2(N-1)\sum_{i=1}^{n} x_{wi}}, \text{ with } \sum_{i=1}^{n} w_i = N \text{ and } \sum_{j=1}^{n} w_j = N. \tag{5.2}$$

If we recall the block of absolute individual differences in (5.2) for variable x as

$$\mathrm{Gsum}(x_{wi}) = \sum_{i=1}^{n}\sum_{j=1}^{n}\left(\left|x_i - x_j\right|\right) * w_i * w_j \tag{5.3}$$

and the total sum of population weighted x in population N as

$$\mathrm{Xsum}(x_{wi}) = \sum_{i=1}^{N} x_i * w_i, \tag{5.4}$$

we can simplify the adjusted standard formulation of the Gini index as

$$\mathrm{GG} = \frac{\mathrm{Gsum}(x_{wi})}{2(N-1) * \mathrm{Xsum}(x_{wi})} = \frac{\mathrm{Gsum}(x_{wi})/(N-1)}{2 * \mathrm{Xsum}(x_{wi})}. \tag{5.5}$$

5.3.4 Within-Group and Between-Group Differentiations of the Gini Coefficient

We split, for each individual, the sums of absolute individual differences in GG according to their current subgroup levels as within-group inequalities and between-group inequalities (details on calculation are presented in Table 5.16). A grouping variable *k* is therefore applied to each individual *i* and a corresponding subgroup level l to each of the individual counterparts *j*. It should be noted that for each individual *i*, only the corresponding subgroup level is considered here as within component. The Gini coefficient can accordingly be fully decomposed as

$$GG = GK + GD. \tag{5.6}$$

The identity within-group component GK summarizes absolute differences between each individual *i* and any other individuals *j* within the same subgroup level $l = k$, whereas GD indicates absolute differences between all individuals *i* and any other individuals *j* for any other subgroup levels l.

$$\mathrm{GK} = \frac{\sum_{ki=1}^{n}\sum_{lj=1}^{n}\left(\left|x_{ki} - x_{lj}\right|\right) * w_i * w_j}{2(N-1)\sum_{i=1}^{n} x_{wi}}, \quad \text{for} \quad l = k. \tag{5.7}$$

The between-group component GD indicates any other inequalities between individual *i* and all other individuals *j* in other subgroups $l \neq k$. Depending on the size of the particular subgroup levels, GK should be usually smaller than GD when describing the overall distribution of inequalities according to within-group and between-group components.

$$\mathrm{GD} = \frac{\sum_{ki=1}^{n}\sum_{lj=1}^{n}\left(\left|x_{ki} - x_{lj}\right|\right) * w_i * w_j}{2(N-1)\sum_{i=1}^{n} x_{wi}}, \quad \text{for} \quad l \neq k. \tag{5.8}$$

Summarizing within-group and between-group differences as

$$\mathrm{GKsum}(x_{wi}) = \sum_{ki=1}^{n}\sum_{lj=1}^{n}\left(\left|x_{ki} - x_{lj}\right|\right) * w_i * w_j \quad \text{for} \quad l = k \tag{5.9}$$

$$\mathrm{GDsum}(x_{wi}) = \sum_{ki=1}^{n}\sum_{lj=1}^{n}\left(\left|x_{ki} - x_{lj}\right|\right) * w_i * w_j \quad \text{for} \quad l \neq k, \tag{5.10}$$

we can simplify within-group term GK and between-group component GD as

$$\mathrm{GK} = \frac{\mathrm{GKsum}(x_{wki})}{2(N-1) * \mathrm{Xsum}(x_{wi})} = \frac{\mathrm{GKsum}(x_{wki})/(N-1)}{2 * \mathrm{Xsum}(x_{wi})} \quad \text{for} \quad l = k \tag{5.11}$$

$$\text{GD} = \frac{\text{GDsum}(x_{wli})}{2(N-1) * \text{Xsum}(x_{wi})} = \frac{\text{Gdsum}(x_{wli})/(N-1)}{2 * \text{Xsum}(x_{wi})} \quad \text{for} \quad l \neq k. \tag{5.12}$$

5.3.5 *Between-Group-Related Decompositions of the Gini Coefficient*

Variations in GD describe any absolute differences in variable x between individual i at group level k $(k = l)$ and any other individual j at any other subgroup levels l $(k \neq l)$. The variations in GD between individuals i and j also include mean group differences between the own subgroup level k and any other subgroup level l. To obtain further information on the sources of between-group differences, we treat mean group differences GMK as an integral part of overall GD components.

$$\text{GMK} = \frac{\sum_{ki=1}^{n} \sum_{lj=1}^{n} \left(\left| \mu_k(x_{ki}) - \mu_l\left(x_{lj}\right) \right| \right) * w_i * w_j}{2(N-1) \sum_{i=1}^{n} x_{wi}},$$
$$\text{with} \sum_{i=1}^{n} w_i = N \text{ and} \sum_{j=1}^{n} w_j = N. \tag{5.13}$$

If we accordingly simplify the absolute mean group differences as

$$\text{GMKsum}\left(x_{wlj}\right) = \sum_{ki=1}^{n} \sum_{lj=1}^{n} \left(\left| \mu_k(x_{ki}) - \mu_l\left(x_{lj}\right) \right| \right) * w_i * w_j, \tag{5.14}$$

inequalities in mean group differences GMK can be expressed as

$$\text{GMK} = \frac{\text{GMKsum}(x_{wli})}{2(N-1) * \text{Xsum}(x_{wi})} = \frac{\text{GMKsum}(x_{wli})/(N-1)}{2 * \text{Xsum}(x_{wi})}. \tag{5.15}$$

Like absolute differences are zero for identical individuals $i = j\{|x_{wi} - x_{wj}| = 0\}$, absolute differences in mean group levels of variable x are zero $\{|\mu_k(x_{ki}) - \mu_l(x_{lj})| = 0\}$ for identical subgroup levels $k = l$. Mean group differences GMK therefore refer only to the between-group index GD (and related subindices), but not to the within-group index GK.

The intended diversity-related between-group decomposition is based on the proposition that any individual differences GD between different group members $\left|x_{ki} - x_{lj}\right|$ can be disentangled using mean group levels by three elements: as within-group difference between individual i and mean group level k $|x_{ki} - \mu_k(x_{wki})|$, as mean group differences between subgroups $\left|\mu_k(x_{wki}) — \mu_l\left(x_{wlj}\right)\right|$, and as within-group differences between individual j and group level l $\left|x_{lj} - \mu_l(x_{wlj})\right|$. This works [for x_{ki}, $x_k = \mu_k(x_{wki})$, $x_l = \mu_l\left(x_{wlj}\right)$, x_{lj}] as long as $x_{ki} \leq x_k \leq x_l \leq x_{lj}$, which is almost by definition not the case for all pairs of group members. If group levels differ the other way around ($x_{ki} > x_l$, $x_k > x_l$, $x_l > x_{lj}$), we obtain additional overlapping components that have to be considered for full decompositions as well.

We use two alternative versions for decomposition of between-group inequalities according to mean subgroup levels. The first version describes between-group differences GD as a net effect of mean between-group differences GMK and within-group differences GDD minus an overlapping term GDQ.

$$\mathrm{GD} = \mathrm{GMK} + \mathrm{GDD} - \mathrm{GDQ}. \tag{5.16}$$

The subindices GDD and GDQ summarize the joint within-group differences to mean group levels as the net effect of absolute differences using $xx_i = \min(x_{ki}, x_{lj})$, $xx_k = \min(x_k, x_l)$, $xx_j = \max(x_{ki}, x_{lj})$, $xx_l = \max(x_k, x_l)$, with GDD for positive differences and GDQ for overlapping terms as negative differences

$$\mathrm{GDD} = \frac{\sum_{i=1}^{n}\sum_{j=1}^{dn}((|xx_i - xx_k|) * w_i * w_j) + \sum_{i=1}^{n}\sum_{j=1}^{dn}((|xx_j - xx_l|) * w_i * w_j)}{2(N-1)\sum_{k=1}^{k}\sum_{i=1}^{n} x_{wki}},$$
$$\text{for } xx_i \le xx_k,\ xx_l \le xx_j, \text{ and } l \ne k \tag{5.17}$$

$$\mathrm{GDQ} = \frac{\sum_{i=1}^{n}\sum_{j=1}^{dn}((|xx_i - xx_k|) * w_i * w_j) + \sum_{i=1}^{n}\sum_{j=1}^{dn}((|xx_j - xx_l|) * w_i * w_j)}{2(N-1)\sum_{k=1}^{k}\sum_{i=1}^{n} x_{wki}},$$
$$\text{for } xx_i > xx_k, xx_l > xx_j, \text{ and } l \ne k. \tag{5.18}$$

Both indices can again be simplified accordingly:

$$\mathrm{GDDsum}(x_{wli}) = \sum_{i=1}^{n}\sum_{j=1}^{dn}((|xx_i - xx_k|) * w_i * w_j) + \sum_{i=1}^{n}\sum_{j=1}^{dn}((|xx_j - xx_l|) * w_i * w_j),$$
$$\text{for } xx_i \le xx_k, xx_l \le xx_j, \text{ and } l \ne k \tag{5.19}$$

$$\mathrm{GDQsum}(x_{wlj}) = \sum_{i=1}^{n}\sum_{j=1}^{dn}((|xx_i - xx_k|) * w_i * w_j) + \sum_{i=1}^{n}\sum_{j=1}^{dn}((|xx_j - xx_l|) * w_i * w_j),$$
$$\text{for } xx_i > xx_k, xx_l > xx_j, \text{ and } l \ne k. \tag{5.20}$$

$$\mathrm{GDD} = \frac{\mathrm{GDDsum}(x_{wli})}{2(N-1) * \mathrm{Xsum}(x_{wi})} = \frac{\mathrm{GDDsum}(x_{wli})/(N-1)}{2 * \mathrm{Xsum}(x_{wi})},$$
$$\text{for } xx_i \le xx_k, xx_l \le xx_j, \text{ and } l \ne k \text{ and} \tag{5.21}$$

$$\mathrm{GDQ} = \frac{\mathrm{GDQsum}(x_{wli})}{2(N-1) * \mathrm{Xsum}(x_{wi})} = \frac{\mathrm{GDQsum}(x_{wli})/(N-1)}{2 * \mathrm{Xsum}(x_{wi})},$$
$$\text{for } xx_i > xx_k, xx_l > xx_j, \text{ and } l \ne k. \tag{5.22}$$

The alternative second version for decomposition of between-group inequalities according to mean subgroup levels considers the full range of individual within- and mean between-group variations as

$$\text{GD} = \text{GMK} + \text{GDDR} - \text{GDQR} - \text{GDQM}. \tag{5.23}$$

This version has to consider two further kinds of operations—for $xk_i > x_{lj}$ at the individual level and for $x_k > x_l$ at the group level—which were both excluded by definition in the net operations due to the min|max option in the first version. For $x_i > x_j$ any relations in $x_{ki} \leq x_k \leq x_l \leq x_{lj}$ are reversed as $x_{ki} > x_l$, $x_k > x_l$, $x_l > x_{lj}$ to avoid negative sums.

$$\text{GDDR} = \frac{\sum_{i=1}^{n}\sum_{j=1}^{dn}((|x_{ki} - x_k|) * w_i * w_j) + \sum_{i=1}^{n}\sum_{j=1}^{dn}((|x_{lj} - x_l|) * w_i * w_j)}{2(N-1)\sum_{k=1}^{k}\sum_{i=1}^{n} x_{wki}},$$
$$\text{for}[x_i \leq x_j \text{ and } x_{ki} \leq x_k,\ x_l \leq x_{lj}] \text{ or } [x_i > x_j \text{ and } x_{ki} > x_k, x_l > x_{lj}],$$
$$\text{and } l \neq k \tag{5.24}$$

$$\text{GDQR} = \frac{\sum_{i=1}^{n}\sum_{j=1}^{dn}((|x_{ki} - x_k|) * w_i * w_j) + \sum_{i=1}^{n}\sum_{j=1}^{dn}((|x_{lj} - x_l|) * w_i * w_j)}{2(N-1)\sum_{k=1}^{k}\sum_{i=1}^{n} x_{ki}},$$
$$\text{for}[x_i \leq x_j \text{ and } x_{ki} > x_k, x_l > x_{lj}] \text{ or } [x_i > x_j \text{ and } x_{ki} \leq x_k,\ x_l \leq x_{lj}],$$
$$\text{and } l \neq k. \tag{5.25}$$

Accordingly simplified:

$$\text{GDDRsum}(x_{wlj}) = \sum_{i=1}^{n}\sum_{j=1}^{dn}((|x_i - x_k|) * w_i * w_j) + \sum_{i=1}^{n}\sum_{j=1}^{dn}((|x_j - x_l|) * w_i * w_j),$$
$$\text{for}[x_i \leq x_j \text{ and } x_{ki} \leq x_k, x_l \leq x_{lj}]$$
$$\text{or}[x_i > x_j \text{ and } x_{ki} > x_k, x_l > x_{lj}], \tag{5.26}$$

$$\text{GDQRsum}(x_{wlj}) = \sum_{i=1}^{n}\sum_{j=1}^{dn}((|x_i - x_k|) * w_i * w_j) + \sum_{i=1}^{n}\sum_{j=1}^{dn}((|x_j - x_l|) * w_i * w_j),$$
$$\text{for}[x_i \leq x_j \text{ and } x_{ki} > x_k, x_l > x_{lj}]$$
$$\text{or}[x_i > x_j \text{ and } x_{ki} \leq x_k, x_l \leq x_{lj}], \text{ with} \tag{5.27}$$

$$\text{GDDR} = \frac{\text{GDDRsum}(x_{wli})}{2(N-1) * \text{Xsum}(x_{wi})} = \frac{\text{GDDRsum}(x_{wli})/(N-1)}{2 * \text{Xsum}(x_{wi})},$$
$$\text{for}[x_i \leq x_j \text{ and } x_{ki} \leq x_k, x_l \leq x_{lj}]$$
$$\text{or}[x_i > x_j \text{ and } x_{ki} > x_k, x_l > x_{lj}], \text{ and } l \neq k \text{ and} \tag{5.28}$$

$$\text{GDQR} = \frac{\text{GDQRsum}(x_{wli})}{2(N-1) * \text{Xsum}(x_{wi})} = \frac{\text{GDQsum}(x_{wli})/(N-1)}{2 * \text{Xsum}(x_{wi})},$$
$$\text{for}[x_i \leq x_j \text{ and } x_{ki} > x_k, x_l > x_{lj}]$$

$$\text{or}\left[x_i > x_j \text{ and } x_{ki} \le x_k, x_l \le x_{lj}\right], \text{ and } l \ne k. \tag{5.29}$$

For $x_k > x_l$ we add—just as we do for GDD(R) and GDQ(R) on the individual level—an overlapping term at group level GDQM as a counterpart to the absolute mean group differences GMK.

$$\text{GDQM} = 2 * \frac{\sum_{i=1}^{n}\sum_{j=1}^{n}\left(\left|\mu_k(x_{ki}) - \mu_l\left(x_{lj}\right)\right|\right) * w_i * w_j}{2(N-1)\sum_{i=1}^{n} x_{wi}},$$
$$\text{for}\left[x_i \le x_j \text{ and } x_k < x_l\right]$$
$$\text{or}\left[x_i > x_j \text{ and } x_k \ge x_l\right], \text{ and } l \ne k. \tag{5.30}$$

Accordingly simplified:

$$\text{GDQMsum}(x_{wlj}) = 2 * \sum_{i=1}^{n}\sum_{j=1}^{n}\left(\left|\mu_k(x_{ki}) - \mu_l\left(x_{lj}\right)\right|\right) * w_i * w_j,$$
$$\text{for}\left[x_i \le x_j \text{ and } x_k < x_l\right]$$
$$\text{or}\left[x_i > x_j \text{ and } x_k \ge x_l\right], \text{ and } l \ne k, \text{ and} \tag{5.31}$$

$$\text{GDQM} = \frac{\text{GDQMsum}(x_{wli})}{2(N-1) * \text{Xsum}(x_{wi})} = \frac{\text{GDQMsum}(x_{wli})/(N-1)}{2 * \text{Xsum}(x_{wi})}. \tag{5.32}$$

5.3.6 Diversity-Related Compositions of Gini Subindices

For decomposition of between-group inequalities according to mean group levels, we mentioned two opposing joint individual within-group differences: GDD versus GDQ for Version 1 and GDDR versus GDQR for Version 2. Each of these components, which are all non-integral parts of the original Gini calculation, are sensitive to joint within-group variations, referring to each element of absolute individual differences across subgroups. GDD and GDQ (as well as GDDR and GDQR) together describe the full amount of joint within-group variation net of differences in mean group levels for each element of Gini calculation in different subgroups. We indicate these inequality-based joint variations within different subgroup levels as diversity measures. For the calculation of Gini-based diversity indices, we use both versions of joint within-group differences to mean group levels (GDD, GDQ and GDDR, GDQR). The resulting Gini-based diversity indices GDIV1 and GDIV2 can thus be regarded as further Gini-based subindices as well.

Diversity indices GDIV1 and GDIV2 are based on different notions of how within-group variations are recognized by individuals from different subgroups. GDIV1 refers to the net effect of inequality-based diversities.

$$\text{GDIV1} = \text{GDD} + \text{GDQ}, \quad \text{for} \quad l \neq k; \tag{5.33}$$

$$\text{GDIV1} = \frac{\sum_{i=1}^{n}\sum_{j=1}^{dn}((|xx_i - xx_k|) * w_i * w_j) + \sum_{i=1}^{n}\sum_{j=1}^{dn}((|xx_j - xx_l|) * w_i * w_j)}{2(N-1)\sum_{k=1}^{k}\sum_{i=1}^{n} x_{wki}}$$
$$\text{for} \quad l \neq k; \tag{5.34}$$

$$\text{GDIV1sum}(x_{wlj}) = \sum_{i=1}^{n}\sum_{j=1}^{dn}((|xx_i - xx_k|) * w_i * w_j) + \sum_{i=1}^{n}\sum_{j=1}^{dn}((|xx_j - xx_l|) * w_i * w_j) \quad \text{for} \quad l \neq k; \tag{5.35}$$

$$\text{GDIV1} = \frac{\text{GDDsum}(xw_{li}) + \text{GDQsum}(x_{wli})}{2(N-1) * \text{Xsum}(x_{wi})} = \frac{\text{GDIV1sum}(x_{wli})/(N-1)}{2 * \text{Xsum}(x_{wi})} \quad \text{for} \quad l \neq k. \tag{5.36}$$

GDIV2 refers instead to the total effect of inequality-based diversities:

$$\text{GDIV2} = \text{GDDR} + \text{GDQR} \tag{5.37}$$

$$\text{GDIV2} = \frac{\sum_{i=1}^{n}\sum_{j=1}^{dn}((|x_{ki} - x_k|) * w_i * w_j) + \sum_{i=1}^{n}\sum_{j=1}^{dn}((|x_j - x_{lj}|) * w_i * w_j)}{2(N-1)\sum_{k=1}^{k}\sum_{i=1}^{n} x_{wi}}$$
$$\text{for} \quad l \neq k; \tag{5.38}$$

$$\text{GDIV2sum}(x_{wlj}) = \sum_{i=1}^{n}\sum_{j=1}^{dn}((|x_{ki} - x_k|) * w_i * w_j) + \sum_{i=1}^{n}\sum_{j=1}^{dn}((|x_{lj} - x_l|) * w_i * w_j), \quad \text{for} \quad l \neq k; \tag{5.39}$$

$$\text{GDIV2} = \frac{\text{GDDRsum}(x_{wli}) + \text{GDQRsum}(x_{wli})}{2(N-1) * \text{Xsum}(x_{wi})} = \frac{\text{GDIV2sum}(x_{wli})/(N-1)}{2 * \text{Xsum}(x_{wi})} \quad \text{for} \quad l \neq k. \tag{5.40}$$

GDIV1 may be regarded as a lower bound in overall diversities, assuming full reports of within-group variation by each individual *i* and *j* from different subgroups *k* and *l*. GDIV1 is conceptualized as the net effect of absolute differences using $xx_i = \min(x_{ki}, x_{lj})$, $xx_k = \min(x_k, x_l)$, $xx_j = \max(x_{ki}, x_{lj})$, and $xx_l = \max(x_k, x_l)$. Differences in individual levels x_{ki} and x_{lj} as well as the related differences in group levels x_k and x_l are fully recognized by each other and therefore interchangeable between any individuals *i* and *j*.

GDIV2 marks upper bounds in inequality-based diversities between each pair of individuals i and j from different subgroups k and l. This specification assumes that each of the individuals refers primarily to their own subgroup level and takes further notice of variations in subgroup-related differences of other individuals to their subgroup level.

We use both Gini-based diversity indices to describe inequality-related diversities across subgroups.

5.3.7 Population Adjustments for Gini-Based Subindices and Diversity Indices

Within-group and between-group-related subindices GK and GD are calibrated according to the total population. Total Gini GG can such be fully described by GK and GD according to their overall contribution in population GG = GK + GD. Depending on the group size of the particular subgroup levels—GK should be usually smaller than GD—describing the overall amount and distribution of inequalities when considering grouping levels. Variations in GK and GD are driven either by differences in distribution of variable x_i or by differences in population size of group levels $\{k|l\}$.

We apply further population-adjusted versions for subindices GK and GD as GGK and GGD to control for differences in population size of individuals j at group levels $\{k|l\}$. GGK and GGD can therefore be directly compared according to their amount of internal inequalities net of population differences in subgroups.

The identity within-group population $\{Nk\}$ for $k = 1$ is $Nk = \sum_{j=1}^{nk} w_j$. The overall Gini coefficient GG has been assigned to population $N - 1$ (instead of N), as absolute differences in $|x_i - x_j|$ for $i = j$ are by definition zero and thus do not contribute to overall absolute differences in population. As all individual identities $i = j$ are only part of identity group levels $k = 1$, the $N - 1$ adjustment refers only to the within-group term GGK, but not to between-group terms GGD.

Any population adjustment of the within-group component GGK is related to $(Nk - 1)$ instead of to the overall population $(N - 1)$, and any between-group-related indices GGD, including each of their related subindices $l \neq k$, are related to population $(N - Nk)$. The population adjustments thus depend on the current population size of subgroup levels k and l and vary across individuals, in contrast to the overall $(N - 1)$ assignment in Gini raw indices. After adjustment, the within-group index GGK and the between-group index GGD do not sum up to GG. Table 5.1 summarizes Gini raw indices [G...] and the population adjusted versions [GG...]. The overall index GG is already, by definition, adjusted to the total population.

Population adjustments can accordingly be processed for any further between-group-related Gini-based subindices. As all terms are adjusted for the same between-group related population $(N - Nk)$. The composition equation also holds after adjustment for GGD (as described in Table 5.2).

Table 5.1 Population adjustments for within- and between-group Gini indices

Gini—raw indices		Gini population-adjusted indices		Total	
GG	$\frac{\text{Gsum}(x_{wi})/(N-1)}{2*\text{Xsum}(x_{wi})}$	GG	$\frac{\text{Gsum}(x_{wi})/(N-1)}{2*\text{Xsum}(x_{wi})}$		Total inequalities
GK	$\frac{\text{GKsum}(x_{wi})/(N-1)}{2*\text{Xsum}(x_{wi})}$	GGK	$\frac{\text{GKsum}(x_{wi})/(Nk-1)}{2*\text{Xsum}(x_{wi})}$	for $l = k$	Within-group I
GD	$\frac{\text{GDsum}(x_{wi})/(N-1)}{2*\text{Xsum}(x_{wi})}$	GGD	$\frac{\text{GDsum}(x_{wi})/(N-Nk)}{2*\text{Xsum}(x_{wi})}$	for $l \neq k$	Between-group I
GG = GK + GD					Decomposition

Source Author

Table 5.2 Population adjustments for between-group-related Gini subindices

Gini—raw indices		Gini population-adjusted indices		Between group	
GD	$\frac{\text{GDsum}(x_{wi})/(N-1)}{2*\text{Xsum}(x_{wi})}$	GGD	$\frac{\text{GDsum}(x_{wi})/(N-Nk)}{2*\text{Xsum}(x_{wi})}$	for $l \neq k$	Between-group I
GMK	$\frac{\text{GMKsum}(x_{wi})/(N-1)}{2*\text{Xsum}(x_{wi})}$	GGMK	$\frac{\text{GMKsum}(x_{wi})/(N-Nk)}{2*\text{Xsum}(x_{wi})}$	for l ≠ k	Mean group I
GDD	$\frac{\text{GDDsum}(x_{wi})/(N-1)}{2*\text{Xsum}(x_{wi})}$	GGDD	$\frac{\text{GDDsum}(x_{wi})/(N-Nk)}{2*\text{Xsum}(x_{wi})}$	for $l \neq k$	Joint-within group I [net effect]
GDQ	$\frac{\text{GDQsum}(x_{wi})/(N-1)}{2*\text{Xsum}(x_{wi})}$	GGDQ	$\frac{\text{GDQsum}(x_{wi})/(N-Nk)}{2*\text{Xsum}(x_{wi})}$	for $l \neq k$	Joint-within group I [net overlaps]
GD = GMK + GDD – GDQ		GGD = GGMK + GGDD – GGDQ		for $l \neq k$	Composition
GMK	$\frac{\text{GMKsum}(x_{wi})/(N-1)}{2*\text{Xsum}(x_{wi})}$	GGMK	$\frac{\text{GMKsum}(x_{wi})/(N-Nk)}{2*\text{Xsum}(x_{wi})}$	for $l \neq k$	Mean group I
GDDR	$\frac{\text{GDDRsum}(x_{wi})/(N-1)}{2*\text{Xsum}(x_{wi})}$	GGDDR	$\frac{\text{GDDRsum}(x_{wi})/(N-Nk)}{2*\text{Xsum}(x_{wi})}$	for $l \neq k$	Joint-within group I [full effect]
GDQR	$\frac{\text{GDQRsum}(x_{wi})/(N-1)}{2*\text{Xsum}(x_{wi})}$	GGDQR	$\frac{\text{GDQRsum}(x_{wi})/(N-Nk)}{2*\text{Xsum}(x_{wi})}$	for $l \neq k$	Joint-within group I [full overlaps]
GDQM	$\frac{\text{GDQMsum}(x_{wi})/(N-1)}{2*\text{Xsum}(x_{wi})}$	GGDQM	$\frac{\text{GDQMsum}(x_{wi})/(N-Nk)}{2*\text{Xsum}(x_{wi})}$	for $l \neq k$	Mean group I [full overlaps]
GD = GMK + GDDR – GDQR – GDQM		GGD = GGMK + GGDDR – GGDQR – GGDQM		for $l \neq k$	Composition

Source Author

As Gini-based diversity indices are derived from between-group components, population adjustments can be also processed, holding decompositions for raw versions GDIV and for population-adjusted version GGDIV. (Between-group related decompositions for Gini-based diversity indices are described in Table 5.3).

Table 5.3 Population adjustments for Gini-based diversity indices

Gini—raw diversity indices		Gini pop. adj. diversity indices		Between group	
GDIV1	$\frac{\text{GDIV1sum}(x_{wi})/(N-1)}{2*\text{Xsum}(x_{wi})}$	GGDIV1	$\frac{\text{GDIV1sum}(x_{wi})/(N-Nk)}{2*\text{Xsum}(x_{wi})}$	for $l \neq k$	Diversity index I [net effect]
GDIV2	$\frac{\text{GDIV2sum}(x_{wi})/(N-1)}{2*\text{Xsum}(x_{wi})}$	GGDIV2	$\frac{\text{GDIV2sum}(x_{wi})/(N-Nk)}{2*\text{Xsum}(x_{wi})}$	for $l \neq k$	Diversity index II [total effect]
GDIV1 = GDD + GDQ		GGDIV1 = GGDD + GGDQ		for $l \neq k$	Decomposition I
GDIV2 = GDDR + GDQR		GGDIV1 = GGDDR + GGDQR		for $l \neq k$	Decomposition II

Source Author

5.3.8 Multidimensional Gini Subindices and Diversity Indices

For multidimensional applications of Gini subindices and diversity indices, we follow the row-first approach by Decanq and Lugo (2012) and others. Any Gini index can, according to this approach, be reassigned as a multidimensional index using dimensional weighting with sum of dimensional weights (dw) = 1. The dimensional weights specify the percentage to which each dimension (or variable) s contributes to the supposed overall inequality across dimensions. A standard approach is to use equal weights or any design-based variable assignments. We also propose the use of welfare-based individual weighting schemes, as dimensional priorities may vary between individuals and may change during the life cycle.

$$\sum_{s=1}^{d} dw_s = 1 \tag{5.41}$$

$$\text{MGG} = \frac{\sum_{s=1}^{d} dw_s \sum_{i=1}^{n} \sum_{j=1}^{n} \left(\left(\left| x_{si} - x_{sj} \right| \right) * w_{si} * w_{sj} \right)}{2(N-1) \sum_{s=1}^{d} dw_s \sum_{i=1}^{n} * x_{wsi}},$$

$$\text{for} \sum_{i=1}^{n} w_i = \sum_{j=1}^{n} w_j = N \text{ and } \sum_{s=1}^{d} dw_s = 1. \tag{5.42}$$

With

$$\text{MGsum}(x_{wsi}) = \sum_{s=1}^{d} dw_s \sum_{i=1}^{n} \sum_{j=1}^{n} \left(\left(\left| x_{si} - x_{sj} \right| \right) * w_{si} * w_{sj} \right) \text{ and} \tag{5.43}$$

$$\text{MXsum}(x_{wsi}) = \sum_{s=1}^{d} dw_s \sum_{i=1}^{N} x_{wsi}, \tag{5.44}$$

we can again simplify also the multidimensional formulation of the Gini index as

$$\mathrm{MGG} = \frac{\mathrm{MGsum}(x_{wsi})}{2(N-1) * \mathrm{MXsum}(x_{wsi})} = \frac{\mathrm{MGsum}(x_{wsi})/(N-1)}{2 * \mathrm{MXsum}(x_{wsi})}. \quad (5.45)$$

Any of the previously differentiated raw [*G*… {GK, GD, GKM, GDD*, GDQ*}] and population-adjusted [GG… {GGK, GGD, GGKM, GGDD*, GGDQ*}] Gini subindices can accordingly be assigned for multidimensional applications as MG… and MGG….

Decanq and Lugo (2012) propose a parameter-driven adaptation to meet further axiomatic requirements for multidimensional Gini applications. We provide a parameter-driven version of MGG (and for any other M-subindices) as an additional option for sensitivity analyses, but use the basic MGG formulation as our standard option, as the standard version can be fully decomposed according to dimensions and subgroups.

Multidimensional assignments are essential for diversity indices, as diversity points not only to variations across groups but also to the dimensional mix of well-being arrangements. It may be questionable whether the diversity indices proposed here, which are derived from and conceptualized as Gini subindices, completely capture all the conceptual requirements which have to be spelled out more explicitly in any case. But at least they may contribute to addressing forms of diversity besides inequalities in multidimensional assessments. Accordingly we use MGDIV{1, 2} and population-adjusted MGGDIV{1, 2} for multidimensional diversity applications.

5.3.9 Decompositions for Gini Subindices and Diversity Indices

We process three types of decompositions: by factors, by subgroups, and—for the multidimensional applications—also by dimensions. Details on (de-)compositions by factors have been described step by step above; Table 5.4 summarizes the options for factor (de)compositions. Decomposition by subgroups can be applied to any of the aforementioned (sub-)indices. Decomposition by dimensions can be applied to any of the multidimensional Gini-based indices; decomposition can also be combined with subgroup differentiations. Decompositions are not available for beta-parameter-driven extensions of multidimensional M-Gini and M-diversity indices.

5.3.10 Rescaling of Indicators

For any Gini-based application to multidimensional inequality and diversity, we suggest the use of harmonized indicators—either normalized at (0, …, 1) ranges or further adjusted. We propose a ct1-scale (1, …, 100)—a kind of percentage

Table 5.4 Factor (de)composition for Gini-based subindices and diversity indices

Raw indices Gini|diversity	Pop. adj. indices Gini|diversity	
GG = GK + GD		
GD = GMK + GDD − GDQ	GGD = GGMK + GGDD − GGDQ	for $l \neq k$
GD = GMK + GDDR − GDQR − GDQM	GGD = GGMK + GGDDR − GGDQR − GGDQM	for $l \neq k$
GDIV1 = GDD + GDQ	GGDIV1 = GGDD + GGDQ	for $l \neq k$
GDIV2 = GDDR + GDQR	GGDIV2 = GGDDR + GGDQR	for $l \neq k$
Raw indices M-Gini|M-diversity	Pop. adj. indices M-Gini|M-diversity	
MGG = MGK + MGD		
MGD = MGMK + MGDD − MGDQ	MGGD = MGGMK + MGGDD − MGGDQ	for $l \neq k$
MGD = MGMK + MGDDR − MGDQR − MGDQM	MGGD = MGGMK + MGGDDR − MGGDQR − MGGDQM	for $l \neq k$
MGDIV1 = MGDD + MGDQ	MGGDIV1 = MGGDD + MGGDQ	for $l \neq k$
MGDIV2 = MGDDR + MGDQR	MGGDIV2 = MGGDDR + MGGDQR	for $l \neq k$

Source Author

scale with positive values ranging from 1 to 100. We further suggest using top- and bottom-coding for continuous variables with high ranges (such as incomes and assets)—to reduce temporary impacts from outliers and provide more robust results for comparative inequality assignments of the entire population, and to provide separate inequality-based analyses when considering the very rich and very poor. We use recoding beyond the first and above the 99th percentile in annual incomes to provide robust inequality measures when monitoring overall population developments over time (detailed inequality analyses on the very top and bottom ends should be done separately).

Any variable x can according to Table 5.5 be rescaled into any other scale in three steps—coding (1), normalization (2), and rescaling (3).

Any rescaled variable y_o (including y_{ct1}) can again be retransformed into the original form of y_c (after coding, if applied).

Table 5.5 Rescaling of indicators (for multidimensional inequality applications)

1. Coding	$y_c = x$
	$y_c = c\min(x)$ if $x < c\min(x)$
	$y_c = c\max(x)$ if $x > c\max(x)$
2. Normalization	$y_n = [y_c—\min(y_c)]/[\max(y_c) - \min(y_c)]$
3. Out-scale (gen.)	$y_o = y_n * (o\max\{y_o\} - o\min\{y_o\}) + o\min(y_o)$
Ct1-scale (1, …, 100)	$y_{\text{ct1}} = y_n * (100 - 1) + 1$

Source Author

5.3.11 Multidimensional Weighting Schemes

Any empirical multidimensional application faces the problem of dimensional weighting—applied relatively as a percentage score so that the dimensional weights sum up to 1 (5.41). A standard practice is to use design-based equal weights—each dimension (or cell containing one or more indicators) receives equal weights. We can thus define dimensional weights by any set of dimensions and sub-dimensions. Within cells, indicators are again considered equally. Indicators from different cells may also be combined by design as joint cells if indicators are directly related (as in the case of employment and unemployment). The merits of this procedure are that it is driven by conceptual settings—independent of the simple number of indicators of each dimension. We therefore use the dimensional design—as outlined above (Figs. 5.2 and 5.3)—as a conceptual starting point.

Equal dimensional weightings are almost by definition constant for all observations. When looking at lifecycle developments, it may be quite likely, however, that individual preferences may vary—and the same may be true of differing or changing economic, social, or regional environments. We therefore propose the use of combined settings of design-driven conceptual and empirical procedures, based on well-being applications for dimensional weightings (Table 5.6 describes the procedure used to derive individual dimensional weights in four steps.). Dimensional weights have to be applied to each indicator in any multidimensional application—for multidimensional deprivation and well-being as well as for multidimensional inequality and diversity.

Table 5.6 Individual multidimensional weighting schemes using individual well-being indicators

		Life-cycle	Source	d_{weights}	Sum
1	Design		{normativ}	weq_var	1.00
2	Living.Std.	{Age groups}	ln_yinc		
	Dim.dWeights	{Controls}	{r2a_var}	wdinc_var	1.00
	Ind.dWeights		{pre_var/mean_pre_c}	wiinc_var	1.00
3	Qual. of Life	{Age groups}	sat_life		
	Dim.dWeights	{Controls}	{r2a_var}	wdsat_var	1.00
	Ind.dWeights		{pre_var/mean_pre_c}	wisat_var	1.00
4	Integration and smoothing		{weq * wdinc * wiinc * wdsat * wisat}	wdi_raw_var	1.00
			{lnweq * lnwdinc * lnwiinc * lnwdsat * lnwisat * 1000}	wdi_var	1.00

Source Author

First (1) we apply conceptual design weights to each indicator. Next we run empirical regressions on two key variables of well-being applications—life satisfaction and (log) equivalized household-net-incomes (2 and 3). These regressions are split by age groups to capture dimensional shifts across the lifecycle. Each regression uses the same basic set of controls (age, sex, regional area, basic differentiation on current municipality, household type and size, migration background, and homeowner) plus the indicator variable. We include two basic pieces of information from each regression—the relative meaning of each indicator to each of the key variables (adjusted r2, r2a_var) and the individual variation in preferences as a ratio of the predicted-to-mean-value of key variables (pre_var/mean_pre_c). The relative meaning is summarized for each key variable as dimensional levels (wdinc_var, wdsat_var), differentiated only by subgroups; the individual variations are summarized for each key variable at the individual level (wiinc_var, wisat_var). All components are integrated and smoothed (using ln-transition) (4) to catch (remaining) outliers. The resulting dimensional weighting schemes provide quite robust dimensional weights—considering relative meanings of dimensions as well as individual preferences, and integrating conceptual and empirical well-being applications.

5.4 Urban Well-Being and Deprivation, Inequality, and Diversity

5.4.1 Urban Population Developments and Trends in Well-Being in Germany

Our empirical analyses of urban developments in Germany refer to the time spans 1985–1989 for West Germany including West Berlin, and 1990–2020 for the overall population of Germany including East Germany in the post-unification period. To describe population developments in the German regions (based on the 16 federal states), we indicate city-states (Hamburg, Bremen, and Berlin) in Table 5.7 as a separate category—in further tables, city-states Hamburg and Bremen are assigned to the Northwest and Berlin to the East. Before unification, the northern and southern parts of West Germany each had about 50% of the population when including the populations of the city-states in the northern region. After unification, the German population increased from 60.5 to between 80 and 82 million. In the city-states that formed large urban centers in the northern and eastern regions, the population has increased in the last decade, whereas in East Germany (outside Berlin), the population has declined steadily over the last three decades. Population developments differ further between the German regions.

With German unification, the overall urban population decreased, as the new Eastern states contained more rural areas. About two thirds of the German population lived in urban areas at that time (as shown in Table 5.8, left), and this has remained relatively stable over the last three decades. These results differ slightly from the

Table 5.7 Population developments in the German regions (states)

	Total	North-West	South-West	City-States	East	
	Mill	pct	pct	pct	pct	Total
1985/1989	*60.5*	*42.9*	*50.6*	*6.6*	*x*	100
1990/1994	79.6	34.1	40.4	6.8	18.7	100
1995/1999	81.1	34.8	41.1	6.7	17.4	100
2000/2004	81.6	35.0	41.6	6.8	16.7	100
2005/2009	81.4	34.9	42.1	6.8	16.3	100
2010/2014	79.9	35.3	41.9	7.0	15.8	100
2015/2020	81.6	35.1	42.0	7.5	15.4	100

SOEPv37. Annual total population in private households

previously mentioned UN results due to some varying definitions, but still present almost the same overall picture.

When looking at population developments by municipalities (as shown in Table 5.8, right), we find diverse developments: an ongoing decrease in the smallest communities of fewer than 2000 inhabitants, with some gains in the small cities up to 20,000 inhabitants and relatively stable populations in medium-sized cities of 20–100,000; slight decreases in larger cities of 100–500,000; and rising populations in the large urban centers with more than 500,000 inhabitants, especially during the last decade.

Distributions and trends in urban areas and municipalities vary between regions (Table 5.9 illustrates the shifts in urban–rural population profiles across German regions.). In the Northwest, we find a higher share of the population in urban centers and low populations in small urban and rural areas. In the Southwest, the population is also concentrated in small urban areas, and we also find higher shares of the population in rural areas than in the North. East Germany is characterized by a

Table 5.8 Population developments in urban areas and municipalities

	Region type		Municipalities [admin. population]					Total
	Urban	Rural	< 2000	2–20,000	20–100,000	100–500,000	> 500,000	
	pct	pct	pct	pct	pct	pct	pct	
1985/89	*75.5*	*24.5*	*4.8*	*34.7*	*27.8*	*16.4*	*16.3*	100
1990/99	68.2	31.8	9.1	32.7	26.3	17.0	14.9	100
2000/09	68.1	31.9	7.8	33.4	27.4	16.7	14.6	100
2010/20	68.1	31.9	5.5	34.3	27.7	15.6	16.9	100

SOEPv37. Annual total population in private households

polarization between large cities, on the one hand, and high population shares of more than 50% in rural areas, on the other—this dichotomy has become even more pronounced over the last three decades since unification.

Overall, median income developments indicate ups and downs in Germany during the 1990s; some economic stagnation without real gains between 2000 and 2015; and a period of some growth between 2015 and 2020. Large cities, other urban areas, and rural areas follow these lines of income developments as well. Between 1992 and 2015, large cities showed slightly higher income levels, and between 2015 and 2020 almost the same mean income as other urban areas. Income levels appear slightly lower for rural areas—following the overall trends—which partly coincides with lower income levels in East Germany (Fig. 5.6 indicates slightly lower income levels in rural areas, Table 5.10 reveals, that this is mainly driven by regional differences).

Developments in mean life satisfaction mirror these economic trends: *u*-shaped turns during the 1990s, a decrease in life satisfaction during the period of economic stagnation (with increasing unemployment) from 2000 to 2006, and a long period of steady recovery afterward from 2006 to 2020. All types of regions display these overall trends. Large cities appear to have had slightly lower values for subjective well-being—despite higher income levels—during the 1990s. Urban areas indicate in general slightly higher levels of subjective well-being than rural areas—which partly coincides again with lower satisfaction levels in East Germany (Fig. 5.7; Table 5.10 [right] show similar regional urban–rural patterns for life satisfaction as before Fig. 5.5; Table 5.10 [left] for median incomes).

These overall trends are again driven by substantial regional differences (Table 5.10 indicates regional urban–rural differences in median incomes [left] and life satisfaction [right]). Income levels are slightly higher in the Southwest than the Northwest, and substantially higher in West German regions than in East German regions. Median income levels are slightly higher in large western cities but different

Table 5.9 Regional developments in urban areas and municipalities

		Urban large cities	Urban cities	Urban small cities	Urban area	Rural cities	Rural area	Total
		> 500,000	100–500.000	20–100,000	< 20,000	> 20,000	< 20,000	
2015/20	North-West	19.0	22.0	27.0	9.5	7.9	14.7	100
	South-West	10.9	12.5	17.8	30.2	5.7	23.0	100
	East	27.8	9.5	2.5	4.2	20.4	35.7	100
	Total	17.3	15.6	18.3	17.1	9.4	22.3	100
1990/94	North-West	17.8	24.8	27.1	9.7	7.2	13.3	100
	South-West	8.6	11.8	15.9	35.5	6.9	21.4	100
	East	24.2	10.4	2.4	6.0	18.5	38.5	100
	Total	15.6	16.3	17.0	19.3	9.7	22.3	100

SOEPv37. Annual total population in private households

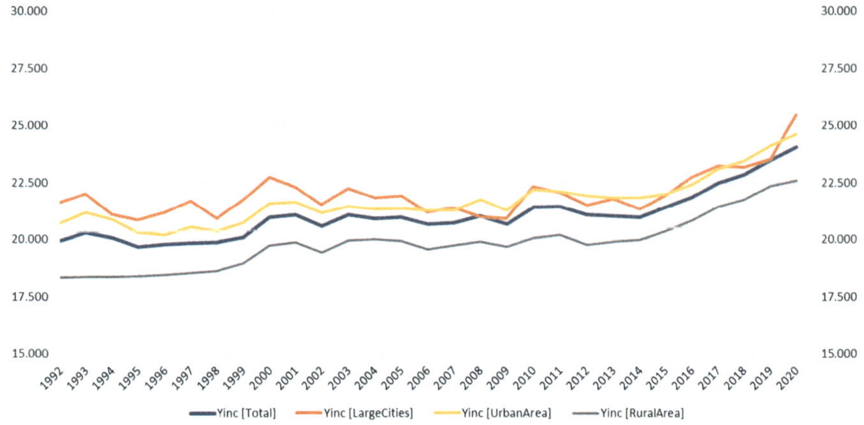

Fig. 5.6 Long-term urban income developments (real household net equivalent incomes), 1992–2020. SOEPv37. Yinc: Med. Annual Eq.Household.Net.Income [€2019; rev.OECD-Scale]

Table 5.10 Regional urban differences in household net income and life satisfaction

	HH-Net-Income				Life Satisfaction			
1992/94	**N-West**	**S-West**	**East**	*Total*	**N-West**	**S-West**	**East**	*Total*
urban large cities	20688	25413	20779	21619	7,0	7,0	6,4	*6,8*
urban cities	21221	21805	16510	20391	7,1	7,2	6,0	*7,0*
urban small cities	20884	21940	14896	21002	7,2	7,1	5,8	*7,1*
urban area	20708	22217	15834	21296	7,0	7,2	6,1	*7,1*
rural cities	20484	20065	16656	18475	7,1	6,9	6,1	*6,6*
rural area	19288	21161	15818	18303	7,1	7,1	6,2	*6,7*
Total	*20694*	*21893*	*16915*	*20120*	*7,1*	*7,1*	*6,2*	*6,9*
2015/20								
urban large cities	23850	26232	21243	*23316*	7,4	7,5	7,1	*7,3*
urban cities	20997	23596	19477	*21597*	7,3	7,4	7,2	*7,3*
urban small cities	23117	24917	20266	*23743*	7,3	7,4	7,3	*7,4*
urban area	23585	24596	20354	*24018*	7,4	7,4	6,9	*7,4*
rural cities	21427	22276	19466	*20626*	7,4	7,3	6,9	*7,2*
rural area	23059	23310	19757	*22006*	7,4	7,3	7,0	*7,2*
Total	*22681*	*24174*	*20079*	*22673*	*7,4*	*7,4*	*7,0*	*7,3*

SOEPv37. Med.Annual.Household-Net-Equivalent Incomes [rev.OECD-scale; €-2019, previous year]; Mean Life satisfaction [0–10-scale; 0 = completely dissatisfied 10 = completely satisfied]

from rural income levels in the western states. The overall lower income levels in rural areas result primarily from the high shares of the East German population with lower incomes. East German incomes increased substantially over time, reducing the difference between eastern and western states. Yet even three decades after unification, the different economic standards are still visible in most types of urban areas.

Mean levels of life satisfaction reflect these trends and relationships: an increase in subjective well-being over the last three decades in all regions at all urban levels, similar values for urban and rural living conditions in all western states, but still some lower scores on subjective evaluations in East Germany.

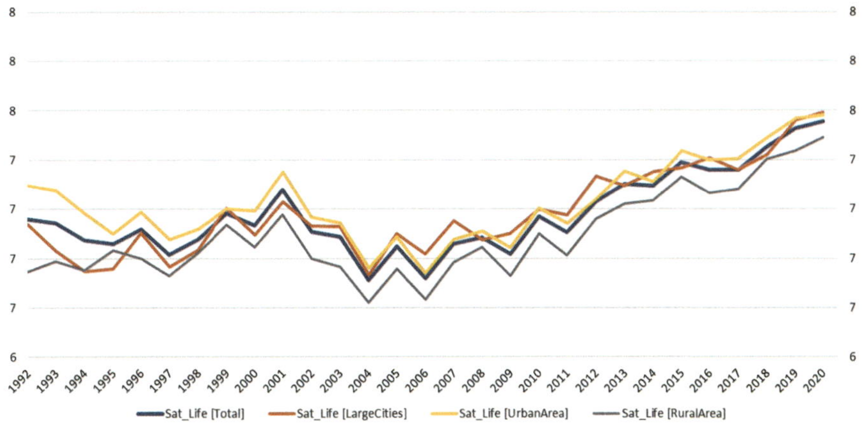

Fig. 5.7 Long-term urban developments in life satisfaction, 1992–2020. SOEPv37. Sat_Life: Mean Life Satisfaction [0–10-scale: 0 = completely dissatisfied 10 = completely satisfied]

5.4.2 Income-Based Urban Well-Being, Deprivation, and Inequality

Median income trends do not reflect variations in distributions. We first look back at national trends in income-based deprivation and well-being, and second, at some further trends in national income inequalities.

Well-Being and Deprivation

Following the European definition of income-based poverty lines at 60% of median-household-net income, we find a decrease over the period 1995–2000 and an almost steady increase over the last two decades from 2000 to 2020 (Fig. 5.8 illustrates long-term developments in income-based poverty rates and partial deprivation as well as in income-based wealth rates and partial well-being). This national trend in income-based deprivation is robust across all kinds of measures according to the standard approach (dotted lines) and the fixed-fuzzy approach, which has a much broader coverage of the population (full lines) according to the 60% line (lines in orange) as well as for the 50% line, indicating higher poverty levels (lines in red). High poverty beyond the 50% median line even doubled between 1992 and 2020 (standard approach). The fixed-fuzzy lines further reveal that these trends of increasing deprivation are not restricted to the population below the poverty line and can be found in income-based partial deprivations above the poverty line as well. The long-term trend toward increasing income-based deprivation has affected the entire bottom half of the income distribution as well and is relatively robust.

When looking at developments in income-based well-being (blue lines), we also find increases in wealth levels at the top end above the 200% median line up to 2015. When looking at the fixed-fuzzy-based well-being trends in incomes, we observe an

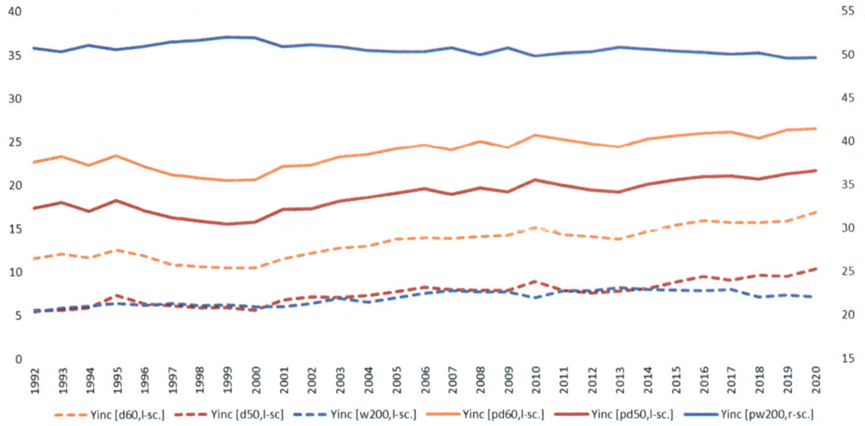

Fig. 5.8 Long-term developments in (partial) deprivation and well-being, 1992–2020. SOEPv37. Yinc: Annual Eq.Household.Net.Income [€2019; rev.OECD-Scale]

almost stagnating trend over the last two decades for the broad majority of the population. When considering both trends in income-based well-being and deprivation, we find increases in deprivation for the lower part of the income distribution and higher rates at the top level of incomes, whereas the overall levels of income-based well-being have remained almost unchanged.

These overall national trends differ between urban regions as well (Table 5.11 indicates regional differences in urban–rural income-based rates for deprivation and well-being). In western regions—especially in the Southwest—deprivation rates in 2020 for large cities are higher than for rural areas—whereas in the eastern cities, rural areas are more affected by income deprivation. High-income levels are more prevalent in large cities across all regions. This also holds for fuzzy-based positive incomes over a much broader share of the population. Tensions between income-based deprivation and well-being seem higher in large cities.

Inequalities and Diversities

Inequalities reflect the tension between rich and poor. We therefore propose a range of Gini-based subindices as simplified factor decompositions of the own group versus other groups.

Gini inequalities are described separately for each region using an urban indicator of subgroup levels (Table 5.12 applies large cities, urban area, and rural area as subgroup levels for Gini within-between-group factor decomposition) as the between-group criterion for factor and subgroup decomposition. The summary table (Table 5.12) illustrates that the main Gini GG can be regarded as the sum of within-group inequalities GK and between-group inequalities GD (GG = GK + GD). Between-group inequalities GD can be further disentangled using their distances from mean group levels and mean group difference GMK using two versions: the net

Table 5.11 Regional urban differences in income-based deprivation and well-being

2020 in percent		Standard approach			Fixed-fuzzy-approach		
		Deprivation	Deprivation	Well-being	Partial deprivation	Partial deprivation	Partial well-being
	[in %-med.]	[z60]	[z50]	[w200]	[z60–100]	[z50–100]	[w200–50]
Total		16.7	10.3	7.0	26.4	21.6	49.7
N-West	(Total)	18.1	11.7	7.5	27.5	22.6	49.1
	Large cities	20.0	12.4	12.9	27.0	23.1	53.0
	Urban area	17.8	11.7	7.0	27.5	22.4	48.5
	Rural area	17.2	11.3	4.6	27.8	22.6	47.6
S-West	(Total)	13.9	7.7	8.2	23.0	18.6	53.7
	Large cities	18.3	10.1	14.2	25.9	22.0	56.7
	Urban area	13.2	7.8	8.3	22.4	18.3	54.7
	Rural area	13.7	6.9	5.8	23.0	18.2	50.6
East	(Total)	20.3	12.9	3.6	31.7	25.9	42.1
	Large cities	18.6	12.3	5.6	29.6	24.4	46.5
	Urban area	18.2	13.1	2.4	30.2	24.7	43.2
	Rural area	21.7	13.1	3.1	33.2	27.0	39.6

SOEPv37. Med.Annual.Household-Net-Equivalent Incomes [rev.OECD-scale; €-2019, previous year]; Standard-A. [z = 60|50%-med.; w = 200%-med.]; Fuzzy-A. [z, u = 60|50–100%med.; w, l = 200–50%-med.]

version GD = GMK + GDD – GDQ and the full version GD = GMK + GDDR – GDDQ – GDQM. Diversity indices indicating joint variations between subgroups can be derived either from net version GDIV1 = GDD + GDQ or full version GDIV2 = GDDR + GDQR. We further propose the use of population adjustments when examining subgroup indices. Any factor decompositions of between-group-related subindices remain [GGD = GGMK + GGDD – GGDQ | GGD = GGMK + GGDDR – GGDQR – GGDQM]; the same holds for population-adjusted diversity indicators [GGDIV1 = GGDD + GGDQ | GGDIV2 = GGDDR + GGDQR].

For all German regions in 2020 (Table 5.13 reveals regional differences in Gini-based urban–rural inequality and diversity subindices), urban differentiations indicate higher income-based inequalities in large cities, followed by urban areas, and some lower inequality levels in rural areas. Population-adjusted within-group

Table 5.12 Gini-based subindices for regional urban income inequalities and diversities

Gini subindices									
2020	GG	GK	GD	GMK	GDD	GDQ	GDDR	GDQR	GDQM
N-West	0.314	0.136	0.178	0.020	0.197	0.039	0.210	0.034	0.019
S-West	0.285	0.127	0.158	0.057	0.150	0.048	0.186	0.033	0.052
East	0.267	0.108	0.159	0.050	0.156	0.047	0.188	0.034	0.045
Gini subindices [GG… population adjusted]									
2020	GG	GGK	GGD	GGMK	GGDD	GGDQ	GGDDR	GGDQR	GGDQM
N-West	0.314	0.313	0.314	0.034	0.349	0.069	0.372	0.060	0.032
S-West	0.285	0.283	0.289	0.101	0.275	0.087	0.341	0.060	0.093
East	0.267	0.266	0.273	0.084	0.268	0.080	0.323	0.058	0.076
Diversity indices [GG… population adjusted]							Subgroup—levels		
2020	GDIV1	GDIV2	GGDIV1	GGDIV2			Urban indicator		
N-West	0.236	0.244	0.418	0.432			*Large cities*		
S-West	0.198	0.219	0.362	0.401			*Urban area*		
East	0.203	0.222	0.348	0.380			*Rural area*		

SOEPv37. Annual.Household-Net-Equivalent Incomes [rev.OECD-scale; €-2019, previous year]
GG… population-adjusted Gini Subindices

inequality GGK is in general the same as main inequality GG and between-group inequality GGD—as overall mean group differences GGMK are quite low. In other words, we find quite similar ranges of income-related inequalities within each of the three subgroups as for the total population. Joint variations in incomes according to the three urban group levels are therefore rather high—as indicated by the diversity indices.

In the Northwest, regional differentiations reveal slightly higher income inequalities (GG). In the Southwest, income inequalities differ most between urban subgroups, and in the East, there are still lower levels of income inequality, as was the case in 1990, before the end of the communist regime. Within-subgroup differentiations (GGK) indicate substantial internal differences, with quite similar lower inequalities in rural areas across all regions and much higher income inequalities in large cities—highest in the richer cities of the Southwest. Mean group levels (GGMK) are relatively low, which means that we have high overlaps in income distributions

Table 5.13 Regional urban differences in income-based inequality and diversity

2020 coeff. {0, …, 1}		Gini subindices				Diversity indices	
		GG	GGK	GGD	GGMK	GGDIV1	GGDIV2
		[Main]	[Within-Gr.]	[Betw.-Gr.]	[Gr-diff]	[Net-effect]	[Full-effect]
Total		0.295	0.294	0.296	0.073	0.380	0.409
N-West	(Total)	0.314	0.313	0.314	0.034	0.418	0.432
	Large cities	0.339	0.360	0.334	0.052	0.437	0.458
	Urban area	0.314	0.314	0.314	0.031	0.419	0.432
	Rural area	0.292	0.270	0.299	0.028	0.400	0.412
S-West	(Total)	0.285	0.283	0.289	0.101	0.362	0.401
	Large cities	0.349	0.403	0.342	0.133	0.439	0.485
	Urban area	0.283	0.280	0.287	0.094	0.361	0.398
	Rural area	0.267	0.247	0.276	0.105	0.337	0.378
East	(Total)	0.267	0.266	0.273	0.084	0.348	0.380
	Large cities	0.291	0.311	0.283	0.119	0.353	0.398
	Urban area	0.262	0.256	0.263	0.051	0.341	0.363
	Rural area	0.258	0.247	0.271	0.077	0.348	0.377

SOEPv37. Med.Annual.Household-Net-Equivalent Incomes [rev.OECD-scale; €-2019, previous year]; GG… population adjusted Gini Sub-Indices

between regions—mean urban group levels differ most within the Southwest and East. Due to the high distributional overlaps, we find quite high rates of income diversity (GGDIV1, 2) that are highest in large cities. It seems that large cities have higher income inequality; urban areas are much more heterogenous, including rich municipalities around cities and low-income suburban areas; and cities of different sizes differ in economic power. Inequalities in the East are also driven by developments in rural areas—it may be that some rural municipalities that are tourist destinations profit more than others.

Long-term developments confirm rising tensions in German income distributions (Fig. 5.9 shows urban–rural developments in income-based Gini inequalitiy and diversity subindices)—main inequalities GG increased between 2000 and 2006 and increased slightly more up to 2020. Mean between-group differences GMK on urban and rural levels decreased in the long run, such that population-adjusted within- and between-group inequalities GGK and GGD almost correspond to the main Gini GG. Diversity indices reveal a steady increase in joint variations by urban and rural levels from 1992 to 2007, with relatively stable further developments in the upper bounds (GGDIV2) and small increases in joint net variations (GGDIV1), indicating slightly increasing income heterogeneities between urban and rural areas since 2010.

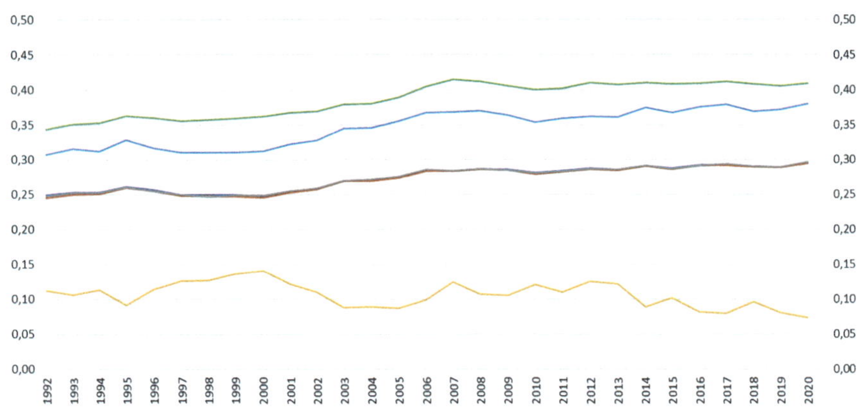

Fig. 5.9 Long-term developments in inequality and diversity, 1992–2020. SOEPv37. Yinc: Annual Eq.Household.Net.Income [€2019; rev.OECD-Scale]

5.5 Multidimensional Urban Well-Being, Deprivation, and Inequalities

We want to at least check multidimensional differences in urban and rural living conditions. We refer first to the dimensions of economic well-being resources (Table 5.14 shows variables and design weights with variations in multidimensional individual weights for each indicator of economic resources) including two sub-dimensions—incomes and assets—broken down by five levels of aggregation—ranging from intra-personal income evaluations, to incomes and assets at individual and household levels, to regional economic performance as a background indicator shaping regional economic living conditions (as mentioned above in Fig. 5.2). The analyses refer to the total population living in private households in 2017, including children (subjective indicators for children are assigned as mean scores of parents or adults who hold financial responsibility within the household).

We apply dimensional weights derived from indicator design and empirical individual schemes of well-being for all applications on multidimensional deprivations and well-being and for multidimensional inequality and diversity. For multidimensional inequality and diversity, all indicators are rescaled according to the ct1-scale (1, …, 100) (as described above in Tables 5.5 and 5.6). All dimensional decompositions are presented as shares of mean multidimensional rates. Figure 5.10 indicates at the top the amount of multidimensional deprivation in economic resources according to the standard approach (*d*) and the fixed-fuzzy-based partial approach (*pd*)—and at the bottom the amount of multidimensional well-being and wealth in economic resources according to the standard approach (*w*) and the fixed-fuzzy-based partial

Table 5.14 Economic resources—indicators and dimensions, design and weighting

Levels	Design weight	D_{wgt} min	D_{wgt} max	Sub-dimension	Indicators	Variables
Regional	0.2	0.131	0.185	reg.income	reg_gdp_pc	Reg.GDP per capita
Social [ass.]	0.1	0.080	0.107	hh.asset	ass_groq	Gross-HH.Assets
	0.1	0.079	0.106	hh.asset	ass_netq	Net-HH.Assets [> = 0]
Social [hinc.]	0.1	0.098	0.128	hh.income	minc_ehc	Monthly eq.Hh-Income EHC
	0.1	0.099	0.140	hh.income	yinc	Annual eq.Hh-Income
Personal	0.2	0.164	0.231	ind.income	pinc	Personal annual income
Intra-personal	0.2	0.211	0.267	sat.income	saty_p	Sat.Hhold-Income

SOEPv37. Design: dimensional design [Sum = 1]; D_{wgt}: dimensional weights [Sum = 1]; Household-Income EHC: household income after eligible housing costs

approach (*pw*)—using individual multidimensional weighting schemes. All multidimensional indices are decomposed by dimensional indicators and for urban–rural subgroups by regions.

In the Northwest, mean multidimensional economic deprivations (Fig. 5.10, top) indicate, according to the standard approach (*d*), lower rates for rural areas than urban areas and large cities. In the Southwest, urban areas seem to have lower risks of multidimensional economic deprivation. In the East, multidimensional economic deprivation is higher in all urban and rural areas, and highest in rural areas. If we consider fixed-fuzzy rates with partial deprivations (*pd*) above the poverty line, we find generally higher degrees of multidimensional economic deprivation in rural areas for all regions. Dimensional decompositions indicate substantial contributions to mean deprivation levels by all indicators. The fixed-fuzzy approach also considers regional differences as partial handicaps, which are higher in rural areas in the Northwest and East.

Mean multidimensional economic well-being (Fig. 5.10, bottom) is lower in eastern than in western regions, and—according to the standard approach (*w*)—mainly driven by positive income evaluations and personal incomes. When considering the fixed-fuzzy extensions (*pw*), we find substantial impacts of other indicators as well. Large cities seem to have slightly higher rates of economic well-being than rural areas in all regions. In the Southwest, urban areas also seem to provide the highest ratings for economic well-being.

Overall multidimensional economic well-being inequalities (MGG) and their within- and between-subgroup differentiations (MGGK, MGGD) are quite similar across regions (Table 5.15 summarizes multidimensional Gini subindices for inequalities and diversities in economic resources). Mean group differences (MGGMK) are quite low, with some higher variations in the East. Multidimensional economic well-being does not vary much within regions according to urban and rural areas. Due to high overlaps, we also find substantial net and full diversity rates (MGGDIV1, 2).

Figure 5.11 describes the amount of Gini-based multidimensional inequalities (MGG) and diversities (MGGDIV1, 2) in economic resources—using individual multidimensional weighting schemes. All multidimensional Gini subindices are decomposed by dimensional indicators and for urban–rural subgroups by regions. Multidimensional economic well-being inequality (MGG) and diversity (MGGDIV1, 2) are driven mainly by differences in individual evaluations and variations in regional performance (Fig. 5.11), with higher impacts from differences in individual evaluations in the eastern regions and higher impacts from variations in regional performance in the western regions. The overall remarkably similar results for mean levels of multidimensional economic well-being inequality and diversity within regions are shaped by underlying conceptual settings on indicators, rescaling, and dimensional weighting, and should be further evaluated using alternative settings.

We observe multidimensional differences in urban and rural living conditions using a selective set of 22 indicators from five basic spheres of well-being at five different levels of aggregation, including the dimensions of economic well-being described above. Empirical results refer to the list of indicators and their conceptual settings for individual dimensional weightings, lines, and rescaling (details in

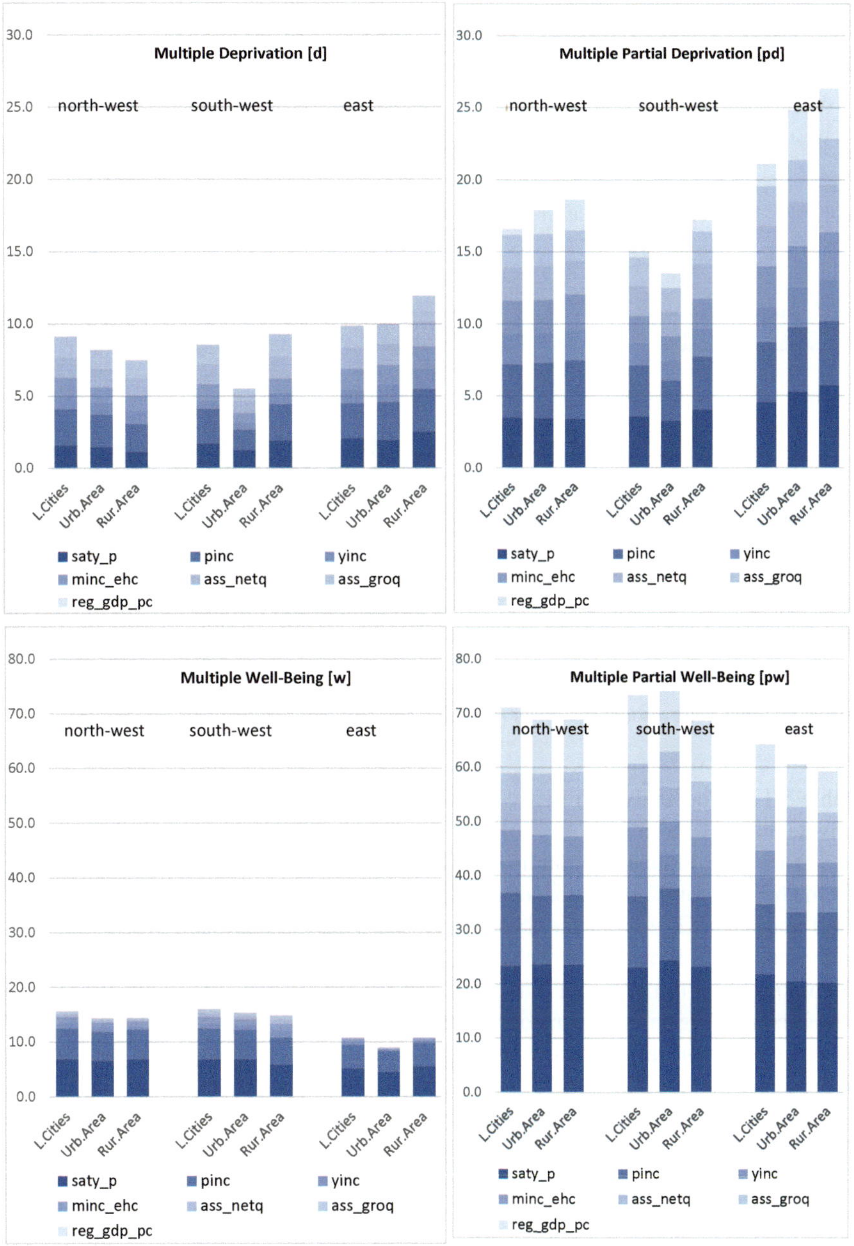

Fig. 5.10 Economic resources – urban deprivation and well-being, Germany 2017. SOEPv37

Table 5.15 Economic well-being resources—regional inequality and diversity

2017	MGG	MGGK	MGGD	MGGMK	MGGDIV1	MGGDIV2
N-West	0.198	0.191	0.208	0.052	0.269	0.280
S-West	0.203	0.195	0.226	0.071	0.283	0.300
East	0.203	0.198	0.206	0.115	0.252	0.297

SOEPv37. Subgroups [within- and between-groups]: Large cities, urban areas, rural areas

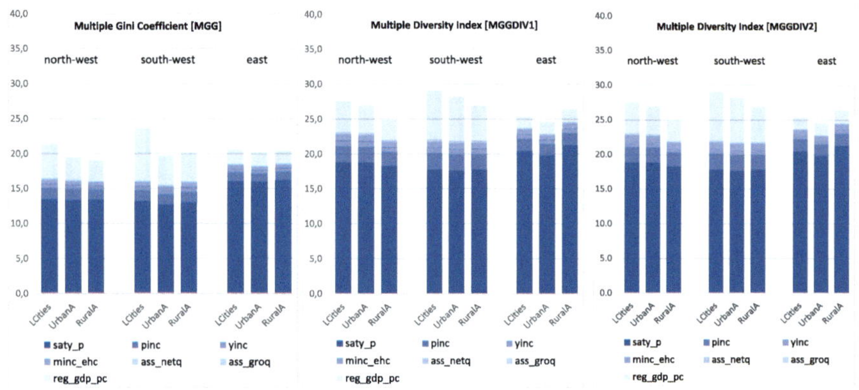

Fig. 5.11 Economic resources—urban inequality and diversity, Germany2017. SOEPv37

Table 5.17). The analyses refer to the adult population aged 20 and older living in private households in 2017.

Figure 5.12 indicates multidimensional deprivation (at the top) and well-being (at the bottom for 22 indicators of five well-being dimensions according to the standard approach (*d*, *w*) and the fixed-fuzzy-based partial approach (*pd*, *pw*)—using individual multidimensional weighting schemes. All multidimensional indices are decomposed by dimensional indicators and for urban–rural subgroups by regions. When comparing multidimensional rates of deprivation and well-being (Fig. 5.12, top), we find not only different levels but also different rankings for the standard approach (*d*, *w*) and the fixed-fuzzy extensions with partial deprivation and partial well-being (*pd*, *pw*). Mean overall pd-levels indicate the lowest deprivation rates in the Southwest—especially in large cities and urban areas. The large cities of the Northwest, other urban areas, and rural areas show almost the same, but slightly higher levels of multidimensional deprivation; in the East, multidimensional deprivation is somewhat higher, especially in rural areas. All dimensions contribute to deprivation levels—with some higher shares from health and social resources across all regions and areas.

Looking at multiple positive indices of well-being (*w*), the overall rates are much higher (Fig. 5.12, bottom): the large majority of individuals share overall high levels of multidimensional well-being across all regions and areas, with some lower levels

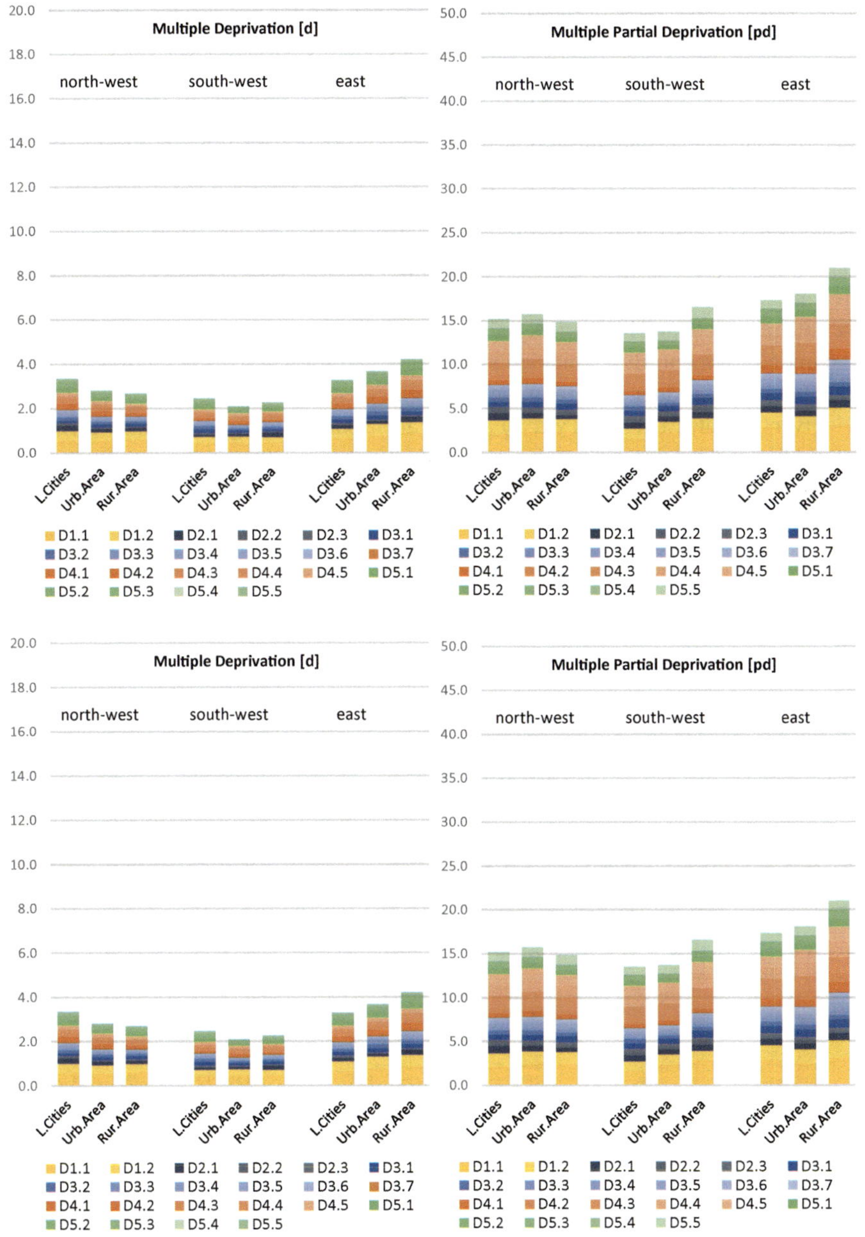

Fig. 5.12 Multidimensional urban well-being and deprivation, Germany 2017. SOEPv37, adult population (age 20+)

in the East. Partial rates of multidimensional well-being (*pd*) rise to around 90% of the adult population within all regions and urban or rural areas.

Figure 5.13 describes the amount of Gini-based multidimensional inequalities (MGG) and diversities (MGGDIV1, 2) for 22 indicators of five well-being dimensions—using individual multidimensional weighting schemes. All multidimensional Gini subindices are decomposed by dimensional indicators and for urban–rural subgroups by regions. Different measures of inequality and diversity underpin the broad range of overlaps in multidimensional well-being between urban and rural areas in different regions (Fig. 5.13, left). Mean multidimensional inequality rates (MGG) are almost the same. In the East, this is a compensating effect of lower inequalities in income resources and higher inequalities in cultural resources including employment and unemployment. Both coincide in higher deprivation and lower well-being when looking at the broader picture of well-being in this region.

Diversity rates (MGGDIV1, 2) are (again) relatively high, similarly driven by internal compensating variations in income resources and cultural resources when looking at the East region (Fig. 5.13, middle and right). These pictures also confirm the high degree of overlap in multidimensional well-being between large cities, other urban areas, and rural areas within each region.

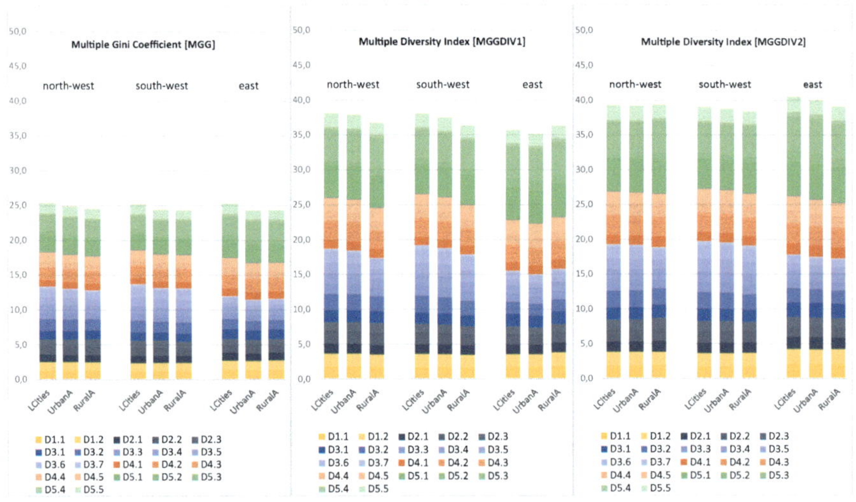

Fig. 5.13 Urban inequalities and diversities in multidimensional well-being, Germany 2017. SOEPv37, adult population (age 20+)

5.6 Analyzing the Emergence of Urban Inequalities in Germany

How did urban inequalities emerge in Germany? This simple question implies at least three types of conceptual questions that must be addressed as a basis for empirical analysis. First, how did urban spaces emerge, and how can urban spaces be defined adequately in contrast to rural spaces today? Second, what kinds of inequalities should be addressed if we consider the urban space as a multifaceted conglomeration of individuals, households, and living arrangements with various backgrounds? And finally, which indicators and conceptual designs should be applied to address urban living arrangements? Each of these three types of conceptual questions offers different perspectives on definitions and measures for empirical research on urban inequalities. We tried to focus on selected aspects of these perspectives.

Due to Germany's historical political and cultural regional fragmentation, urban developments are regionally embedded in the federal states. When examining the emergence of cities in Germany as traditional centers of national cultural development, we find three different regional profiles of urban development. In the Northwest, several cities emerged as large commercial, political, and cultural centers. Examples are Bremen and Hamburg as well as Berlin (in the Northwest of Germany): each of them is now an independent federal city-state. In the Southwest, the rising urban population has led to more multifaceted urban developments: Cologne, as one of the major traditional cultural urban centers in the West, belongs to the state of North Rhine-Westphalia (with Düsseldorf as its state capital), and Frankfurt/Main, as a major urban financial hub in the middle of Germany, belongs to the state of Hesse (with Wiesbaden as its state capital). East Germany differs due to its political history before unification as an independent communist country, and as a region with higher shares of the population living in rural areas (and in large cities as well). Urbanization considers not only large cities but also urban agglomerations around and between urban centers. We address urban developments in Germany by distinguishing between large cities, other types of urban areas, and rural areas in each of the three regional types to separate urban from embedded regional developments.

Inequalities are commonly associated with tensions between rich and poor. We apply unidimensional and multidimensional poverty measures according to (ordinal) FGT measures and the Alkire-Foster measurement framework, extended by a fixed-fuzzy approach to also capture partial degrees of deprivation above the poverty line. These measurement frameworks have been further assigned to the positive part of well-being, including fixed-fuzzy-based partial degrees of well-being below wealth lines.

Urban living conditions differ from those in rural areas not only in terms of the higher population density but also in terms of wide (multi-cultural) variation—urban inequalities are thus also closely related to urban diversity. While inequalities reflect differences in living conditions primarily along (vertical) dimensions of rich versus poor, diversity reflects (horizontal) between-group variations in ways of living. We use the Gini coefficient as a common measure of inequality that is applicable in unidimensional and multidimensional settings, and propose new Gini-based subindices of joint subgroup-variations as diversity indicators, which, of course, require further conceptual and empirical justification. Urban inequalities are thus addressed in our empirical settings as tensions—between deprivation and well-being on the one hand, and between (vertical) inequality and (horizontal) diversity on the other.

The selection of indicators is driven by a strong multidimensional quality-of-life perspective. We use equivalized-household-net incomes as an established key indicator of economic well-being, and satisfaction with life as an established key indicator of subjective well-being. We further address multidimensional settings of urban inequalities. Multiple indicator designs are derived from several quality-of-life-related conceptions. Dimensional differentiations refer to a combination of two dimensions of basic needs—life (health, nutrition) and housing (home, environment)—and three additional dimensions—economic resources (incomes and assets), social resources (participation, qol), and cultural resources (skills, work), adapted from differentiations of economic, social, and cultural capital by Bourdieu and the having-loving-being differentiation of Allardt. Each of the five basic dimensions of well-being can be further differentiated by sub-dimensions. We further propose a differentiation of well-being dimensions for multidimensional applications by aggregation levels (Zapf, Noll), ranging from intra-personal and individual indicators to household and regional levels. For dimensional weighting, we propose a combination of design weights and empirical weights including individual variations, as individual needs and preferences may differ and change during life cycle.

For our empirical analyses, we proceed in three steps: first we refer to unidimensional key variables—focusing on incomes. We then extend the perspective to multidimensional economic well-being—also considering individual income evaluations and regional background indicators as further levels. Finally, we run a full multidimensional setting using a selected range of indicators from all five basic dimensions of well-being, disentangled by aggregation levels ranging from internal and subjective to regional.

Unidimensional results reveal that long-term developments in median incomes and (mean) life satisfaction within large cities, other urban areas, and rural areas coincide with national trends. Income inequality and diversity have increased in the long run. As East German regions had lower living standards after unification—including lower income and life satisfaction—urban developments are partly dominated by regional adaptation processes. Regional East–West differences have diminished over time but still remain visible three decades later.

In 2020, we find higher rates of top incomes (above wealth lines) in large cities in all regions. For deprivation rates, this is less clear. If we use stronger poverty lines or look at fixed-fuzzy results (covering a much broader range of population), the differences in income deprivation do not differ much between large cities, other urban areas, and rural areas within regions. Income-based inequality and diversity are higher in large cities within each region, but differences between other urban and rural areas are somewhat greater between regions than within regions.

Results for multidimensional economic well-being resources in 2017 confirm in general the previous unidimensional differences in income deprivation and economic well-being between regions—higher deprivation rates and lower well-being rates in the East, and slightly higher economic standards in the Southwest and Northwest. Differences between large cities, other urban areas, and rural areas within regions are less clear. Multidimensional inequality and diversity indicate compensating effects across regions—impacts of education levels seem to vary more in western regions, whereas in the East, variations in subjective income evaluations are greater.

Multidimensional results in 2017 for overall well-being across several dimensions and aggregation levels further confirm regional differences in deprivation—overall, well-being is also still lower according to the standard approach, but these differences diminish when considering fixed-fuzzy extensions of well-being, covering almost the entire population of each region. Multiple differences in inequality and diversity in overall well-beings are strikingly similar across regions—with some internal compensating effects and moderate variations in levels between large cities, other urban areas, and rural areas.

5.7 Conclusion

In summary, our results confirm several regional differences, especially when considering deprivation and economic well-being in East Germany, but show almost no systematic variations in inequality or diversity in our urban categories—large cities, other urban areas, and rural areas—in and across regions. Several Gini-based subindices indicate instead, for all unidimensional and multidimensional settings, high diversity-related overlaps in distributions, with similar pictures of within- and between-group variation and low mean group differences. Living conditions in Germany are thus not really divided by differences between large cities, other urban areas, and rural areas.

Instead, we observe higher degrees of inequality over time—also within each kind of urban and rural area: large cities display substantial internal variation between rich and poor parts of the city; economic levels differ between urban and suburban areas; and rural towns and municipalities are very heterogenous (economically).

The differentiation between large cities, other urban areas, and rural areas is more meaningful than ever when considering urbanization processes, approaches to land use regulation, and efforts to stop the dramatic decrease in biodiversity, which may affect the future design and planning of cities and urban areas. Our results cast some doubt, however, on whether the differentiation between large cities, other urban areas, and rural areas can still be applied adequately to characterize differences in current living standards and overall well-being in Germany.

Acknowledgements I am grateful to the editor, Graciela Haydee Tonon, for her trust, help, and incredible patience, to Carsten Schoeder, Jan Goebel, and Markus Grabka for their conceptual and technical support, and to Deborah Anne Bowen for her careful comments and suggestions.

Appendix

See Tables 5.16 and 5.17.

Table 5.16 Calculating the Gini index and subindices as sums of absolute differences

							n_j	1	2	3	4	5	6	7	8	9	10				
							l	1	1	1	2	2	2	2	3	3	3				
							x_j	0	10	20	0	10	20	30	10	20	30				
							w_j	0.5	1.0	1.5	0.5	1.0	1.0	1.5	0.5	1.0	1.5				
							wx_j	0	10	30	0	10	20	45	5	20	45				
							wx_l	13	13	13	19	19	19	19	23	23	23				
n_i	k	k_n	x_i	w_i	wx_i	wx_k	GG														
1	1	3	0	0.5	0	13.33		0	5	15	0	5	10	23	2.5	10	23	92.5	20	72.5	25.8
2	1	3	10	1.0	10	13.33		5	0	15	5	0	10	30	0	10	30	105	20	85	51.7
3	1	3	20	1.5	30	13.33		15	15	0	15	15	0	23	7.5	0	23	113	30	82.5	77.5
4	2	4	0	0.5	0	18.75		0	5	15	0	5	10	23	2.5	10	23	92.5	37.5	55	15
5	2	4	10	1.0	10	18.75		5	0	15	5	0	10	30	0	10	30	105	45	60	30
6	2	4	20	1.0	20	18.75		10	10	0	10	10	0	15	5	0	15	75	35	40	30
7	2	4	30	1.5	45	18.75		23	30	23	23	30	15	0	15	15	0	173	67.5	105	45
8	3	3	10	0.5	5	23.33		2.5	0	7.5	2.5	0	5	15	0	5	15	52.5	20	32.5	24.2
9	3	3	20	1.0	20	23.33		10	10	0	10	10	0	15	5	0	15	75	20	55	48.3
10	3	3	30	1.5	45	23.33		23	30	23	23	30	15	0	15	15	0	173	30	143	72.5
							GMK														
10	3		150	10	185	185		0	0	0	1.4	2.7	2.7	4.1	2.5	5	7.5	1055	325	730	420
								0	0	0	2.7	5.4	5.4	8.1	5	10	15				
n	Gini		G		0.285			0	0	0	4.1	8.1	8.1	12	7.5	15	23	GG	GK	GD	GMK
n-1	Gini		G		0.317			1.4	2.7	4.1	0	0	0	0	1.1	2.3	3.4				
								2.7	5.4	8.1	0	0	0	0	2.3	4.6	6.9				
GG	GK	GD	GMK					2.7	5.4	8.1	0	0	0	0	2.3	4.6	6.9				
GG	GGK	GGD	GGMK					4.1	8.1	12	0	0	0	0	3.4	6.9	10				
								2.5	5	7.5	1.1	2.3	2.3	3.4	0	0	0				
0.32	0.10	0.22	0.13					5	10	15	2.3	4.6	4.6	6.9	0	0	0				
0.32	0.36	0.30	0.17					7.5	15	23	3.4	6.9	6.9	10	0	0	0				

Source Author

Table 5.17 Multidimensional well-being—indicators, design, and dimensional weights

	Dimension	SubDimension	Indicators	Mean	SD	Design	Dwgt(min)	Dwgt(min)	Dwgt(max)	Variables
D1.1	Life	Health	sat_heal	6.6	2.2	0.1000	0.1211	0.1007	0.1466	Sat.Health
D1.2		Health	health	3.3	1.0	0.1000	0.1179	0.1023	0.1408	Health Status
D2.1	Home	Housing	sat_home	7.8	1.9	0.1000	0.1044	0.0890	0.1281	Sat.Home
D2.2		Housing	hroom1_pc	1.8	1.0	0.0500	0.0390	0.0343	0.0481	Rooms per capita
D2.3		Housing	hroom2_pc	45.8	26.6	0.0500	0.0399	0.0345	0.0484	Area (sqm) per capita
D3.1	Econ.Res	Income	sat_hinc	6.8	2.1	0.0400	0.0512	0.0417	0.0688	Sat.Hhold-Income
D3.2		Income	pinc	34008.9	33420.5	0.0400	0.0403	0.0357	0.0541	Personal Annual Income
D3.3		Income	yinc	25696.3	18306.8	0.0200	0.0215	0.0186	0.0295	Annual eq.Hh-Income
D3.4		Income	minc_ehc	1409.0	1043.8	0.0200	0.0213	0.0187	0.0289	Monthly eq.Hh-Income EHC
D3.5		Assets	ass_netq	134313.3	587812.2	0.0200	0.0162	0.0145	0.0204	Net-HH.Assets [>=0]
D3.6		Assets	ass_groq	154615.7	603994.7	0.0200	0.0164	0.0146	0.0205	Gross-HH.Assets
D3.7		Income	reg_gdp_pc	37.9	17.0	0.0400	0.0301	0.0239	0.0376	Reg.GDP per capita
D4.1	Soc.Res	Qual.of.life	sat_life	7.3	1.7	0.0667	0.1017	0.0000	0.1231	Sat.Life
D4.2		Participation	pol_int	2.3	0.9	0.0333	0.0265	0.0237	0.0334	Pol.Interest
D4.3		Participation	engage	1.1	0.5	0.0333	0.0250	0.0224	0.0303	Engagement
D4.4		participation	cul_int	1.9	0.7	0.0333	0.0304	0.0271	0.0429	Cult.Interest
D4.5		Participation	cul_act	1.8	1.0	0.0333	0.0256	0.0222	0.0326	Cult.Engagem.
D5.1	Cult.Res	Work	lab_emp	7.9	5.5	0.0400	0.0356	0.0274	0.0473	Employment
D5.2		Handicap	lab_uem	11.4	2.4	0.0400	0.0348	0.0234	0.0466	UnEmployment
D5.3		Work	lab_wi	61.5	44.7	0.0400	0.0355	0.0241	0.0486	HH-WorkIntensity
D5.4		Work	reg_emp	59.5	4.3	0.0400	0.0303	0.0245	0.0380	Reg.Empl. per capita
D5.5		Skills	educy	12.4	2.8	0.0400	0.0351	0.0316	0.0429	Years-of-Education

SOEPv37. Design: dimensional design [Sum=1]; Dwgt: dimensional weights [Sum=1];
Household-Income EHC: Household Income after eligible housing costs.

References

Aaberge R, Brandolini A (2015) Multidimensional poverty and inequality. In: Handbook of income distribution, vol 2. Elsevier, pp 141–216

Alkire S, Ballon P, Foster J, Roche JM, Santos ME, Seth S (2015) Multidimensional poverty measurement and analysis: a counting approach. Oxford University Press

Alkire S, Foster J (2011a) Counting and multidimensional poverty measurement. J Publ Econ 95(7–8):476–487

Alkire S, Foster J (2011b) Understandings and misunderstandings of multidimensional poverty measurement. J Econ Inequality 9(2):289–314. https://doi.org/10.1007/s10888-011-9181-4

Allardt E (1993) Having, loving, being: an alternative to the Swedish model of welfare research. In: Nussbaum MC, Sen AK (eds) The quality of life. Clarendon Press, Oxford, pp 88–94

Atkinson AB (2015) Inequality: what can be done? Harvard University Press, London

Atkinson AB (2019) Measuring poverty around the world. Princeton University Press

Bartels C, Jaeger S, Obergruber N (2020) Long-term effects of equal sharing: evidence from inheritance rules for land. IZA DP no. 13665

Betti G, Verma V (2007) Fuzzy measures of the incidence of relative poverty and deprivation: a multi-dimensional perspective. Stat Methods Appl 17(2):225–250

Blotevogel HH (O.J.) Kulturelle Stadtfunktionen und Urbanisierung. Interdependente Beziehungen im Rahmen der Entwicklung des deutschen Städtesystems im Industriezeitalter. Sonderdruck aus: Teuteberg, Hans Jürgen (Hg): Urbanisierung im 19. und 20. Jahrhundert. Historische und Geographische Aspekte

Bohmann S (2021) Whether you think you can, or you can't—it's true. Locus of control in status reproduction. Dissertation. Humboldt-University Berlin, Social Sciences

Bourdieu P (1984) Distinction: a social critique of the judgment of taste. Routledge & Kegan Paul, London

Bronfenbrenner U (1979) The ecology of human development: experiments by nature and design. Harvard University Press, Cambridge, Mass

Decancq K, Lugo MA (2012) Inequality of wellbeing: a multidimensional approach. Economica 79:721–746. https://doi.org/10.1111/j.1468-0335.2012.00929.x

Foster J, Greer J, Thorbecke E (1984) A class of decomposable poverty measures. Econometrica 52(3):761–766

Foster J, Greer J, Thorbecke E (2010) The Foster–Greer–Thorbecke (FGT) poverty measures: 25 years later. J Econ Inequality 8(4):491–524

Frick JR, Goebel J, Schechtman E, Wagner GG, Yitzhaki S (2006) Using analysis of Gini (ANOGI) for detecting whether two subsamples represent the same universe. Sociol Methods Res 34(4):427–468. https://doi.org/10.1177/0049124105283109

Gini C (1936) On the measure of concentration with special reference to income and statistics. Colorado College Publication, general series no. 208, 73–79

Goebel J, Grabka MM, Liebig S, Kroh M, Richter D, Schröder C, Schupp J (2018) The German socio-economic panel study (SOEP). J Econ Stat. https://doi.org/10.1515/jbnst-2018-0022

Glatzer W, Zapf W (Hrsg) (1984) Lebensqualität in der Bundesrepublik. Objektive Lebensbedingungen und subjektives Wohlbefinden. Frankfurt/Main New York, Campus

Giesselmann M, Bohmann S, Goebel J, Krause P, Liebau E, Richter D, Schacht D, Schröder C, Schupp J, Liebig S (2019) The individual in context(s): research potentials of the socio-economic panel study (SOEP) in sociology. Euro Sociol Rev 35(5):738–755

Hawes J (2018) The shortest history of Germany. CPI Group (UK) Ltd., Croydon

Heuer H (1977) Sozioökonomische Bestimmungsfaktoren der Stadtentwicklung. Deutsches Institut für Urbanistik, Kohlhammer

Krause P (1984) Verstädterung und Wandel der Siedlungsstruktur im Industrialisierungsprozess. Eine empirische Analyse der deutschen Entwicklung. Diplomarbeit, Universität Mannheim

Krause P (2016) Quality of life and inequality. In: Bruni L, Porta PL (eds) Handbook of research methods and application in happiness and quality of life. Edward Elgar Publishing, pp 111–147

Krause P (2019) Wealth and partial wealth. In: Wealth(s) and subjective well-being. Springer, pp 87–128
Maasoumi E (1986) The measurement and decomposition of multi-dimensional inequality. Econometrica 54(4):991–997
MacGregor N (2014) Germany. Memories of a nation. British Museum, Allan Lane, Penguin Books
Maslow AH (1943) A theory of human motivation. Psychol Rev 50(4):370–396
Mukhopadhyay N, Sengupta PP (eds) (2021) Gini inequality index. Methods and applications. CRC Press, Boca Raton
Mueller GP (2019) Item nonresponse and fuzzy logic. Faculty of Economics and Social Sciences, University of Fribourg, Lausanne. FORS Working Paper 01-2019
Noll HH (2018) Social monitoring and reporting: a success story in applied research on social indicators and quality of life. Soc Indic Res 135(3):951–964
Nussbaum M (2000) Women and human development: the capabilities approach. Cambridge University Press, Cambridge New York. ISBN 9780521003858
Peichl A, Pestel N (2013) Multidimensional affluence: theory and applications to Germany and the US. Appl Econ 45(32):4591–4601. https://doi.org/10.1080/00036846.2013.795280
Piketty T, Zucman G (2015) Wealth and inheritance in the long run. In: Atkinson (ed) Handbook of income distribution, vol 2B. Elsevier B.V., pp 1303–1368. https://doi.org/10.1016/B978-0-444-59429-7.00016-9
Rippin N (2014) Considerations of efficiency and distributive justice in multidimensional poverty measurement. Universität Göttingen, eDiss, http://hdl.handle.net/11858/00-1735-0000-0022-5E2E-B
Rippin N (2015) Multidimensional poverty in Germany: a capability approach. In: Forum for social economics (special issue: capability approach and multidimensional well-being in high-income countries). https://doi.org/10.1080/07360932.2014.995199
Sen A (1976) Poverty: an ordinal approach to measurement. Econometrica 44(2):219–231
Sen A (1979) Issues in the measurement of poverty. Scand J Econ 81(2):285–307
Sen A (1999) Development as freedom. Oxford University Press, New York
Socio-economic Panel (SOEP) (2021) Data 1984–2020, version v37. https://doi.org/10.5684/soep.v37
Wagner GG, Frick JR, Schupp J (2007) The German socio-economic panel study (SOEP): scope, evolution and enhancements. Schmollers Jahrbuch 127(1):139–169
Yitzhaki S (1994) Economic distance and overlapping of distributions. J Econ 61:147–159

Chapter 6
Future Tripping: What Might Inequality Look Like in a Warmer World? A Human Development Perspective

Fernanda Pavez Esbry, Rehana Mohammed, and Jon Hall

Abstract Dangerous planetary change and social imbalances are interlinked, and often reinforce each other. Climate change for instance, is expected to exacerbate inequalities between people, communities, and countries. This chapter considers how climate change and inequalities interact and reinforce each other through two main channels: unequal exposure and unequal vulnerability to climate change hazards. We pay special attention to urban environments, where many of the risks to people from climate change are compounded. And we consider inequalities between and within countries in terms of the three channels: exposure to climate change hazards, vulnerability to hazards, and the ability to recover from hazards. The effects of climate change on people will depend on the dynamic interaction between these channels. The human development levels of those affected also plays a part. We argue that climate change is an "inequality bomb" that seems set to drive existing inequalities wider within and between many countries and threatens to redraw the human development map. And we illustrate this argument using projected changes in mortality rates, just one of the very many facets of inequality that will be affected by global warming.

Keywords Climate change · Human development · Inequality · Urban

---

The views expressed in this publication are those of the authors and do not necessarily represent those of the United Nations, including UNDP, or the UN Member States.

F. P. Esbry (✉) · R. Mohammed · J. Hall
Human Development Report Office at the United Nations Development Program, New York, US
e-mail: fernanda.pavezesbry@un.org

R. Mohammed
e-mail: rehana.mohammed@undp.org

J. Hall
e-mail: jonathan.hall@undp.org

G. H. Tonon (ed.), *Urban Inequalities*, The Urban Book Series,
https://doi.org/10.1007/978-3-031-59746-6_6

6.1 Introduction

This chapter considers what the climate crisis means for inequalities in human development. It pays particular attention to what might happen in cities around the world.

Global warming threatens to redraw the development map within and between countries. The impacts of a changing climate on development outcomes are manifold. More frequent flooding and wildfires could be catastrophic in some areas. Warmer temperatures will lead to heatwaves that could impact mortality rates and lead to areas becoming virtually uninhabitable. Changing rainfall patterns will affect food and water security. In other places, however, warmer temperatures might be beneficial to human health for some groups, for example, there might be fewer very cold days in winter.

Climate change has been described as an inequality bomb, pushing existing inequalities wider and creating new ones. The impacts will fall unequally on the world, with some countries, predominantly in the global South, bearing the brunt of the negative effects while also being least able to adapt to them. While within countries too inequalities will widen. We consider here what this might mean for human development.

We begin by describing human development, before considering why inequalities matter both in themselves and to humanity's efforts to tackle climate change. Next, we consider how climate change might affect inequalities in human development and whether cities will fare differently to other areas.

6.2 The Human Development Approach: What, Why, and How

Discussions about development, progress, or the good life, have been raging for at least the past 2000 years. By the end of the twentieth century many economists argued that economic growth and development were synonymous, either directly or indirectly. There was "an implicit assumption that economic growth was synonymous with progress: an assumption that a growing gross domestic product (GDP) meant life must be getting better" (OECD 2008).

Yet GDP was never intended as a proxy for well-being. Simon Kuznets, one of the fathers of the system of national accounts, warned the US Congress in 1934 of GDP's limitations noting that "the welfare of a nation can scarcely be inferred from a measurement of national income" (Kuznets 1934).

UNDP's Human Development Report and Index were introduced in 1990 to encourage a rethink about the relationship between economic growth and development. As Amartya Sen, a co-creator of the Human Development Approach, said

"Human development, as an approach, is concerned with what I take to be the basic development idea: namely, advancing the richness of human life, rather than the richness of the economy in which human beings live, which is only a part of it" (Sen 2004).

Human development has the capabilities approach at its center. This argues that well-being should be understood in terms of people's "doings" and "beings"—the various activities, states of being, and opportunities that people might pursue if they so choose. Human development is the process of enlarging people's choices and freedoms, which range from living a long and healthy life and being educated, to enjoying political freedom and a decent standard of living.

6.2.1 What is Human Development?

The first *Human Development Report*, published in 1990 by the United Nations Development Program, introduced a new approach for advancing human well-being. It is about:

- People: improving the lives people lead rather than assuming that economic growth will lead, automatically, to greater well-being for all. Income growth is seen as a means to development, rather than an end in itself.
- Opportunities: giving people more freedom to live lives they value. In effect this means developing people's abilities and giving them a chance to use them. For example, educating a girl would build her skills, but it is of little use if she is denied access to jobs or does not have the right skills for the local labor market.
- Choice: providing people with opportunities, not insisting that they make use of them. No one can guarantee human happiness, and the choices people make are their own concern. The process of development—human development—should at least create an environment for people, individually and collectively, to develop to their full potential and to have a reasonable chance of leading productive and creative lives that they value.

The Human Development Index (HDI) ranks countries by their level of human development. The report recognizes that the concept of human development is much broader than the HDI, making it impossible to come up with a comprehensive measure, or even a comprehensive set of indicators, because many areas of human development are non-quantifiable.

6.3 Inequality and the Anthropocene

Individual people are the fundamental building block and unit of analysis for human development. Therefore, the distribution of human development outcomes, or inequalities, between people are fundamental to assessing human development.

Yet look below the surface and inequalities have other effects. For instance, the 2019 Human Development Report (UNDP 2019) argued that:

- Inequalities in socioeconomic status and income have a profound influence on key areas of human development such as education and health.
- Income and wealth inequalities often translate into political inequalities, allowing some groups greater influence over policymaking and a greater ability to shape policies in their favor.
- Beyond their distributive effects, inequalities also contribute to oppressive social conditions, in that they often foster and compound relations of domination and standing between groups.

In other words, economic, political, and social inequalities diminish people's choices and freedoms, making inequality a core concern for human development. Inequalities are also driving—and being driven by–dangerous planetary change. A characteristic of our era that has been described as the Anthropocene.

Never have so many of the planet's systems been knowingly affected by a single species. Humans are driving climate change (Masson-Delmotte et al. 2021; Pörtner et al. 2022) and harming the integrity of many of the ecosystems that sustain human lives and other species. Our choices are shaping the evolution of life on Earth through legacies that will unfold over millions of years to come (Malhi 2017). Climate change, biodiversity loss and many other environmental challenges—from air pollution to plastics use—are receiving individual attention. But the way these and other planetary pressures are interlinked—and the speed, scale, and scope of the unprecedented planetary changes unfolding as a result—has motivated a name for this era: the Anthropocene—the age of humans, where humans' impact on the planet is so stark that it is driving dangerous planetary change.[1] And human influence has unequivocally impacted the planet. Greenhouse gas emissions are causing global surface temperature and sea levels to rise, Arctic ice area to diminish, extreme weather events to multiply and intensify and annual global precipitation to increase (Boulton et al. 2022). Biodiversity loss is staggering up to 1 million species are estimated to be facing extinction, and the population wild mammals, fish, birds, reptiles, and amphibians fell by an average of 68% between 1970 and 2016 (Almond et al. 2020). Cities, particularly in developed countries, are among the biggest contributors to planetary pressures. Urbanization has direct and indirect impacts on ecosystems, both within and around cities. As cities expand, more land is converted for urban development. Urban areas exert greater pressure on resources, including energy and water. Weak urban infrastructure, such as inadequate wastewater treatment facilities, contributes to pollution of surrounding ecosystems and contamination of freshwater supplies. Urban areas account for a majority of greenhouse gas emissions from final energy

[1] See the 2020 Human Development Report (UNDP 2019) for a review of the evidence, as well as the debate from multiple perspectives from the natural sciences, social sciences, and humanities that the Anthropocene frame has motivated.

use (Mahendra et al. 2021). Dangerous planetary change and social imbalances are interlinked, and often reinforce each other. Climate change for instance, is expected to exacerbate inequalities between people, societies, and countries.

This chapter considers how climate change and inequalities interact and reinforce each other through two main channels: unequal exposure and unequal vulnerability to climate change hazards. Some countries, communities and people are more exposed to climate change-related hazards, such as extreme weather events and high temperatures, than others. In addition, inequalities shape the vulnerability of countries and people to hazards. For instance, while coastal settlements in many countries may be exposed to the sea level rise, some countries will be better equipped to prepare and adapt to this hazard than others. In a similar vein, with the same exposure to hazards such as storms, floods, or wildfires, some people or groups will be better able to protect themselves than others.

Inequalities in exposure and vulnerability to hazards are themselves shaped by existing inequalities. It is because of the significant existing disparities that climate change is potentially an "inequality bomb". No country or community will be immune to the effects of climate change, but some will be considerably more impacted than others.

The rest of this chapter looks ahead to investigate what impacts climate change might have on human development and inequality. It considers inequalities between countries at different levels of human development and within-country inequalities in terms of the three channels outlined above: exposure to climate change hazards, vulnerability to hazards, and ability to recover from hazards. The effects of climate change on people will depend on the dynamic interaction between these channels. And as noted by the Intergovernmental Panel on Climate Change (Pörtner et al. 2022), in the near term, the risks to natural and human systems from climate change depend largely on exposure and vulnerability rather than on differences in hazards between emissions scenarios.

This chapter pays special attention to urban environments, where many of the risks to people from climate change are compounded. Any assessment of the impact climate change will have on inequalities needs to pay attention to cities, if only because of the sheer number of people inhabiting them—estimated to reach to 68% of the global population by 2050, with an additional 2.5 billion people projected to be living in urban areas by 2050, with up to 90% of this increase concentrated in the regions of Asia and Africa (Pörtner et al. 2022, p. 916).

Patterns of urbanization today are different from those of the past. Today the highest rates of urbanization are in Africa and South and Southeast Asia, and urbanization is taking place in more low-income countries than before. In the past, urbanization was usually associated with income growth. But today urbanization is coming with "an explosion of poor megacities" in countries where income growth has remained slow or stagnant (Glaeser 2014). And while the vast majority of the world's poor live in rural areas, a rising share of the poor are moving to and living in urban areas (see Ravallion et al. 2007; Mahendra et al. 2021).

This chapter argues that urbanization has the potential to increase the risks to people arising from climate change, and so it comprises an additional layer to analysis of inequalities. Cities are often constructed in ways that exacerbate the impacts of extreme weather. For example, greenery and vegetation, that can moderate temperature, are often built over. While areas with impermeable surfaces that can exacerbate flooding are often built. People living in informal settlements, a hallmark of urbanization in less developed countries, are especially vulnerable to climate change-related shocks and risks. Looking at urban environments highlights different exposure to hazards, vulnerability, and ability to recover among the communities and people, where most climate change impacts are ultimately felt.

6.3.1 Looking Ahead: The Changing Face of Inequality in a Changing Climate

In 2022 UNDP and the Climate Impact Lab released a Human Climate Horizons (HCH) platform at https://horizons.hdr.undp.org/. It allows users to generate projections—under different scenarios—that consider how climate change may affect several aspects of human development at a fine granular level of geography. A disaggregated view—which the HCH offers—shows how some cities in developed countries will face higher mortality risks from global warming just as some cities in developing countries will face lower mortality risks. Moreover, territories such as Florida in the USA or the West Coast of India, which are likely to have relatively low exposure to some climate change risks, such as mortality, are likely to be more vulnerable to other threats, such as flooding and rising sea levels. The analysis that follows comes from that data set.

The Human Climate Horizons platform provides localized information on some of the future impacts of climate change across several dimensions of human development. It is an open access and scalable digital public good that offers a window to possible futures and is the result of joint work of the Climate Impact Lab and the UNDP's Human Development Report Office. Covering more than 24,000 regions worldwide, two different policy scenarios and time horizons through the end of twenty-first century, HCH provides empirically grounded data and information on the potential human costs of climate change. In the first release, HCH looks at the effects of temperature changes on mortality, labor, and energy demand. Future updates will include information on coastal impacts, damage to infrastructure and climate change's potential to disrupt food production systems (UNDP 2019).

6.4 Unequal Exposure to Climate Change Hazards

There is evidence of differentiated exposure to climate change hazards between countries and people. Climate change presents many interacting hazards to human and natural systems, including from more frequent and more intense weather extremes, including high temperatures, heavy precipitation events, and droughts. In addition are hazards from slow-onset processes, such as ocean acidification and sea level rise.

The HCH captures climate change hazards in terms of average annual temperatures (see Fig. 6.1) and days of extreme hot weather (see Fig. 6.2). These hazards are projected to be significantly lower in countries at higher levels of human development than low human development countries. This is because many developing countries are in tropical and sub-tropical regions, which have greater exposure to warmer temperatures (Watson et al. 1998).

We have already shown that rapid urbanization is occurring in many developing countries, and so it is worth considering how living in an urban settlement influences exposure to hazards.

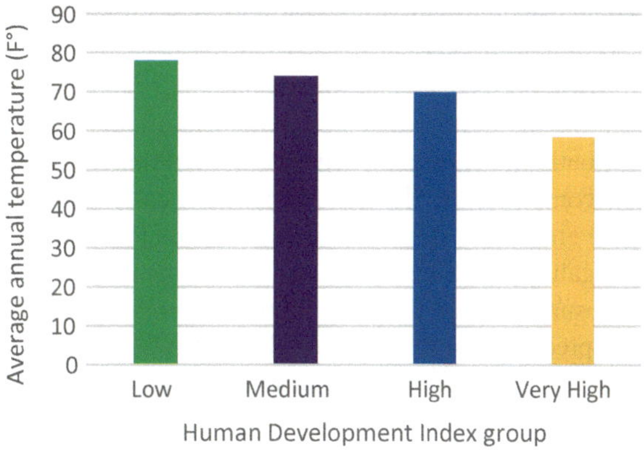

Fig. 6.1 Lower HDI countries are expected to experience higher average annual temperatures due to climate change. *Source* Elaborated by authors with data from UNDP 2022 and UNDP and Climate Impact Lab 2022. Note: Estimates are for the 2020–2039 period, under RCP 4.5 scenario

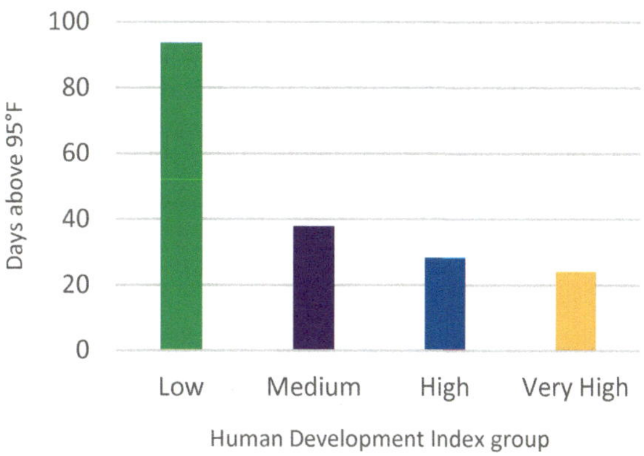

Fig. 6.2 Lower HDI countries are expected to experience more days of extreme hot weather due to climate change. *Source* Elaborated by authors with data from UNDP 2022 and UNDP and Climate Impact Lab 2022. Note: Estimates are for the 2020–2039 period, under RCP 4.5 scenario

Urbanization is known to increase exposure to climate change hazards (Pörtner et al. 2022). Many cities are in areas exposed to climate hazards. Urban areas face greater flooding hazards, have greater exposure to heatwaves, and greater exposure to water scarcity from severe droughts (Pörtner et al. 2022), Rosenzweig et al. 2018).

Flooding is the most pervasive risk to people associated with climate change and is expected to worsen as rainfall and sea level rise intensifies (McDermott 2022). One analysis of global exposure to floods estimates that 1.8 billion people, or almost a quarter of the world's population, are directly exposed to inundation depths of over 0.15 m. Almost 90% of flood-exposed people live in low- and middle-income countries (Rentschler et al. 2022).

Urbanization is also contributing to increasing exposure to hazards over time. For example, urbanization might increase exposure to flooding because economic development and population growth often occur in coastal cities. Several expanding megacities—Bangkok, Mumbai, Ho Chi Minh City, Manila—are in low-lying coastal areas and vulnerable to flooding, sea level rise and more hazards (Ahmed et al. 2010).

There are also considerable within-country inequalities between people in their exposure to climate hazards. These inequalities in exposure to hazards often reflect existing patterns of inequality, such as long-term discrimination and poverty, both of which can contribute to the formation of informal settlements in places with greater exposure to floods.

An analysis of 52 countries found that poor people are disproportionately exposed to flooding and droughts, especially in urban areas (Winsemius et al. 2018). While in South Africa, spatial segregation and discrimination against black communities, has contributed to the growth of chronically underserved informal settlements that have greater exposure to climate hazards (Williams et al. 2019). In the Indian cities Bengaluru and Surat, migrant settlements in low-lying areas result in high levels of exposure to urban flooding, compounding migrant communities' existing insecurity and precariousness (Chu and Michael 2019).

In many Latin American cities, the urban poor are concentrated in hazard-prone areas, including locations close to mountain ranges and floodplains (Chu and Michael 2019). In the United States, there is evidence that historically discriminatory housing policies contribute to unequal exposure to extreme heat in urban areas (Hoffman et al. 2020).

Inequalities in exposure to extreme heat have been documented at the neighborhood level: people residing in neighborhoods with higher racial diversity, extreme poverty, and lower levels of education are more likely to be exposed to extreme heat (Voelkel et al. 2018).

6.5 Unequal Impacts of Climate Change Hazards

No country will be spared the negative impacts of climate change. But there are significant inequalities between countries in terms of how adversely impacted they will be. This means that at the same level of exposure to climate change hazards, some groups fare worse than others.

Overall, climate change is poised to reduce global economic output in the long run (UNDP 2019). Economic losses from disasters have been rising, primarily due to population and economic growth in areas prone to disasters (Botzen et al. 2019). While inequality between countries has declined over the past decades, global warming has inhibited this decline (Diffenbaugh and Burke 2019). In short, the impacts of climate change and structural inequalities are mutually reinforcing. Existing inequalities shape how vulnerable societies and people are to the impacts of climate change, and how the wider societal impacts of climate change will be felt.

There is growing evidence that vulnerability is shaped primarily by development processes rather than by individual hazardous events (Birkmann et al. 2022). Some countries and regions of the world are more vulnerable than others: Africa, Asia, and Central America have been described as global vulnerability hotspots, with socioeconomic, demographic, and environmental conditions that heighten the adverse impacts of climate change (Birkmann et al. 2022). In many countries, vulnerability to climate change is shaped by historical conditions, and climate change itself is only one of many long-term development challenges (For instance see Douglass and Cooper 2020).

The regions that are more vulnerable to the adverse impacts from climate change are also home to more people: there are nearly twice as many people living in the most vulnerable countries compared to those in less vulnerable countries (Birkmann et al. 2021).

The inequalities in the impacts of climate change between countries also reflects the fact developing countries have fewer resources for adaptation relative to wealthier countries. The HCH's estimations account for the varying ability of countries to adapt to climate change, considering for instance that wealthier countries are better able to adapt to increased temperatures with cooling measures than poorer countries. This adaptation gap between countries is overlayed on the existing variation in exposure to climate change hazards particularly that many poorer countries are in warmer regions of the world and so are more vulnerable to climate change.

The HCH depicts climate change impact in terms of mortality, using changes in death rates due to global warming. Temperature-related mortality is projected to decrease in colder, wealthier countries, as days of extreme cold decrease, while increasing in warmer, poorer countries, as days of extreme heat increase.

Figure 6.3 shows that while death rates are set to worsen in low human development countries, they are projected to improve at higher levels of human development.

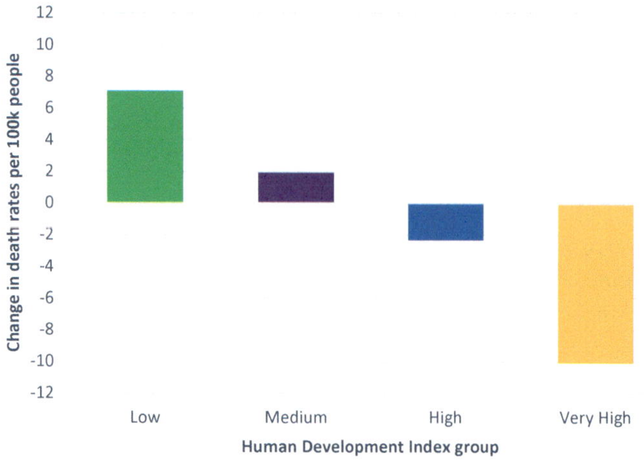

Fig. 6.3 Projected change in mortality rates due to global warming per 100k people at the country level, by Human Development Index group. *Source* Elaborated by authors with data from UNDP 2022 and UNDP and Climate Impact Lab 2022. Note: Estimates are for the 2020–2039 period, under RCP 4.5 scenario

Using city level estimates, HCH data shows that cities in low HDI countries are set to see higher death rates than those in high HDI countries. Within countries, some areas will face considerably more impacts from climate change than others.

While wealthier countries may experience higher direct property losses due to disasters, economic growth impacts in wealthier countries are found to be smaller compared to developing countries (Hoeppe 2016; Kahn 2005). Indeed, disasters have the strongest negative impacts in developing countries (Klomp and Valckx 2014). Global warming is expected to diminish economic output in hotter, poorer countries while increasing output in cooler, wealthier countries (Diffenbaugh and Burke 2019).

Climate change also makes it harder for people escape poverty. The 2022 IPCC report documents five channels through which climate change impedes poverty alleviation: (i) it worsens living conditions, (ii) it threatens food security and nutrition, (iii) it disrupts access to ecosystem services, (iv) it facilitates the spread of vector-borne disease, and (v) it reinforces inequalities between gender, age groups, social class, and race (Pörtner et al. 2022).

Urbanization will play a large role in determining how the impacts of the climate change on people and societies will unfold. Other things equal, urban environments experience greater temperate-related mortality impacts (Calleja-Agius et al. 2021). In part because of the urban heat island effect, by which cities experience elevated temperatures compared to their surrounding areas.

Within cities, patterns of poverty and informal settlements also contribute to the inequality because some groups are more at risk from the heat than others (Milan and Creutzig 2015). Poorer people in particular face greater risks from urban heat exposure due to inadequate housing, limited access to air conditioning and cooler indoor environments, and often undertake occupations with higher heat exposure

(Chu and Michael 2019). Informal settlements often tend to be in areas that are highly exposed to climate change hazards while having relatively poor-quality infrastructure and services, leading to greater risks to people.

In addition to its direct impacts on people's health, urban heat also leads to unequal economic stresses on people and households via energy demands (e.g., for cooling in warm periods), higher medical costs due to heat-related illness or missing work and other labor impacts (Dodman et al. 2022). Higher temperatures are not the only way in which urban societies will be impacted: there is evidence of several dynamic interactions between climate change and flooding, water scarcity, pollution, drought, and infrastructure damage (Dodman et al. 2022).

6.6 The Future

Climate change will have very different impacts across some of the most populous cities in the world. These different impacts will affect both inequality within a country and between countries. A localized analysis of projected climate change hazards and impacts, using data published by UNDP's Human Climate Horizons Platform, paints a fascinating picture.

We focus here on the Brazil, Russian Federation, India, China, and South Africa countries (BRICS) and the United States, HCH's moderate (RCP4.5) warming scenario.[2]

The average projected number of additional days above 35 °C per annum in the 2020–2039 period, compared to a world without climate change varies enormously. Across India there is enormous additional variation between cities. On one hand, the city of Srinagar in the Himalayas, is projected to experience just one more day a year above 35C (and just one day each year in total). On the other, Ahmedabad, in India's dry west, is projected to experience an additional 22 days above 35C each year and 166 days above 35C in total.

Across the whole country, over the period 2020–2039 India is projected to experience an average 95 days a year above 35C, some 17 more days each year than it would have experienced without climate change.

There is also considerable variation across cities in Brazil, the US, and to a lesser but still considerable extent, China, and South Africa.

These additional inequalities in temperature will manifest themselves in other ways. Perhaps most notably, from a human development perspective, is their implications for mortality. At the risk of oversimplifying, an increase of hot days per annum in places where temperatures are already high will lead to additional deaths. But an increase in temperatures that moderate very cold winters can lead to fewer deaths.

---

[2] Representative Concentration Pathway (RCP) 4.5 is a scenario that stabilizes radiative forcing at 4.5 W m^{-2} in the year 2100 without ever exceeding that value.

The patterns are, therefore, complex. But the following tables summarizes the HCH projections on changes in mortality and days above 35C among the 6 countries we considered, see Table 6.1.

6.7 Conclusions

This chapter has argued that climate change is projected to have a significant and very varied impact on urban inequalities in human development because of the existing inequalities in people's exposure to both climate hazards and the projected impacts of changing climate.

There are many ways to consider urban inequality and we have used just three broad lenses to consider how climate change might impact urban areas compared to: non-urban areas; other urban areas within a nation; and other urban areas worldwide.

Using all three geographic lenses we have demonstrated that climate change is likely to have a significant impact on inequality across many aspects of human development, not least because many of the very physical characteristics of a city mean urban areas in hotter climates are particularly exposed to climate hazards.

Using data from UNDP's Human Climate Horizons we analyze just one particular facet of human development inequality, to look at how mortality rates across and within cities of the BRICS countries plus the USA are projected to change by 2040 under a moderate global warming scenario.

The modeled estimates show that there will be considerable change within and across countries and that climate change will redraw the urban inequality map. In Russia, for example mortality rates are project to decline by 25 deaths per 100,000 people per annum. In India they are projected to rise by 3 deaths. Within countries patterns of mortality will also change quite dramatically. In the United States of America for instance, mortality rates are projected to rise in the city of Phoenix by 16 deaths per 100,000 people per annum, yet and decline in Minneapolis by more than 37 deaths.

Table 6.1 Projected global warming and its impacts on people: numbers of days above 35C each year, and mortality rates

Brazil

	Average number of days above 35C in 2020–2039		Of which are additional from climate change	Mortality changes in deaths per 100k people in 2020–2039	
Country level	23 days		12 days	– 1 deaths	
Brasilia	6 days		4 days	– 2.3 deaths	
City level	Manaus 65 days, 44 additional	Sao Paulo 0 days		Manaus + 5.4 deaths	Brasilia – 4.3 deaths

Russia

	Average number of days above 35C in 2020–2039		Of which are additional from climate change	Mortality change in deaths per 100k people in 2020–2039	
Country level	1 day		1 day	– 25 deaths	
Moscow	0 days		0 days	– 26.4 deaths	
City level	Rostov-na-Donu 7 days, 6 additional	Saint Petersburg 0 days		Rostov-na-Donu – 8.4 deaths	Novosibirsk – 35.5 deaths

India

	Average number of days above 35C in 2020–2039		Of which are additional from climate change	Mortality change in deaths per 100k people in 2020–2039	
Country level	97 days		17 days	+ 3 deaths	
New Delhi	125 days		20 days	+ 7.3 deaths	
City level	Ahmadabad 171 days, 27 additional	Srinagar 1 day, 1 additional		Amritsar + 10.9 deaths	Srinagar – 5.5 deaths

China

	Days above 35C		Of which are additional from climate change	Mortality changes in deaths per 100k people in 2020–2039	
Country level	7 days		3 days	– 2 deaths	
Beijing	15 days		5 days	– 2.8 deaths	
City level	Nanchang 30 days, 11 additional	Dalian 0 days		Luoyang + 2.8 deaths	Xining – 20.1 deaths

(continued)

Table 6.1 (continued)

South Africa					
	Average number of days above 35C in 2020–2039		Of which are additional from climate change	Mortality changes in deaths per 100k people in 2020–2039	
Country level	13 days		6 days	– 2 deaths	
Johannesburg	2 days		1 day	– 2 deaths	
City level	Pretoria 20 days, 8 additional	Cape Town 0 days		Cape Town – 1.7 deaths	Vereeniging – 2.3 deaths

United States					
	Average number of days above 35C in 2020–2039		Of which are additional from climate change	Mortality changes in deaths per 100k people in 2020–2039	
Country level	23 days		12 days	– 3 deaths	
Washington DC	16 days		9 days	+ 3.2 deaths	
City level	Phoenix 142 days, 18 additional	San Francisco 0 days		Phoenix + 16 deaths	Minneapolis – 37.2 deaths

Six country comparison			
	Average number of days above 35C in 2020–2039	Of which are additional from climate change	Mortality change in deaths per 100k people in 2020–2039
Brazil	23 days	12 days	– 1 death
China	7 days	3 days	– 2 deaths
India	97 days	17 days	+ 3 deaths
Russia	1 day	1 day	– 25 deaths
South Africa	13 days	6 days	– 2 deaths
United States of America	23 days	12 days	– 3 deaths

Source Authors' own calculations from the HCH data, 2023

References

Ahmed K, Philips BR, Pillai P, Shyamsundar P, Wang L (2010) Climate risks and adaptation in Asian coastal megacities: a synthesis report. World Bank Group. https://documents.worldbank.org/en/publication/documents-reports/documentdetail/866821468339644916/climate-risks-and-adaptation-in-asian-coastal-megacities-a-synthesis-report

Almond R, Grooten M, Petersen T (eds) (2020) Living Planet Report 2020—bending the curve of biodiversity loss. World Wildlife Fund

Birkmann J, Feldmeyer D, McMillan J, Solecki W, Totin E, Roberts D, Trisos C et al (2021) Regional clusters of vulnerability show the need for transboundary cooperation. Environ Res Lett 16(9). https://doi.org/10.1088/1748-9326/ac1f43

Birkmann J, Liwenga E, Pandey R, Boyd E, Djalante R, Gemenne F, Leal Filho W, Pinho PF, Stringer L, Wrathall D (2022) Poverty, livelihoods and sustainable development. In: Pörtner H-O, Roberts DC, Tignor M, Poloczanska ES, Mintenbeck K, Alegría A, Craig M, Langsdorf S, Löschke S, Möller V, Okem A, Rama B (eds) Climate change 2022: impacts, adaptation and vulnerability. Contribution of Working Group II to the Sixth Assessment Report of the Intergovernmental Panel on Climate Change. Cambridge University Press, pp 1171–1274. https://doi.org/10.1017/9781009325844.010

Botzen WJW, Deschenes O, Sanders M (2019) The economic impacts of natural disasters: a review of models and empirical studies. Rev Environ Econ Policy 13(2). https://doi.org/10.1093/reep/rez004

Boulton CA, Lenton TM, Boers N (2022) Pronounced loss of Amazon rainforest resilience since the early 2000s. Nat Clim Change 12(3):271–278. https://doi.org/10.1038/s41558-022-01287-8

Calleja-Agius J, England K, Calleja N (2021) The effect of global warming on mortality. Early Hum Dev 155:105222. https://doi.org/10.1016/j.earlhumdev.2020.105222

Chu E, Michael K (2019) Recognition in urban climate justice: marginality and exclusion of migrants in Indian cities. Environ Urban 31(1):139–156. https://doi.org/10.1177/0956247818814449

Diffenbaugh NS, Burke M (2019) Global warming has increased global economic inequality. Proc Natl Acad Sci 116(20):9808–9813. https://doi.org/10.1073/pnas.1816020116

Dodman D, Hayward B, Pelling M, Castan Broto V, Chow W, Chu E, Dawson R, Khirfan L, McPhearson T, Prakash A, Zheng Y, Ziervogel G (2022) Cities, settlements and key infrastructure. In: Pörtner H-O, Roberts DC, Tignor M, Poloczanska ES, Mintenbeck K, Alegría A, Craig M, Langsdorf S, Löschke S, Möller V, Okem A, Rama B (eds) Climate change 2022: impacts, adaptation and vulnerability. Contribution of Working Group II to the Sixth Assessment Report of the Intergovernmental Panel on Climate Change. Cambridge University Press, pp 907–1040. https://doi.org/10.1017/9781009325844.008

Douglass K, Cooper J (2020) Archaeology, environmental justice, and climate change on islands of the Caribbean and southwestern Indian Ocean. Proc Natl Acad Sci 117(15):8254–8262. https://doi.org/10.1073/pnas.1914211117

Glaeser EL (2014) A world of cities: the causes and consequences of urbanization in poorer countries. J Eur Econ Assoc 12(5):1154–1199. https://doi.org/10.1111/jeea.12100

Hoeppe P (2016) Trends in weather-related disasters—consequences for insurers and society. Weather Clim Extremes 11:70–79. https://doi.org/10.1016/j.wace.2015.10.002

Hoffman J, Shandas V, Pendleton N (2020) The effects of historical housing policies on resident exposure to intra-urban heat: a study of 108 US urban areas. Climate 8(1). https://doi.org/10.3390/cli8010012

Kahn ME (2005) The death toll from natural disasters: the role of income, geography and institutions. Rev Econ Stat 87(2):271–284. https://doi.org/10.1162/0034653053970339

Klomp J, Valckx K (2014) Natural disasters and economic growth: a meta-analysis. Glob Environ Change 26:183–195. https://doi.org/10.1016/j.gloenvcha.2014.02.006

Kuznets S (1934) National income, 1929–1932. In National income, 1929–1932. National Bureau of Economic Research, pp 1–12. https://www.nber.org/system/files/chapters/c2258/c2258.pdf

Mahendra A, King R, Du J, Dasgupta A, Beard VA, Kallergis A, Schalch K (2021) Seven transformations for more equitable and sustainable cities. World resources report. Towards a more equal city. World Resources Institute. https://doi.org/10.46830/wrirpt.19.00124

Malhi Y (2017) The concept of the Anthropocene. Annu Rev Environ Resour 42:77–104. https://doi.org/10.1146/annurev-environ-102016-060854

Masson-Delmotte V, Zhai P, Pirani A, Connors SL, Péan C, Berger S, Caud N, Chen Y, Goldfarb L, Gomis MI, Huang M, Leitzell K, Lonnoy E, Matthews JBR, Maycock T, Waterfield TK, Yelekçi O, Yu R, Zhou B (eds) (2021) Climate change 2021: the physical science basis. In: Contribution of Working Group I to the sixth Assessment Report of the Intergovernmental Panel on Climate Change, vol 2. Cambridge University Press, pp 3–32. https://doi.org/10.1017/9781009157896.001

McDermott TK (2022) Global exposure to flood risk and poverty. Nat Commun 13(1):3529. https://doi.org/10.1038/s41467-022-30725-6

Milan BF, Creutzig F (2015) Reducing urban heat wave risk in the 21st century. Curr Opin Environ Sustain 14:221–231. https://doi.org/10.1016/j.cosust.2015.08.002

Organisation for Economic Cooperation and Development (2008) Statistics, knowledge and policy 2007: measuring and fostering the progress of societies [Preface]. https://doi.org/10.1787/978 9264043244-en

Pörtner H-O, Roberts DC, Tignor M, Poloczanska ES, Mintenbeck K, Alegría A, Craig M, Langsdorf S, Löschke S, Möller V, Okem A, Rama R (eds) (2022) Climate change 2022: impacts, adaptation, and vulnerability. Contribution of Working Group II to the Sixth Assessment Report of the Intergovernmental Panel on Climate Change. Cambridge University Press. https://doi.org/10.1017/9781009325844

Ravallion M, Chen S, Sangraula P (2007) New evidence on the urbanization of global poverty. Popul Dev Rev 33(4):667–701. https://doi.org/10.1111/j.1728-4457.2007.00193.x

Rentschler J, Salhab M, Jafino BA (2022) Flood exposure and poverty in 188 countries. Nat Commun 13(1):3527. https://doi.org/10.1038/s41467-022-30727-4

Rosenzweig C, Solecki WD, Romero-Lankao P, Mehrotra S, Dhakal S, Ibrahim SA (eds) (2018) Climate change and cities: second assessment report of the urban climate change research network. Cambridge University Press

Sen A (2004) Amartya Sen: a more human theory of development. Interview with Amartya Sen. Asia Society, 6 Dec. https://asiasociety.org/amartya-sen-more-human-theory-development

United Nations Development Programme (2019) Human Development Report 2019. Beyond income, beyond averages, beyond today: inequalities in human development in the 21st century. https://hdr.undp.org/content/human-development-report-2019

United Nations Development Programme (2022) What is human climate horizons? https://horizons.hdr.undp.org/#/what-is-hch

Voelkel J, Hellman D, Sakuma R, Shandas V (2018) Assessing vulnerability to urban heat: a study of disproportionate heat exposure and access to refuge by socio-demographic status in Portland, Oregon. Int J Environ Res Public Health 15(4):640. https://doi.org/10.3390/ijerph15040640

Watson RT, Zinyowera MC, Moss RH (1998) The regional impacts of climate change: an assessment of vulnerability. Cambridge University Press

Williams DS, Costa MM, Sutherland C, Celliers L, Scheffran J (2019) Vulnerability of informal settlements in the context of rapid urbanization and climate change. Environ Urban 31(1):157–176. https://doi.org/10.1177/0956247818819694

Winsemius H, Jongman B, Veldkamp T, Hallegatte S, Bangalore M, Ward P (2018) Disaster risk, climate change, and poverty: assessing the global exposure of poor people to floods and droughts. Environ Dev Econ 23(3):328–348. https://doi.org/10.1017/S1355770X17000444

Chapter 7
Urban Inequalities and Climate Change: Contributions to the Study of Resilience in Argentine Cities

Marìa Laura Zulaica

Abstract Latin American cities are exposed to threats triggered by climate change that generate environmental impacts. In addition, there is evidence of an accelerated process of urbanization which consequences are intensified by the absence of comprehensive planning policies, deepening urban inequalities. In Argentina, the effects of climate change are visible and have differential regional scope, depending on the response capacity of cities in the face of environmental dangers. This ability to respond to extreme events can be interpreted in terms of resilience. In this context, the present work analyzes resilience in 32 Argentine urban agglomerates through the construction of a synthetic index, the Resilience Index, which integrates three variables: exposure, sensitivity, and adaptation. The results are spatially represented using a Geographic Information System. The response capacity in the cities analyzed accounts for the existing inequalities in the national territory. Although the concept of resilience was incorporated into the urban planning and management processes of some cities, the importance of focusing on adaptation measures that involve everything from works and infrastructure to training actions, awareness, and action in emergencies is highlighted. The opening of new questions stimulates to continue deepening the analysis of inequalities, to advance toward sustainability and urban resilience from its inclusion in the local political agenda.

Keywords Environmental change · Inequalities · Planning policies · Resilience indicators · Urban sustainability

M. L. Zulaica (✉)
Institute of Habitat and Environment (IHAM), Faculty of Architecture, Urbanism and Design (FAUD), National University of Mar del Plata (UNMdP)—National Scientific and Technical Research Council (CONICET), Mar del Plata, Buenos Aires Province, Argentina
e-mail: laurazulaica@conicet.gov.ar

G. H. Tonon (ed.), *Urban Inequalities*, The Urban Book Series,
https://doi.org/10.1007/978-3-031-59746-6_7

7.1 Introduction

In latest decades, the climate of our planet accelerated its process of change. This change mainly generated an increase in the average temperature of the planet (global warming) and alterations in the frequency and magnitude of floods, droughts, and extreme storms. These consequences impacted the quality of life of present and future generations on a regional and global scale with differential environmental impacts.

A recent statement from the Intergovernmental Panel on Climate Change (IPCC 2022) states that, over the next two decades, the planet will face several inevitable climate hazards with global warming of 1.5 °C (2.7°F). For their part, specialists also argue that, if this level of warming is temporarily exceeded, additional serious impacts will be generated, some of which will be irreversible. Among them, risks to infrastructure and low-lying coastal settlements stand out. These impacts are enhanced as a result of the urbanization process, and are of concern, given that more than 90% of all urban areas are coastal.

Latin America shows a process of urbanization so fast that is described as an urban explosion, being today the most urbanized developing region in the world with 80% of its population residing in cities (Montero and García 2017). In this context, cities are key actors in the climate crisis, as they are also responsible for between 40 and 70% of greenhouse gases (Vera and Sordi 2021).

The largest share of urban growth is taking place in developing countries, where large portions of the urban population live in poverty, and in areas vulnerable to extreme weather events. In this sense, floods are exacerbated by extreme weather events such as heavy rainfall, hurricanes, and gradual sea level rise, which constitute a latent risk for urban areas. On a regional scale, Latin America and the Caribbean are particularly vulnerable to the impacts of climate change. By 2050, sea level rise, temperature, and changes in rainfall patterns are predicted to reach an estimated annual cost of approximately 2–4% of the Gross Domestic Product (Vera and Sordi 2021).

Within this scenario, cities and other human settlements are exposed to environmental risks, many of them triggered by climate change, which implies strong transformations in environmental dynamics.

However, the current trend of development poses serious constraints to achieving a more inclusive and spatially balanced urbanization; these cities, compared to international standards, exhibit higher levels of inequality, a phenomenon that bears implications for a set of social problems such as the growth of informal settlements, lack of services, residential segregation, gentrification, gaps related to mobility and time use, insecurity and environmental impacts, among others (Jordán et al. 2017). As a result, the effects of climate crisis translate into urban inequalities.

Inequality has long been a major issue in social sciences and has taken on renewed relevance over the past two decades. Nijman and Wei (2020) argue that inequalities in cities are simply a reflection of rapid urbanization, where social issues are increasingly urban issues.

In turn, the study of current urban inequalities requires the analysis of global processes associated with modes of production and consumption that give rise to environmental consequences that impact on cities. In that sense, although the effects of climate change on economic output and financial stability received considerable attention, less relevance was given to the relationship between climate change, income inequality (Cevik and Jalles 2022) and the ability to cope with impacts. This is no small problem considering that climate change could undermine poverty eradication efforts, disproportionately affect poorer regions, and worsen income inequality within countries (World Bank 2020).

Thus, environmental risks in a context of climate change are variable and their study is at the center of political agendas to face transformations and respond to emerging urban problems. According to Zulaica and Vazquez (2021), although the conceptual discussion of risk is extensive (Wynne 1992; Wilches Chaux 1993; Beck 1998), from the analytical point of view, in principle two constitutive dimensions can be identified: dangerousness, defined as the probability of occurrence of a natural or technological phenomenon with potential to generate damage (Natenzon 1998); and vulnerability, which refers to the social, economic, cultural, etc., conditions that expose a population and make it prone to be affected and harmed by dangerousness (Lavell 1996). According to Moreno Crossley (2008), vulnerability denotes a condition of defenselessness: the possibility of suffering some type of damage or uncertainty, and it is situational, since it is conditioned by the context in which it develops (Mackenzie et al. 2014). The discussion of risk is approached from different disciplinary lines that generally involve the concepts of danger, threat, exposure, vulnerability, and more recently resilience.

Cities in Latin America and the Caribbean are the most unequal in the world (ONU-Hábitat 2012) and many policies aimed at reducing them have slowed down since 2015 (CEPAL 2016a). As a consequence of these dissimilar characteristics, the capacity to respond to climatic phenomena differs between cities and toward their interior.

In Argentina, the effects of climate change are visible, they have different scopes in different regions of the country (CIMA 2014; Camillone 2018) and pose a challenge for the management of cities. Schejtman et al. (2015) point out that local governments must face disasters caused by rains, floods, droughts, fires, and that high temperatures and most extreme phenomena produce changes in the way of inhabiting a city, demanding new goods and services. While some aspects related to climate phenomena will show gradual changes allowing urban environments to adapt, it should be observed that the increased probability and frequency of extreme weather events will require an immediate response when trying to reduce risks (Salimi and Al-Ghamdi 2020).

In this framework, this paper proposes to analyze resilience in 32 Argentine urban agglomerates. To this end, three constitutive dimensions of resilience (exposure, sensitivity and adaptation) are addressed, and a synthetic index, the Resilience Index, is constructed. Previously, the threat is addressed as a constituent element of the risk that expresses differential scope in the Argentine territory.

7.2 Constituent Elements of Risk, Some Conceptualizations

The threat, alluding to the dangerous dimension of the risk, refers to potentially hazardous phenomena of natural or anthropic origin that can cause social, environmental, and economic impacts (Turnbull et al. 2013). When analyzing threats in the context of climate change, the concept refers to both extreme and sudden weather events (storms, frosts) and situations of gradual and slow manifestation (changes in average precipitation or temperature regimes), as long as they have the potential to generate damage (SAyDS 2015a).

Although the concept of resilience is more frequently used to refer to disaster risk, it is increasingly introduced to analyze the capacity to respond and adapt to climate change, proposing a rereading of vulnerability in proactive terms. In this sense, although there are numerous antecedents of vulnerability studies associated with environmental change (Adger 2006; Cutter et al. 2009), more recent research has focused on resilience analysis (Santiago-Verat et al. 2018).

As expressed by Zulaica and Vazquez (2021), resilience is a complex concept with numerous meanings (Hallegatte and Engle 2019). According to Asadzadeh et al. (2017), although the term resilience was first formulated in the field of ecology by Holling (1973), it has been used since the sixteenth century and covers various theoretical-conceptual frameworks: from socio-ecological systems to later developments in the field of sustainability, mitigation and adaptation, and more recently, disaster risk reduction. Following the authors, resilience is currently a priority issue in academia and on international agendas and acquires significant relevance in a context of environmental change and in urban planning policies. Unlike the concept of vulnerability, resilience is generally considered a positive and desirable attribute to face dangerous events (Scherzer et al. 2019).

7.3 Complexity, Environmental Problems, Inequality and Resilience

Analyzing interactions among natural systems, such as those associated with climate change, and the social systems, is a complex issue. In fact, research on environmental problems revealed the difficulties of the traditional conceptual and methodological bases to do so (García 2011). On the other hand, although there are controversies around the meaning of complexity (Becerra 2020), there is a broad consensus in considering that, to address complex problems such as environmental ones, it is necessary to attain an association of disciplines. In this way, complexity introduces a scientific paradigm into the field of science, which involves a new way of doing and understanding science (Rodríguez Zoya and Leónidas Aguirre 2011) from the comprehension of the world as an entity where everything is intertwined.

The Complex Systems Theory proposed by García (2000, 2006) constitutes an integral framework to face "complex problems," enabling the integration of disciplinary perspectives. A complex system is a representation of a fragment of reality, conceptualized as an organized whole (system), in which the elements are not "separable" and, therefore, cannot be studied in isolation (García 2006). Complexity is an inherent quality of systems (Bustamante-Ubilla and Arenas-Torres 2022) and is associated with the impossibility of considering particular aspects of a phenomenon, process or situation from a unique specific discipline since, in reality, the situations and processes are not presented in such a way that they can be classified by their correspondence with any particular discipline. Along these lines, Morin (1994) points out that complexity is multidimensional, bringing together order, disorder, organization, and diversity.

From the conception of a Complex Systems Theory, interdisciplinarity is not only of a theoretical or methodological nature, but rather of a political-evaluative nature (Becerra and Castorina 2016). Its innovation and relevance lie in the fact that it is an approach from Latin America, built for interdisciplinary research on specific complex problems (Rodríguez Zoya 2017). Among the complex problems that were analyzed under this approach, Becerra (2014) mentions hunger, environmental degradation, social marginalization in urban situations, technological-productive development, among others. He also adds that all of them pose the difficulty of involving multiple processes of a diverse nature whose interrelationships cannot be ignored, nor can their elements be separated or treated in isolation.

Environmental problems, such as those derived from climate change, emerge from complex interactions between society and nature and are not exempt from inequalities, since they emerge from the current development model (Leff 2000). Within this framework, the Complex Systems Theory establishes a theoretical and methodological perspective for the construction of interdisciplinary knowledge that allows addressing inequalities. Still, as García (2011) points out, this theory emerges from the alarm generated by catastrophes attributed to climate change, which affected various regions of the world in the 1960–1970s. This led to the creation of the "Drought and Man" program within the scope of the International Federation of Institutes for Advanced Studies, to address the social, ethical, cultural, and political consequences of climate change, with unequal impacts in different regions of the planet.

Although the concept of inequality has traditionally been considered in the field of income (Sen, 2000), it is currently incorporated into new collective spheres, relating to other aspects such as socioeconomic level, gender, ethnic-racial condition, stages of the life cycle, territorial conditions (CEPAL 2016b). From a social perspective, Altschuler (2016) states that inequality is currently a crucial problem in Latin American societies, mostly affected by the implementation of neoliberal models in the last quarter of the last century.

When referring to socio-environmental inequalities, Gobert and León-Bärecke (2017) point out that they are interpreted through the standpoints of different conceptions of justice and different social representations of equity. This approach supposes considering multiple interests and problems around the imbalances that are manifested as a phenomenon of unequal access to environmental benefits and to the costs of environmental degradation. The authors establish a typology of socio-environmental inequalities, the analysis of which requires integrated approaches such as the one proposed here, based on the Complex Systems Theory: differentiated exposure to environmental impacts (pollution); unequal access to natural resources; differences in the abilities of communities to influence decision-making processes; differential effects of environmental policies; inadequate management of externalities; among others.

Inequality is a historical and structural characteristic of Latin American societies (CEPAL 2016b). In this context, the concept of sustainable development, widely discussed in the literature (Reboratti 2000; Guimarães 2003; Gallopín 2010; Gallopín et al. 2014), is proposed as a development model to address inequalities. In this line, the concept of resilience assumed in this research makes it possible to operationalize environmental inequalities from a comprehensive perspective, taking into account that they represent a multidimensional and multidetermined phenomenon that characterizes different territorial areas, especially the urban ones.

7.4 Cities as Scenarios of Climate Change with Unequal Effects

Climate change interacts with ongoing global trends in urbanization to create regionally specific impacts and risk profiles; however, the considerable capacity of settlements to cope with challenges of climate change is also recognized (Dodman et al. 2022).

The inclusion of urban environmental problems derived from climate change in the political agenda is a growing fact in local, national, and international governments. Indeed, the 2030 Agenda for Sustainable Development (Naciones Unidas 2015), adopted by the United Nations General Assembly in September 2015, proposes among the Sustainable Development Goals (SDGs) "to make cities and human settlements inclusive, safe, resilient, and sustainable" (SDG 11). Likewise, this concern was part of the discussion at the third United Nations conference on Housing and Sustainable Urban Development held in Quito in 2016—Habitat III—(Naciones Unidas 2017).

In this context, cities are key territories to address the effects of climate change, since it is there where its impacts are most significantly manifested and, at the same time, where most of its causes are found; although it is a global phenomenon, the approach at the local level takes on a transformative meaning by linking urban planning and development to the responses necessary to face environmental changes (Herrero et al. 2018). Hence, the importance of exploring resilience of Argentine cities in the context of environmental changes taking place in different regions.

Studies of quality of life and well-being conducted in Latin America and Argentina (Lucero 2008; Tonon and Castro Solano 2012; Tonon and Rodriguez de la Vega 2016; Velazquez and Celemín 2020), account for territorial inequalities at different scales of analysis, including urban inequalities (Mikkelsen et al. 2020). On the other hand, there are relevant studies that contribute to the operationalization of the concept of resilience (Fraser et al. 1999; Suárez Ojeda 2001; Folke 2006; Simonovic, and Peck 2013; Southwick et al. 2014; López Bracamonte et al. 2017; Quiroga 2019; Tonon and Zulaica 2022). Based on the conceptual bases and methodologies used in these studies, it is possible to advance in the analysis of resilience of Argentine cities to account for urban inequities, in this case focused on climate change.

7.5 The Urban Resilience Approach

Climate change induces environmental changes to which local governments must respond. In recent years, the concept of urban resilience emerged as a central theme of the development of cities and human settlements, as a basis for a wide range of interventions. It is defined as a feature of sustainable urban development that recognizes the urban area as a complex and dynamic system that must continuously adapt to various challenges in an integrated and holistic manner, through its functional, organizational, physical, and spatial structure, including different scales (ONU-Hábitat 2015).

Based on the aforementioned, the importance of regulating territorial housing policies and planning, and the construction of infrastructure that cares about climate risk in the face of current and projected hazards are highlighted, as well as the design of adequate infrastructure, strategically located and resistant to events, using territorial planning standards based on risk assessments, and including early warning and emergency management systems at the municipal level.

In line with the above, urban resilience is determined by the positive response capacity of governments and communities to organize and generate strategies to prevent, mitigate, and adapt to adversity, reducing inequalities. In this regard, Simonovic and Peck (2013) argue that it is necessary to focus on urban resilience planning with actions aimed at strengthening local capacity including community participation. This will allow a better understanding of the diversity, needs, strengths, and vulnerabilities within communities since threats do not affect everyone in the same way.

7.6 Methodology

The territory of the Argentine Republic is characterized by a wide ecological diversity that is reflected in a diverse productive matrix. According to the provisional data of the national census carried out in May 2022 by the National Institute of Statistics and Censuses, the population reaches 47,327,407, being more than 90% urban, and unequally distributed in the territory occupied by the different provinces. The most populated is Buenos Aires, which gathers approximately 40% of the total population, concentrated especially in the Metropolitan Area of Buenos Aires autonomous city.

To respond to the proposed objective of this study, it was decided to include the 32 urban agglomerates in which INDEC examines the Permanent Survey of Households (EPH). These urban agglomerates are displayed in Fig. 7.1.

Firstly, threats faced by urban agglomerates in a context of climate change are analyzed. In this case, based on previous studies (Zulaica and Vazquez 2021), and information generated and processed by CIMA (2014), available on the official website of the National Ministry of Environment and Sustainable Development, through the Climate Change Risk Maps System—SIMARCC—(SAyDS 2015b), estimates were obtained corresponding to a scenario of average emissions in the near future (projection 2015–2039). Based on this information, indicators referring to the change in average temperature and annual precipitation are examined.

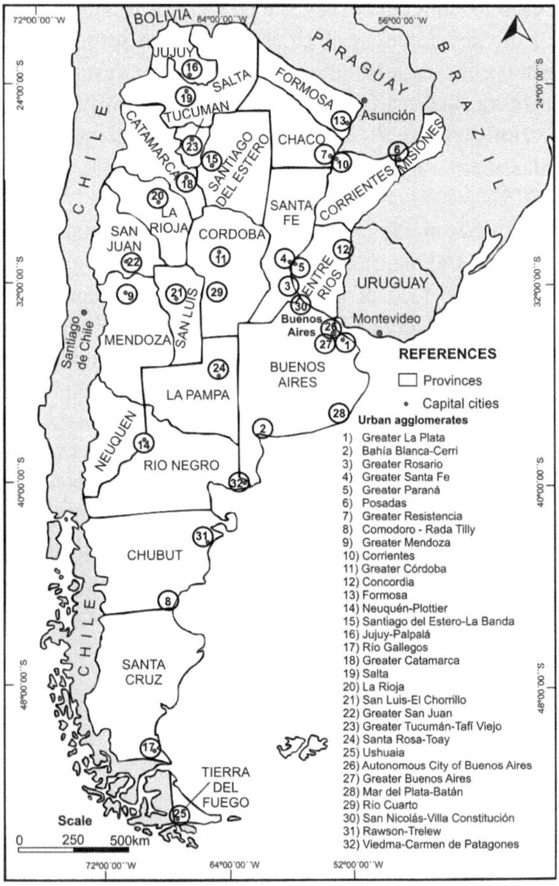

Fig. 7.1 Urban agglomerates of Argentina. *Source* Author's elaboration based on Mikkelsen et al. (2020) and Zulaica and Vazquez (2021)

Subsequently, to achieve the proposed objective, resilience of Argentine cities is evaluated considering three main dimensions: exposure, sensitivity, and adaptation. The analysis of resilience of the agglomerates, where the EPH was surveyed, focused on the construction of a synthetic index, the Resilience Index (ResI), which allowed to integrate different aspects that affect the concept, trying to respond to social objectives of improving communication and establishing comparisons. To this end, based on the available and feasible information to be processed, indicators that characterized the dimensions considered were selected.

7.6.1 Selection of Indicators

In general terms, exposure assesses the level at which an area may be affected as a result of environmental changes. In this case, it was only possible to consider the population density indicator calculated from data of the population estimates made by the National Institute of Statistics and Censuses for the year 2020 (INDEC 2015), considering the spatial units (departments or districts in the case of Buenos Aires province) that group the agglomerates.

Sensitivity refers to the intrinsic characteristics of a territory and includes indicators that express general aspects of the population of each agglomerate. EPH indicators corresponding to the first half of 2020 were selected, referring mainly to housing conditions—obtained from the pre-census of housing carried out by INDEC in 2020—, sanitary—extracted from the data of the EPH of 2020[1]—, and related to the vulnerable population in the cities, according to the information from the National Registry of Popular Neighborhoods (RENABAP), updated to 2022.[2]

Finally, adaptation considers in this case the capacity of each agglomerate to respond to environmental changes. At this point, we included indicators related to the educational and labor characteristics of the population surveyed in the first 2020 semester of EPH, and also the integration of the municipalities that made up the agglomerates of the Argentine Network of Municipalities Facing Climate Change (RAMCC). This Network was formed in 2010 during the First International Conference "Municipalities and Climate Change," held in the city of Monte Caseros, Corrientes, and was established as an instance of coordination and promotion of local public policies to face climate change in the cities and towns of Argentina.[3] The list of indicators contained in the dimensions is summarized in Table 7.1.

---

[1] https://precensodeviviendas.indec.gob.ar/.

[2] https://datos.gob.ar/dataset/desarrollo-social-registro-nacional-barrios-populares.

[3] The RAMFCC (https://www.ramcc.net/) is a coalition of Argentine municipalities that coordinates and promotes strategic plans to address climate change. This commitment to climate action is framed by the objectives of the Global Covenant of Mayors for Climate and Energy, which addresses three aspects of climate action: climate change mitigation, adaptation to adverse effects, and universal access to safe, clean, and affordable energy.

Table 7.1 Dimensions and indicators that make up the Resilience Index (ResI)

Dimensions	Indicators
Exposure	Population density of the administrative units comprising the agglomerates
Sensitivity	Percentage of population with health coverage
	Percentage of inadequate housing over total housing
	Percentage of dwellings in slums and popular settlements out of total dwellings
Adaptation	Percentage of unemployed population aged 10 and over
	Percentage of population with complete or higher secondary education aged 10 years or over
	Percentage of population of municipalities that make up the agglomerates, which do not belong to the Network of Municipalities facing climate change

Source Author's elaboration

7.6.2 Standardization of Indicators

The construction of the ResI comes from background studies in which indices of quality of life and well-being were built (Lucero 2008; Velazquez 2016; Mikkelsen et al. 2020), habitability (Zulaica and Oriolani 2019), vulnerability (Zulaica and Ferraro 2016), and resilience (Zulaica and Vazquez 2021), adjusting and updating the indicators according to the proposed objectives.

In this sense, to facilitate the comparison of the indicators used, it is necessary to define a common structure, since each of them has a different classification system and their terminologies also differ (Braulio-Gonzalo et al. 2015). As a result, standardization procedures were used to transform data into dimensionless units. Based on previous studies, the Omega Score technique (Buzai 2003; Buzai and Baxendale 2006) was used, which converts indicator data to a measurement range between 0 and 1. This procedure transforms indicators data into a measuring range that runs from 0 (worst situation) to 1 (best situation). These values correspond to the minimum and maximum data, respectively.

The formulas used for standardization are shown below according to their positive or negative sense:

Indicators which increase implies a worse relative situation:

$$\mathrm{SV} = (H - d)/(H - l).$$

Indicators which increase implies a better relative situation:

$$\mathrm{SV} = [1 - (H - d)/(H - l)],$$

where SV: standardized indicator value; d: original data to be standardized; H: highest indicator value; l: lowest indicator value.

7.7 Construction of the Resilience Index (ResI) and Comparative Analysis of Spatial Distribution

The ResI is obtained by the addition of the standardized values of the exposure, sensitivity, and adaptation indicators. In this way, as King (2016) argues, the development of indices allows simplifying the information, facilitating the understanding, and interpretation of the emergent findings.

The results obtained for each of the variables were spatially represented in maps elaborated in a Geographic Information System (GIS) of free access, QGIS (version 3.22), and the particular situation of each agglomerate was analyzed. The integration of the results in an index allowed to distinguish five categories of the ResI, which reflect the favorable, intermediate, and unfavorable situations (Resilience: very low, low, medium, high, and very high).

The spatial configuration was obtained from a classification in intervals by natural cuts. This method identifies the breakpoints between classes using a statistical formula (Jenk optimization), which minimizes the sum of the variance within each of the classes.

7.8 Results

Based on the methodological procedure described above, the results were organized into three sections. The first section corresponds to the analysis of the threats that are a constitutive part of the concept of risk, and that were characterized from previous studies. The second section is the analysis of resilience with the results obtained from the indicators and their integration into the synthetic index. Finally, the last section discusses the concept of urban resilience in light of the results obtained.

7.8.1 The Threats of Climate Change

Threats derived from climate change can be defined as the possibility, probability, or potential for climatic phenomena to impact specific territories affecting their population and activities (Chavarro Pinzon et al. 2008).

The review of previous studies systematized by Zulaica and Vazquez (2021) indicates that, according to the report of the Sea and Atmosphere Research Center (CIMA 2014) between 1960 and 2010 changes that took place in Argentina were fewer than those observed in other regions. Such report highlights that in the middle of the country there were smaller increases in temperature, with reductions in some areas, and more significant increases in the minimum temperature than in the maximum temperature. Moreover, in general, temperatures showed decreases in the middle of the country. The same source highlights that in Patagonia temperature increase was greater than in the rest of the country, and that the changes in the east and north reflected the warming observed in the average and minimum temperatures. Regarding rainfall, within this period they increased in most of the country, with inter-annual variations. The largest increases were recorded in the east and, in contrast, over the Patagonian and Cuyo Andes, rainfall showed negative trends. Also, in much of the country, rainfall was more intense and frequent (CIMA 2014).

In compliance with the above, Camilloni (2018) points out that temperature increased 0.5 °C in the north-central region of the country while minimum temperatures increased approximately 1 °C, and maximum temperatures were reduced by practically the same proportion during the same period of time. The Cuyo region shows warming tendencies in average temperatures, probably causing the generalized retreat of glaciers. According to the same source, at the seasonal level, summers tend to be longer and winters more moderate, recording a decrease in the occurrence of frost and an increase in the frequency of heat waves. On the other hand, the most important rivers show a reduction in their annual flows since the 1980s; these trends highlight possible increases in water deficit in the aforementioned regions (SAyDS 2015a).

Since 1960, the increase in annual rainfall and the frequency of extreme incidents in regions of the center-east of the country stand out. Among the consequences of these trends, Camilloni (2018) points to the occurrence of frequent flooding and overflows in depressed areas. The provinces of Misiones, Santa Fe, and Entre Ríos show the largest increases. On the other hand, and accompanying the previous process, floods occurred on the banks of the Paraná and Uruguay rivers, and in lowland areas that mainly affected the provinces of Buenos Aires, La Pampa, Santa Fe, and Córdoba (Barros and Camilloni 2016). Toward the center-northwest, the province of Tucumán evidences changes associated with increased rainfall, intense precipitation, increases in cloudiness derived from decreases in daytime temperatures, and increase in nighttime temperatures (Minetti and González 2006).

Regarding the Metropolitan Area of Buenos Aires, Ludueña et al. (2012) highlight increases in the average annual temperature around 0.5 °C and a probable increase of 0.6 m in mean river level that could affect, in the long term, the banks of the Matanza-Riachuelo River and the streams of the north of the city and Reconquista River, also foreseeing the increase of extreme temperatures, heat waves, and the frequency of storms/southeast strong winds, with threats of flooding on the coast of the Río de la Plata.

In terms of average temperatures, between 1960 and 2010, there was an increase of less than 1 °C in most of the country (CIMA 2014). According to Zulaica and Vazquez (2021), data obtained from SIMARCC for the administrative units that make up the agglomerates indicate, in a scenario of moderate emissions and in the near future (2015–2039), more significant changes (greater than 1 °C) in Cuyo (Greater San Juan and Greater Mendoza), toward the Argentine northwest (Salta and Jujuy-Palpalá) and in the center-east of the country (Greater Santa Fe, Corrientes, and Concordia).

On the other hand, in the period 1960–2010, precipitation increased in most of the country, with more significant increases in the east, although the most relevant took place in semi-arid areas (CIMA 2014). The analysis carried out by Zulaica and Vazquez (2021) from the scenario considered allows detecting changes greater than 10 mm in the northeast of the country (Corrientes, Greater Posadas, Greater Resistencia), in the center-east (Greater La Plata, Autonomous City of Buenos Aires, San Nicolás-Villa Constitución) and toward the center (Santiago del Estero-La Banda, San Luis-El Chorrillo, Santa Rosa-Toay). Likewise, there were negative changes in the average rainfall in Tierra del Fuego.

On the national scale, a decrease in the maximum length of days in the year with practically no precipitation is observed in the humid Pampas and in Patagonia, consistent with the increase in annual rainfall; however, to the north and west, longer dry periods are observed (CIMA 2014). In the scenario considered, the maximum change of dry days with increases between 5 and 10 days was registered toward the center and northwest of the country in districts that make up the agglomerates of Greater Tucumán-Tafí Viejo, Jujuy-Palpalá, Santiago de Estero-La Banda, and Salta (Zulaica and Vazquez 2021). This indicator is related to the changes in the occurrence of heat waves which maximums are estimated in the Argentine northwest (Salta, Jujuy-Palpalá) with increases also toward the northeast (Greater Posadas).

7.8.2 Resilience in Argentine Cities

The analysis of the distribution of the Resilience Index (ResI) deepens the differentiations between agglomerates. Thus, the most critical situations from the analyzed set of data (ResI very low: 0.376–0.4) are presented in the districts that make up Greater Buenos Aires and Formosa. In second place (low ResI: 0.4–0.5) are the Autonomous City of Buenos Aires and Santiago del Estero-La Banda. In this condition, almost 52% of the total population of the agglomerates is concentrated. The high percentage of population in this category is directly related to the fact that the agglomerates that concentrate most of the country's population (Autonomous City of Buenos Aires and districts of Greater Buenos Aires) are included.

At the other end, the best conditions in the analyzed context (very high ResI: 0.7–0.776) are identified in La Rioja, San Luis-El Chorrillo, Greater Paraná and Rawson-Trelew, grouping less than 4% of inhabitants. The rest of the agglomerates are concentrated among intermediate conditions of the index. The results obtained are shown in Fig. 7.2.

The results recorded for this set of sensitivity indicators reveal the most unfavorable situations in Formosa, and after it, in Jujuy-Palpalá, Greater Resistencia, and Concordia. In this condition, for exposure the only feasible indicator to use was density, the most critical situations stand out in Greater Córdoba, Greater Rosario, Mar del Plata-Batán, Autonomous City of Buenos Aires, Greater Buenos Aires, Greater Mendoza, and Greater Santa Fe. On the other hand, the most unfavorable conditions, related to adaptation, are presented in the districts of Greater Buenos Aires, followed by Greater La Plata, Formosa, Santiago del Estero-La Banda, Greater

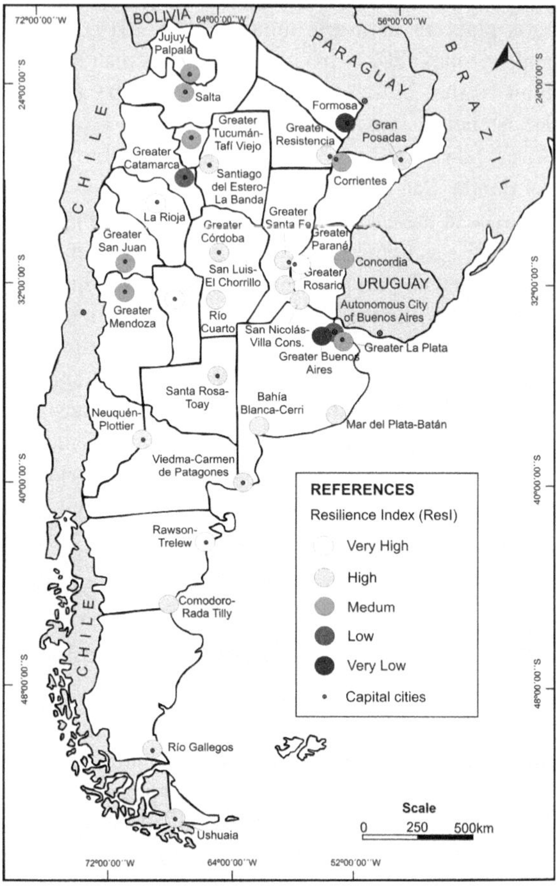

Fig. 7.2 Resilience Index (ResI) in Argentine urban agglomerates. *Source* Author's elaboration

Mendoza, Greater San Juan, Greater Tucumán-Tafí Viejo, Santa Rosa-Toay, and Río Gallegos.

In relation to exposure, the analysis should include infrastructures as key elements of this variable. However, the systematized information available did not allow to account for this aspect to establish comparisons. Therefore, density data were included that let us establish, in general terms, the number of people potentially exposed per unit area.

Housing is par excellence the daily environment of families living in the agglomerates. With regard to sensitivity indicators, data on inadequate housing generally show unfavorable conditions in the Autonomous City of Buenos Aires, Jujuy-Palpalá, Salta, Formosa, and Corrientes.

On the other hand, the deficiencies in infrastructure, the precariousness in the land tenure regime, and the population residing in informal settlements allow us to distinguish conditions of spatial segregation and social marginality, often associated with the unplanned occupation of cities. In this sense, the indicator referring to the percentage of dwellings surveyed in slums and settlements over the total number of dwellings reflects more critical situations in Concordia, Formosa, Greater Resistencia, and Posadas.

Whether or not having health coverage reflects precarious working conditions that make families more sensitive to the need to deal with threats. When the percentage of the population with health coverage is analyzed, it is observed that Formosa, Greater Resistencia, Santiago del Estero-La Banda, Jujuy-Palpalá, and Greater San Juan record the lowest values.

In relation to adaptation indicators, the percentage of the unemployed population, which to some extent represents the economic response capacity of families, exhibits more critical conditions in Mar del Plata-Batán, Greater Córdoba, Ushuaia, and Greater Rosario, followed by the districts of Greater Buenos Aires, the Autonomous City of Buenos Aires, Greater Mendoza, Río Cuarto, Salta, and Greater Santa Fe.

In turn, to a greater or lesser extent, the level of education constrains the access to the labor market and can be reflected in unemployment. In this way, completing the secondary level and advancing in higher levels means having greater possibilities of access to the labor market. Thus, the highest percentages of the population that reached secondary or higher education are presented in the Autonomous City of Buenos Aires and Ushuaia.

The RAMCC is defined as an instrument for coordinating and promoting local public policies to address climate change. Its main objective lies in municipal, regional, or national projects or programs, related to climate change mitigation and/or adaptation, based on the mobilization of local, national, and international resources. The participation of municipalities in the Network represents, to some extent, the concern for the problem and the preparation to adapt to new contexts. A growing group of agglomerates integrated into this network stand out (Jujuy-Palpalá, Salta, Corrientes, Posadas, Greater Resistencia, Greater Catamarca, La Rioja, Greater Córdoba, Greater Santa Fe, Greater Paraná, Concordia, San Luis-El Chorrillo, Río Cuarto, Greater Rosario, Mar del Plata-Batán, Neuquén Plottier, Rawson-Trelew, and Comodoro Rivadavia–Rada Tilly).

As anticipated, there are changes in risk management approaches focused first on vulnerability and then on resilience, the latter perspective seen as a more proactive and positive expression to face threats (Simonovic and Peck 2013) derived from disaster risk, including those derived from climate change. The results achieved from the ResI obtained from the analyzed data show heterogeneous situations in the urban agglomerates of Argentina, expressing inequalities. The interpretation of these inequalities will require the analysis of their historical, contextual and situated evolution, which involve diverse and interrelated temporal, sociopolitical, and spatial scales. In this sense, the Complex Systems Theory provides an interesting approach to achieve it.

7.9 Final Comments

Environmental problems, such as those arising from climate change, are complex and pose inequalities both in their causes and in the effects they generate. Consequently, addressing these problems and the inequalities they imply requires approaches capable of assuming this complexity. The Complex Systems Theory constitutes an interesting alternative to address the inequalities that are due to a series of vulnerabilities. In this framework, the concept of resilience is assumed as an interesting alternative to deal with environmental inequalities.

Given the dissimilar characteristics of the territories and societies that inhabit them, the evaluation of resilience acquires importance for the design of planning policies. Although resilience is not a directly observable phenomenon, indicators and indices are analytical, communicative, and collaborative tools that contribute to raising awareness and supporting decision-making, facilitating planning and policy development through a better understanding of complex and multidimensional problems (Raghavan Sathyan et al. 2018), such as those that come from environmental changes.

Without considering hazardous conditions implicit in risk management, the assessment of the resilience of urban agglomerates in Argentina reveals more critical situations in Greater Buenos Aires, Formosa, the Autonomous City of Buenos Aires, and Santiago del Estero-La Banda. The spatial distribution of the ResI shows relatively random situations, although the lowest values are concentrated in the north of the country. In general, the best conditions can be visualized in the center.

The ResI involves three dimensions in its construction: exposure, sensitivity, and adaptation. The indicators considered are linked to the availability of systematized data or to the possibility of constructing them from pre-existing information common to the agglomerates. In this sense, according to Agol et al. (2014), the selection of indicators is often subjective, generally depending on factors such as cost, ease of understanding, reliability, and comparability. The exposure dimension has only one indicator (population density), which in this instance provides interesting, although not sufficient, data. Although the importance of deepening the analysis from the incorporation of new indicators is considered, the results of this study constitute useful inputs for the understanding of the problem in a general way, providing an interesting diagnostic basis to generate strategies for action in the face of environmental phenomena, which effects are enhanced in the areas of lower response capacity.

Cities are the scene of the main manifestations of climate change and, therefore, investigating the capacity of adaptation and response of society in the face of threats is imminent. Faced with this, the concept of resilience provides elements to deepen the new challenges today's cities face, while allowing the analysis of urban inequalities.

In that sense, although there are critical views on the conception of urban resilience (Borie et al. 2019), this perspective applied to risk management has become a systematic approach to reduce the impact of climate change on the built environment (Fakhruddin et al. 2019). In this regard, Salimi and Al-Ghamdi (2020) state that the characteristics of urban resilience should be considered in the planning phase of urban projects and not when problems occur.

The opening of new questions arising within the framework of the analysis carried out inspires to continue developing the study of inequalities, to advance toward sustainability and urban resilience from their inclusion in the local political agenda.

References

Adger N (2006) Vulnerability. Glob Environ Chang 3(16):268–281. https://doi.org/10.1016/j.gloenvcha.2006.02.006

Agol D, Latawiec A, Strassburg B (2014) Evaluating impacts of development and conservation projects using sustainability indicators: opportunities and challenges. Environ Impact Assess 48:1–9. https://doi.org/10.1016/j.eiar.2014.04.001

Altschuler B (2016) Desigualdades sociales desde el enfoque de la complejidad: integrando disciplinas para pensar nuestra realidad. Divulgatio 1(1). http://ridaa.unq.edu.ar/handle/20.500.11807/263

Asadzadeh A, Kötter T, Salehi P, Birkmann J (2017) Operationalizing a concept: the systematic review of composite indicator building for measuring community disaster resilience. Int J Disaster Risk Reduct 25:147–162. https://doi.org/10.1016/j.ijdrr.2017.09.015

Barros V, Camilloni I (2016) La Argentina y el cambio climático, de la física a la política. Eudeba, Buenos Aires

Becerra G (2014) Interdisciplina y Sistemas Complejos. Un Enfoque Para Abordar Problemáticas Sociales Complejas. Psocial 1(1):34–43

Becerra G (2020) La Teoría de los Sistemas Complejos y la Teoría de los Sistemas Sociales en las controversias de la complejidad. Convergencia Revista de Ciencias Sociales 27:1–23. https://doi.org/10.29101/crcs.v27i83.12148

Becerra G, Castorina JA (2016) Acerca de la noción de "marco epistémico" del constructivismo. Una comparación con la noción de "paradigma" de Kuhn. Revista Iberoamericana de Ciencia, Tecnología y Sociedad 11(31):9–28

Beck U (1998) La sociedad del riesgo. Hacia una nueva modernidad. Paidós, Barcelona

Borie M, Pelling M, Ziervogel G, Hyams K (2019) Mapping narratives of urban resilience in the global south. Glob Environ Chang 54:203–213. https://doi.org/10.1016/j.gloenvcha.2019.01.001

Braulio-Gonzalo M, Bovea MD, Ruá MJ (2015) Sustainability on the urban scale: proposal of a structure of indicators for the Spanish context. Environ Impact Assess Rev 53:16–30. https://doi.org/10.1016/j.eiar.2015.03.002

Bustamante-Ubilla M, Arenas-Torres F (2022) Epistemological foundations of complexity theory. Sustainability 14:13316. https://doi.org/10.3390/su142013316

Buzai G (2003) Mapas sociales urbanos. Lugar Editorial, Buenos Aires

Buzai G, Baxendale C (2006) Análisis socioespacial con sistemas de información geográfica. Buenos Aires: Grupo de Ecología del Paisaje y Medio Ambiente (GEPAMA), Lugar Editorial

Camilloni I (2018) Argentina y el cambio climático. Cienc Invest 68(5):5–10

CEPAL—Comisión Económica para América Latina y el Caribe (2016a) Horizontes 2030: La Igualdad en el Centro del Desarrollo Sostenible. Trigésimo sexto período de sesiones de la Cepal. Naciones Unidas, Santiago de Chile

CEPAL—Comisión Económica para América Latina y el Caribe (2016b) Matriz de la desigualdad en América Latina. Naciones Unidas, Santiago de Chile

Cevik S, Jalles JT (2022) For whom the bell tolls: climate change and income inequality. Working Paper 22/103. International Monetary Fund

Chavarro Pinzón M, García Guerrero A, García Portilla J, Pabón J, Prieto Rozo A, Ulloa Cubillos A (2008) Amenazas, riesgos, vulnerabilidad y adaptación frente al cambio climático. UNODC, Naciones Unidas, Bogotá

CIMA—Centro de Investigaciones del Mar y la Atmósfera (2014) Cambio Climático en Argentina; Tendencias y Proyecciones, Tercera Comunicación Nacional sobre Cambio Climático. Secretaría de Ambiente y Desarrollo Sustentable de la Nación, Buenos Aires

Cutter S, Emrich C, Webb J, Morath D (2009) Social vulnerability to climate variability hazards: a review of the literature. Final report to Oxfam America. Hazards and Vulnerability Research Institute—University of South Carolina, Columbia

Dodman D, Hayward B, Pelling M, Castan Broto V, Chow W, Chu E, Dawson R, Khirfan L, McPhearson T, Prakash A, Zheng Y, Ziervogel G (2022) Cities, settlements and key infrastructure. In: Climate change 2022: impacts, adaptation and vulnerability. In: Pörtner H-O, Roberts DC, Tignor M, Poloczanska ES, Mintenbeck K, Alegría A, Craig M, Langsdorf S, Löschke S, Möller V, Okem A, Rama B (eds) Contribution of working group II to the sixth assessment report of the intergovernmental panel on climate change. Cambridge University Press, Cambridge, UK and New York, NY, USA, pp 907–1040

Fakhruddin B, Reinen-Hamill R, Robertson R (2019) Extent and evaluation of vulnerability for disaster risk reduction of urban Nuku'alofa, Tonga. Prog Disaster Sci 2:100017. https://doi.org/10.1016/j.pdisas.2019.100017

Folke C (2006) Resilience: the emergence of a perspective for social–ecological systems analyses. Glob Environ Chang 16(3):253–267. https://doi.org/10.1016/j.gloenvcha.2006.04.002

Fraser MW, Richman JM, Galinsky MJ (1999) Risk, protection, and resilience: toward a conceptual framework for social work practice. Social Work Res 23:131–143. https://doi.org/10.1093/swr/23.3.131

Gallopín G (2010) El desarrollo sostenible desde una perspectiva sistémica. Sostenible? 11:19–35

Gallopín G, Jiménez Herrero L, Rocuts A (2014) Conceptual frameworks and visual interpretations of sustainability. Int J Sustain Dev 17(3):298–326

García R (2000) El conocimiento en construcción. De las formulaciones de Jean Piaget a la teoría de los sistemas complejos. Gedisa, Barcelona

García R (2006) Sistemas complejos: conceptos, método y fundamentación epistemológica de la investigación interdisciplinaria. Editorial Gedisa, Barcelona

García R (2011) Interdisciplinariedad y sistemas complejos. Revista Latinoamericana de Metodología de las Ciencias Sociales 1(1)

Gobert J, León-Bärecke N (2017) Desigualdades socioambientales: desafios para los urbanistas y sus territorios. Ciudades. hal-01841277

Guimarães R (2003) Tierra de sombras: desafíos de la sustentabilidad y del desarrollo territorial y local ante la globalización corporativa. Serie Medio Ambiente, División de Desarrollo Sostenible y Asentamientos Humanos No 67, CEPAL, Santiago de Chile

Hallegatte S, Engle N (2019) The search for the perfect indicator: reflections on monitoring and evaluation of resilience for improved climate risk management. Clim Risk Manag 23:1–6. https://doi.org/10.1016/j.crm.2018.12.001

Herrero AC, Natenzon C, Miño M (2018) Vulnerabilidad social, amenazas y riesgos frente al cambio climático en el Aglomerado Gran Buenos Aires, Documento de Trabajo No 172, Programa Ciudades. CIPPEC, Ciudad Autónoma de Buenos Aires

Holling CS (1973) Resilience and stability of ecological systems. Annu Rev Ecol Syst 4:1–23

INDEC (2015) Estimaciones de población por sexo, departamento y año calendario 2010–2025. No 38 Serie de Análisis Demográfico. Instituto Nacional de Estadísticas y Censos, Buenos Aires

IPCC (2022) Climate change 2022: impacts, adaptation and vulnerability. In: Pörtner H-O, Roberts DC, Tignor M, Poloczanska ES, Mintenbeck K, Alegría A, Craig M, Langsdorf S, Löschke S, Möller V, Okem A, Rama B (eds) Contribution of working group II to the sixth assessment report of the intergovernmental panel on climate change. Cambridge University Press, Cambridge, UK and New York, NY, USA

Jordán R, Riffo L, Prado A et al (2017) Desarrollo sostenible, urbanización y desigualdad en América Latina y el Caribe. Dinámicas y desafíos para el cambio estructural. CEPAL, Naciones Unidas, Santiago de Chile
King L (2016) Functional sustainability indicators. Ecol Ind 66:121–131. https://doi.org/10.1016/j.ecolind.2016.01.027
Lavell A (1996). Degradación Ambiental, Riesgo y Desastre Urbano. Problemas y Conceptos: Hacia la Definición de una Agenda de Investigación. In: Fernández MA (Comp.) Ciudades en riesgo. Degradación Ambiental, Riesgos urbanos y Desastres en América Latina compilado. Lima: Red de Estudios en Prevención de desastres en América Latina, pp 21–60
Leff E (2000) Tiempo de Sustentabilidad. Ambiente Sociedade III(6/7)
López Bracamonte FM, Limón Aguirre F (2017) Componentes del proceso de resiliencia comunitaria: conocimientos culturales, capacidades sociales y estrategias organizativas. Psiencia 9. https://doi.org/10.5872/psiencia.v9i3.236
Lucero P (2008) Territorio y Calidad de Vida, una mirada desde la Geografía Local, Mar del Plata y Partido de General Pueyrredon. Eudem, Universidad Nacional de Mar del Plata, Mar del Plata
Ludueña C, Wilk D, Quiroga R (2012) Argentina: Mitigación y Adaptación al Cambio Climático; Marco de la preparación de la Estrategia 2012–2016 del BID en Argentina. Banco Interamericano de Desarrollo, Washington, D.C
Mackenzie C, Rogers W, Dodds S (2014) New essays in ethics and feminist philosophy. Oxford UP, Oxford
Mikkelsen C, Zulaica L, Ares S (2020) Aglomerados Urbanos Argentinos: construcción de un índice de bienestar en tres momentos (2003, 2008 y 2014). Acta Geográfica 14(35):140–164
Minetti J, González J (2006) El cambio climático en la provincia de Tucumán. Sus impactos. Serie Conservación de la Naturaleza No 17. Fundación Miguel Lillo, San Miguel de Tucumán
Montero L, García J (2017) Panorama multidimensional del desarrollo urbano en América Latina y el Caribe. CEPAL, Naciones Unidas, Santiago de Chile
Moreno Crossley JC (2008) El concepto de vulnerabilidad social en el debate en torno a la desigualdad: problemas, alcances y perspectivas. Center for Latin American Studies, University of Miami, Working Paper Series No 9, Miami
Morin E (1994) Introducción al pensamiento complejo. Gedisa, Barcelona
Naciones Unidas (2015) Resolución A/RES/70/1 de la Asamblea General el 25 de septiembre de 2015 que aprueba el documento final de la cumbre de las Naciones Unidas para la aprobación de la agenda para el desarrollo después de 2015 Transformar nuestro mundo: la Agenda 2030 para el Desarrollo Sostenible
Naciones Unidas (2017) Nueva Agenda Urbana. Hábitat III-Naciones Unidas, Quito
Natenzon C (1998) Riesgo, vulnerabilidad e incertidumbre. Desastres por inundaciones en Argentina. Seminario sobre Problemas ambientales e vulnerabilidade. Abordagens integradoras para o campo da Saude Publica, Ministerio de la Salud, Río de Janeiro—Brasil
Nijman J, Wei YD (2020) Urban inequalities in the 21st century economy. Appl Geogr 117:102188. https://doi.org/10.1016/j.apgeog.2020.102188
ONU-Hábitat (2012) Estado de las ciudades de América Latina y el Caribe: Rumbo a una nueva transición urbana. Programa de las Naciones Unidas para los Asentamientos Humanos, Río de Janeiro
ONU-Hábitat (2015) Resiliencia urbana, Conferencia de las Naciones Unidas sobre la Vivienda y el Desarrollo Urbano Sostenible. Organización de Naciones Unidas, Nueva York
Quiroga R (2019) Cómo medir la resiliencia: algunas ideas. Misión exploratoria para la asistencia técnica sobre indicadores asociados a la gestión del riesgo de desastres y resiliencia en CDMX. CEPAL, México
Raghavan Sathyan A, Funk C, Aenis T, Winker P, Breuer L (2018) Sensitivity analysis of a climate vulnerability index—a case study from Indian watershed development programs. Clim Change Responses 5(1)
Reboratti C (2000) Ambiente y sociedad: conceptos y relaciones. Editorial Planeta Argentina, Buenos Aires

Rodríguez Zoya L (2017) Complejidad, interdisciplina y política en la teoría de los sistemas complejos, de Rolando García. Civilizar Ciencias Sociales y Humanas 17(33):221–242
Rodríguez Zoya L, Leónidas Aguirre J (2011) Teorías de la complejidad y ciencias sociales. Nuevas estrategias epistemológicas y metodológicas. Nómadas, Revista Crítica de Ciencias Sociales y Jurídicas 30(2):147–166
Salimi M, Al-Ghamdi S (2020) Climate change impacts on critical urban infrastructure and urban resiliency strategies for the Middle East. Sustain Cities Soc 54:1–17. https://doi.org/10.1016/j.scs.2019.101948
Santiago-Vera T, García-Millán M, Michael-Rosset P (2018) Enfoques de la resiliencia ante el cambio climático. Agricultura, Sociedad y Desarrollo 15(4):531–539
SAyDS—Secretaría de Ambiente y Desarrollo Sustentable de la Nación (2015a) Tercera comunicación nacional de la República Argentina a la Convención Marco de las Naciones Unidas sobre el Cambio Climático. SAyDS, Jefatura de Ministros, Presidencia de la Nación, Buenos Aires
SAyDS—Secretaría de Ambiente y Desarrollo Sustentable de la Nación (2015b) Vulnerabilidad y adaptación al cambio climático para la gestión y planificación local. SAyDS, Jefatura de Ministros, Presidencia de la Nación, Buenos Aires
Schejtman L, Bidart M, Deleersnyder D (2015) Hacia ciudades sustentables. CIPPEC-Programa Ciudades, Buenos Aires
Scherzer S, Lujala P, Rod J (2019) A community resilience index for Norway: an adaptation of the baseline resilience indicators for communities (BRIC). Int J Disaster Risk Reduct 36:101107. https://doi.org/10.1016/j.ijdrr.2019.101107
Sen A (2000) Desarrollo y libertad. Planeta, Barcelona
Simonovic S, Peck A (2013) Dynamic resilience to climate change caused natural disasters in coastal megacities quantification framework. Br J Environ Clim Change 3(3):378–401. https://doi.org/10.9734/BJECC/2013/2504
Southwick SM, Bonanno GA, Masten AS, Panter-Brick C, Yehuda R (2014) Resilience definitions, theory, and challenges: interdisciplinary perspectives. Eur J Psych Traumatol 5:1–14. https://doi.org/10.3402/ejpt.v5.25338
Suárez Ojeda E (2001) Resiliencia descubriendo las propias fortalezas. Paidós, Buenos Aires
Tonon G, Castro Solano A (2012) Calidad de vida en Argentina: percepciones macro y micro sociales. Revista Estudios Políticos, No 27 (septiembre-2012), pp 157–171
Tonon G, Rodríguez de la Vega L (2016) Measuring quality of life and inequalities in South America. In: Tonon G (ed) Indicators of quality of life in Latin America, social indicators research series, vol 62. Springer, Heilderberg, Dordrecht, New York, London, pp 3–17
Tonon G, Zulaica L (2022) Community well-being and community resilience to achieve sustainability goals. In: ISQOLS 2022 quality of life for futures resilient: sustainability, equity & wellbeing, International Society for Quality of Life Studies. Burlington, Vermont USA
Turnbull M, Sterrett C, Hilleboe A (2013) Hacia la Resiliencia Una Guía para la Reducción del Riesgo de Desastres y Adaptación al Cambio Climático. Warwickshire, Reino Unido: Catholic Relief Services—United States Conference of Catholic Bishops
Velázquez GA, Celemín JP (2020) Calidad de vida por departamentos y regiones en la Argentina de finales del siglo XIX. Travesía 22(1):135–165
Vera F, Sordi J (2021) Ecological design: strategies for the vulnerable city: adapting precarious areas in Latin America and the Caribbean to climate change. In: Housing and urban development division. Inter-American Development Bank
Wilches Chaux G (1993) La Vulnerabilidad Global. In: Maskrey A (Comp.) Los desastres no son naturales. Red de Estudios Sociales en Prevención de Desastres en América Latina, Bogotá, pp 11–44
World Bank (2020) Poverty and shared prosperity: reversals of fortune. World Bank, Washington, DC
Wynne B (1992) Uncertainty and environmental learning. Reconceiving science and policy in the preventive paradigm. Global Environ Change 2(2):111–127. https://doi.org/10.1016/0959-3780(92)90017-2

Zulaica L, Ferraro R (2016) Vulnerabilidad socioambiental en Monte Caseros, Corrientes: aportes al estudio de las problemáticas socio-espaciales de la localidad. Journal De Ciencias Sociales 4(7):93–114

Zulaica L, Oriolani F (2019) Quality of life and habitability conditions in peri-urban areas of southern Mar del Plata, Argentina: a multimethod study. Appl Res Qual Life 14:659–683. https://doi.org/10.1007/s11482-018-9618-z

Zulaica L, Vazquez P (2021) Ciudades argentinas en el contexto del cambio climático: exploraciones para el análisis del riesgo y la resiliencia urbana. Cuadernos de Geografía 30(2):396–417. https://doi.org/10.15446/rcdg.v30n2.87584

Chapter 8
The Role of Inequality for Digital Learning in Pandemic Times: The Case of Mexico City's Civil Society

Arturo Flores-López and Pedro Flores-Crespo

Abstract This chapter shows how inequalities in digital learning in urban spaces were addressed. The key research question is to determine the role inequality played during the pandemic as part of the technological learning process faced by civil society organizations in Mexico City. It presents how different types of civil society organizations actively engaged in the use of digital tools during the COVID-19 crisis in the Mexico City context. This process covered such aspects as downloading a document, setting up a WhatsApp group, or using Zoom. Although these tools were known before the pandemic, during this challenging period, they were intensely used. Since in some cases, it is believed that there was a direct correlation between such aspects as Wi-Fi access or technological literacy with the city zone where these organizations operate. This work expects to be in the interest of academics and practitioners in the field of education and those who analyze civil society organizations in general.

Keywords Digital learning · Inequalities · COVID-19 · Mexico City · Urban studies

8.1 Introduction

The COVID-19 pandemic became a time when the human race, for the first time in many years, was sharing the same problem. It was a time full of uncertainty, and the main challenge became coping with the social distancing suggested avoiding the

A. Flores-López (✉)
Anahuac Research Center for International Relations, School of Global Studies, Anahuac Mexico University, Mexico, Mexico
e-mail: arturo.floresl@anahuac.mx

P. Flores-Crespo
School of Social Sciences and Politics, Universidad Autónoma de Querétaro (FCPS), Querétaro, Mexico
e-mail: pedro.florescrespo@uaq.mx

G. H. Tonon (ed.), *Urban Inequalities*, The Urban Book Series,
https://doi.org/10.1007/978-3-031-59746-6_8

proliferation of the virus. Stopping the whole world from doing its daily activities was a significant task for every country. We can say that until then, it became relevant for everyone to look at the quality of the technology available since it was going to be the only way to keep in contact with each other, work, or perform the activities that the whole world carried on regularly. Technological access and digital literacy gave a considerable advantage to those used to dealing with these issues over those who had never thought about it. Mexico as a nation was one of the countries with the worst results in terms of COVID-19 crisis management. However, Mexico City took specific measures which had positive results when the crisis hit one of the largest metropolitan areas globally. Nevertheless, most measures appeared not to be directed at alleviating civil society organizations' situation.

In this chapter, we look at how some civil society organizations in Mexico City used technology during the COVID-19 crisis. The key research question is to determine the role inequality played during the pandemic as part of the technological learning process faced by civil society organizations in Mexico City. In order to estimate the impact of inequality, we collected information via a questionnaire from 19 civil society organizations based in Mexico City. Some of the results state that there are certain city zones where technological resources are inaccessible or difficult to access. Furthermore, they identified that the Mexico City government could have made a more significant effort to reduce the prevailing economic and social inequalities.

The chapter is divided into seven sections. The first one is dedicated to socio-digital inequalities. Developed by Helsper (2021), this approach goes beyond the "traditional" inequality approaches. The second section deals with inequality and the COVID-19 context in Mexico (CONEVAL 2020). In this section, we offer data about how inequality increased due to the COVID-19 pandemic. The third part deals with the digital strategy for learning. A plan that was launched before the pandemic and that emphasizes the importance that technology has for the learning processes implemented by the federal office in charge of public education in Mexico. The fourth unit speaks about Internet habits in Mexico. Results from surveys carried out in 2021 and 2022 by the Internet Association of Mexico are presented. The fifth part introduces an overview of Mexico City's situation. The sixth section is dedicated to the methodology used to collect the information to develop this work, followed by the survey results with 19 civil society organizations. In the last part, we present the conclusions of the chapter. We believe this work could interest experts, academics, implementers, and civil society organizations in Mexico City and other latitudes.

8.2 Socio-digital Inequalities

One of the most interesting debates which are taking place nowadays has to do with the way inequalities have permeated the "traditional" perspectives of Bourdieu (1986), Sen (1997), and others. At present times, inequalities are also existing in the digital aspect of our lives. One of the authors who recently manifested in this sense is Helsper (2021). She starts highlighting inequalities as follows:

- Compound and intersectional, with objective, economic, functional, subjective, social, and well-being dimensions.
- Systematic and structural, rather than as differences based on individual preferences or merit.
- Evaluated in terms of inequalities in outcomes rather than merely on the basis of inequalities in opportunities.
- Resisted only when there is a collective consciousness that inequalities are unjust (Helsper 2021, pp. 23–24).

Furthermore, Helsper (2021) identifies what she calls *socio-digital inequalities.* "I use the term 'socio-digital inequalities' instead of digital divides or inequalities. I am concerned with inequalities of outcomes in digital societies, not solely digital connection, or engagement" (Helsper 2021, p. 28). In this regard, she identifies three scopes under which socio-digital inequalities operate: "first- (access), second- (skills and use), and third level (outcomes) digital divides" (Helsper 2021, p. 29). Firstly, about access, Helsper mentions:

> definitions of digital inequalities started by describing differences between people who did and did not have access to desktop computers and the internet… at a more macro level, poor digital infrastructure in lower-income neighborhoods and rural areas was another cause of digital exclusion. The discourse and the issues around digital inequalities during this period resonate with traditional inequalities literature's emphasis on wealth and income as the most important aspect of inequalities (Helsper 2021, p. 29).

Dealing with the second scope (skills and use), Helsper establishes that "digital literacy is a broad, multidimensional, and dynamic concept. In an ever-changing world that is increasingly digital, the continuous upgrading of our skills becomes necessary. To stay 'equally included' we need to commit to lifelong learning that allows us to undertake (or reject) the next new activity in ways that benefit us" (Helsper 2021, p. 31). This second realm remarks learning processes, which are frequently missed from the discussions on inequality. According to Sen (1999), process matters not only what a person can achieve.

Finally, regarding the third level (outcomes) or digital divides, Helsper states that "the implicit assumption in most research, policy, and digital inclusion initiatives is that increased and broad uptake of ICTs automatically leads to universally positive outcomes…as in traditional inequalities literature, the digital inequalities literature tends to situate particular types of outcome as more 'worthy' than others" (Helsper 2021, pp. 31–32). In order to have a better understanding of the approach developed by Helsper, see Fig. 8.1 which summarizes these three different scopes of inequalities in the digital context.

The interesting part to identify is whether, in the cases introduced here, which type of socio-digital inequalities developed by Helsper is present in the Mexico City civil society organizations under scrutiny.

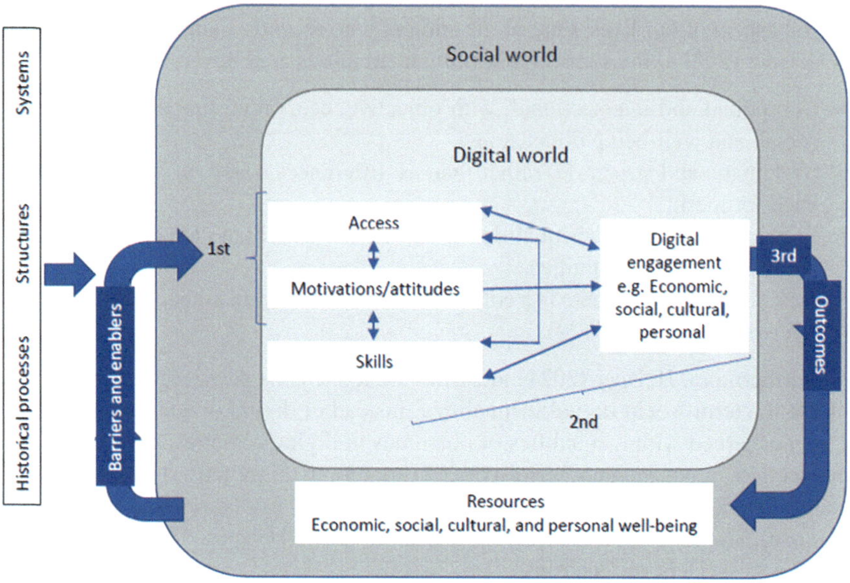

Adapted from (Helsper 2021, p.33)

Fig. 8.1 Scopes of socio-digital inequalities. Adapted from Helsper (2021, p.33)

8.3 Inequality and the COVID-19 Context in Mexico

One of the main aspects that called global attention was how the Mexican federal administration deals with the pandemic. It was ambivalent with the messages sent to the population, and this behavior had adverse effects. Mexico became one of the countries with the deadliest toll rates in the world. In August 2020, Mexico was among the top ten countries in terms of infections and the top five in terms of deaths per one hundred thousand inhabitants (Lustig et al. 2020). Due to the depth of the problem, extreme measures had to be taken. Governments implemented lockdown policies of various degrees to contain the spread of the virus. Although the federal government did not implement full-lockdown measures in Mexico, "stay-at-home" recommendations were put in place. Inevitably, these measures caused a sharp reduction in activity, a fall in employment and income, and a rise in poverty and inequality (Lustig et al. 2020).

Despite introducing extreme measures, the Mexican government did not provide economic support as in other countries. During the lockdown and recovery phase, the Mexican government did not introduce new safety nets. This contrasts with most high-income and Latin American countries (Hale et al. 2020). Undoubtedly the most tragic effects of the COVID-19 pandemic had to do with the devastating economic consequences it had, especially for the poor. For example, Lustig et al. (2020) estimate the increase in poverty due to the COVID-19-induced economic dislocation in

Argentina, Brazil, Colombia, and Mexico. Their estimates suggest that Brazil and Mexico could face the most significant increase in the poor among these countries. However, they also find that the expanded social assistance governments have introduced in response to the crisis could have a sizeable offsetting effect in Brazil and Argentina. In Colombia, the mitigation is relatively modest. In Mexico, it is nil: the federal government has provided no (at least, not as of this article) additional social assistance in the wake of the crisis (Lustig et al. 2020). The way the government in Mexico dealt with the problem was not as expected since most Latin American countries speedily approved economic measures to support the population. For instance, Peru and Uruguay quickly passed income support legislation among Latin American countries, with Argentina, Bolivia, and Paraguay following. Then by April, Ecuador, Chile, Honduras, Colombia, and Guatemala had also passed income support policies (Hoehn-Velasco et al. 2022). The only aid received by the population was in the shape of advanced payments of a pension scheme, which already existed before the pandemic, and secondly, via microcredits with a preferential rate. First, individuals could receive an advanced two-month payment from the non-contributory pension system. Second, small and medium-sized enterprises in the formal and informal sectors were given credits. For the formal sector, these credits were capped at 25,000 MXN (1100 USD) with a maturity of three years at a 6.5% annual interest rate. México Evalúa suggests that the advanced payment is equivalent to 0.2% of GDP and the credits to 0.1% of GDP (cited in Hoehn-Velasco et al. 2022).

In its 2020 report, CONEVAL (*Consejo Nacional de Evaluación de la Política de Desarrollo Social*) established that the number of people it classified as poor rose to 55.7 million, or 43.9% of the population, up from 41.9% in 2018. Hence, the conditions of inequality between different sectors of the population increased.

8.4 The Digital Strategy for Learning: An Effective Way to Address Inequality?

Sustainable development goal four aims to "ensure inclusive and equitable quality education and promote lifelong learning opportunities for all." Target 4.5 focuses on eliminating disparities in education and ensuring equal access to vulnerable groups, including persons with disabilities, indigenous peoples, women, and those who "require particular attention" (see UNESCO https://sdg4education2030.org/the-goal). Hence, providing access to education to all groups in society becomes a crucial element in achieving sustainable development goals.

Despite the perception that the Mexican state was not providing the necessary assistance during the COVID-19 crisis, previous to its start, there was a strategy in practice related to the use of technology for learning. Since 2019 the Secretariat of Education, known as *Secretaría de Educación Pública* (SEP), has installed a national digital strategy for learning. According to the 2019–2023 Educational Digital Agenda (SEP 2020), Mexico's efforts to include ICTs in education began in the 1980s with the production and delivery of computers to primary and lower-secondary school establishments and have evolved.

The 2019–2023 Educational Digital Agenda (SEP 2020) took note of previous programs and required to come up with five working lines: teacher training and updating; digital literacy, inclusion, and citizenship; production, promotion, access, and social use of digital learning resources; better connectivity and infrastructure of ICTs together with Learning and Knowledge Technologies (LTKs) and communication; and research, innovation, and creativity in digital education. These axes are embedded in the National Agreement on Education (Acuerdo Educativo Nacional) and its New Mexican School; they respond to the General Law on Education (Arts, 84, 85, and 86) and, as of 2019, to the country's Constitution (Arts. 3, 6).

It can be argued that great hope was given to social networks as digital learning instruments during the pandemic. The role of Facebook and WhatsApp was crucial since, in many cases, these were the only channels available to share information. Furthermore, people not used to communicate or exchanging information through these means became acquainted with these tools. It was evident the role that online education was going to play in the coming years, as commented by Lloyd:

> with online education likely to become the norm in the coming months or years, there is an urgent need to scale up these initiatives and find new solutions. Some examples are using digital platforms to upload videos and documents by teachers and students when counting on the Internet. Another option, which is being used by some intercultural universities, is the creation of forums on WhatsApp and Facebook that can be accessed from cell phones (Lloyd 2020, p. 119).

Findings suggest that the digital gap in Mexico cannot be merely understood as the access to devices and the Internet that the US National Telecommunications and Information Administration (NTIA) defined in 1999 but as a holistic set of factors that intervene in the use of technologies: Internet and mobile, social class, economic capital, emotions and perceptions, and public policy (Robinson et al. 2015). Finding access to technology in Mexico City would be challenging if this were true. However, access would not be the only factor; there would also be aspects related to technological literacy. Recent research has shown that within the same city, in this case, Mexico City, there can be examples of the existence of a two or three-speed city, characterized by dramatic differences related to access or the population's literacy on technology (see Pansera et al. 2022). There is also a direct correlation between the type of instrument from which people communicate and the technology available, as shown in the reports of the Internet Association of Mexico presented next. Digital strategies are limited to address long-standing inequalities.

8.5 About Internet Habits in Mexico

According to a report presented by FORBES in March 2021, Facebook continues to be crowned as the most-used social platform in the world, as the applications it owns represent four of the five most-used social platforms globally, followed by YouTube and WhatsApp. In the case of Mexico, this was enhanced in a country with more than 100 million social network users, representing a growth of 12.4% compared to 2019. The report also sustains that Mexicans stated that they spend an average of 9 h surfing the net, two hours above the global average, with the most visited sites being Google, YouTube, and Facebook (see Garduño in FORBES 2021).

In the last years, there has been an increasing interest in measuring the Internet habits of the Mexican population. This is how the Internet Association of Mexico has been researching to know more about the prevailing trends. The 17th study on Internet habits in Mexico was published in May 2021 by the Internet Association of Mexico; it was research carried out between April and May 2021. Seventy percent was done via phone, and thirty percent through the Internet. The sample size was 2375 interviews, and 1905 were Internet users (on a national scale), with a margin of error of 2.25% and a level of trust of 95%. As of 2020, there were 84.1 million Internet users in Mexico, representing 72.0% of the population aged six or over. During 2020 and as a result of the COVID-19 confinement, Internet users had the highest growth observed in the last five years. Interestingly, 77% of non-users planned to acquire a connection within the next six months after the survey. During the confinement, the performance of the fixed and mobile Internet operators was positive since only 9.4% of users were not satisfied with their current Internet service. The most crucial reason was linked to the fact that they perceived that the speed obtained was lower than the contracted. Also, Internet consumption habits by users suffered impacts during the pandemic. Mainly in the time users spend connected before migration from work, school, and entertainment to the home. 1 in 10 users decided to increase the download speed they had contracted prior to the pandemic. Despite the negative economic impact, users reported decreased usage and bill payment (source https://www.asociaciondeinternet.mx/).

The 18th study on Internet habits in Mexico was published in May 2022 by the Internet Association from Mexico. It was research carried out between the 12th and the 25th of April 2022. The sample size was 1761 interviews, and 1430 were Internet users (on a national scale), with a margin of error of 2.25% and a level of trust of 95% (see https://www.asociaciondeinternet.mx/). By 2021, there will be 88.6 million Internet users in Me; this represents 75.6% of the population aged six and over. The growth of 3.60% (compared to 2021) is explained by the recovery of purchasing power, reconfiguration of household spending, and new digital needs derived from COVID-19 confinement. The pandemic certainly encouraged and increased use of digital tools at all socioeconomic levels. The study also found an increase in Internet users between age, school, and work online. There was also an amplified penetration of mobile devices and access to social media and messaging snapshots.

There were also two interesting pieces of information released with the 2022 report. Firstly, the socioeconomic level is the main factor explaining the lack of connectivity, and secondly, age is a relevant factor for Internet use. Since almost half of the people defined as not users are older than 55 years old. The primary means of connection are through fixed networks; 91.7% of users connect. Users who hire the service at home spend an average of 401.70 pesos per month. 74.4% of Internet users connect through mobile networks, spending, on average, 176 pesos per month. Home remains the central place of connection, despite the end of confinement. There is a low Internet adoption for connecting to Wi-Fi in public squares and parks. In Mexico, the smartphone has become the main point of access to the Internet, according to 95.4% of Internet users interviewed. This device has the highest adoption and allows many children to study during confinement. There is also a variety of devices that persist in the face of purchasing power barriers. The Smart TV and video game consoles projected an increase over the year due to entertainment needs in the face of confinement.

The Internet is a transversal tool for our daily activities; for this reason, 74% of Internet users declare that they do not have a specific day to spend more time connected than the rest of the week. 26% have a favorite day, 58.8% connect more on weekends, and 41.2% have a greater connection during the week, with Monday being the most important day (13.8%). 25.2% of Internet users reported connecting to the Internet all day equally. Half of the users prefer to connect between 4 and 9 pm.

Video calls show a decrease in use as a result of deconfinement. While many offices continue online or in hybrid formats, the vast majority have returned to work in the office.

5 out of 10 Internet users have made video calls during the last year. WhatsApp remains the most-used one with 63.3%, followed by Zoom with 46.9%, and in third place, Google Meet with 27.3%.

95% of Internet users have at least one social network or instant messaging platform. 16.4% of users do not have an active account, and 83.6% have never had one. WhatsApp, Facebook, and YouTube are the most used among users who maintain an active account. Facebook and TikTok show the highest growth during the year.

8.6 An Overview of Mexico City's Situation. Between Progress and Inequality

Jaramillo's article in March of 2021 provided recommendations for how Mexico City could deal with the pandemic since it was perceived as one of the places on a global scale where inequality produced more significant hassle. According to Jaramillo, "both the national and Mexico City governments must strengthen their test-and-trace strategy and widen the distribution of social protection payments, but they must also target these initiatives better by using a spatial approach and taking social stratum (and the realities of the poorest households) into account" (Jaramillo 2021). Furthermore, he claimed that the high death toll faced in Mexico City was directly related to the inequalities that prevailed before the pandemic in one of the largest metropolitan areas worldwide.

> Although the CDMX (capital region) government has been one of the few state-level governments in the country to pursue a neighborhood-focused response (increasing testing in the most affected neighborhoods, for example), the terrible truth is that Mexico City's administration has not managed to adapt to the deadly truth whose warning signs were evident even at the very beginning of the pandemic: the virus does not affect everyone equally. It infects

> (and kills) the poorest because of the extreme inequalities that were a fact of life long before the arrival of COVID-19 (Jaramillo 2021, p. 1).

Even though Mexico City implemented some initiatives to reduce the impact of COVID-19, as will be introduced further in this chapter, the truth is that it became one of the devastated areas by the pandemic in the whole of Mexico.

Several things happened in the Mexico City context, which inevitably affected how civil society organizations (CSOs), or *colectivos*, operate. A change of government translated into a tense relationship between the government of Mexico City and how these groups used to relate. Before the 2018 local government took office, civil society organizations had a relatively smooth relationship, they even developed joint projects, and some received funding from the local authority. However, once the fourth transformation (4 T) took office, a discourse of austerity and rejection of the third sector became a central aspect of the 2018–2024 local administration. Previous efforts that we can mention as an example of the close relationship that CSOs had with the government of Mexico City was the initiative called *El Laboratorio*. In 2012, the mayor of Mexico City, Miguel Angel Mancera, created this space. The Laboratorio's purpose was to create an experimental space in which citizens, civil society, academia, business, and government could meet to reflect on the problems facing the city and then take joint actions, including utilizing new technologies for digital governance (Pansera et al. 2022).

The territory of Mexico City appears to be a privileged site to look at the penetration of technology, as expressed by some authors, "socio-territorial marginalization is an ideal context to observe the social penetration of ICTs for several reasons, including their availability, since they are an expensive and scarce resource, in an area where there is no market" (Ali et al. 2020).

Furthermore, one of the main aspects for the reduction of technological inequality is increasing access to technology within the population, especially within its poorest sectors; "the primary task should be social integration by providing infrastructure—from broadband access to electronic devices—to increase their possibilities of reaching better jobs, therefore, an improved life quality" (García-Díaz 2021, p. 38). However, enhancing access to infrastructure will not automatically improve the users' conditions since there might be other aspects to consider.

> The city has two speeds; there is a cosmopolitan city, there is a city of knowledge, access, law, opportunities, work, and services, and there is a great city that is not that. So, the probability of it being a smart city is high for this city that is already encrypted, that has services, that has traffic lights, to call it that stupidly, smart or self-regulated, that an ambulance has a route with a GPS phone to take, here depending on where you live the emergency services, that help you or not, to die or live; it is that simple (social activist) (cited in Pansera et al. 2022).

8.7 Our Research

8.7.1 Methodology

In order to find out the way technological access permeated throughout Mexico City, we developed an instrument (questionnaire) consisting of ten questions, which were asked to 19 civil society organizations in November 2022. In this effort, we covered 10 of the 16 localities that compose Mexico City; hence, we could interact with organizations from different social, economic, and cultural backgrounds and with various purposes. The sample was selected on a non-probabilistic basis.

Each of the organizations is based in the following localities of Mexico City. Three are in Benito Juaréz, three in Gustavo A. Madero, two in Venustiano Carranza, two in Iztacalco, two in Coyoacan, two in Iztapalapa, one in Azcapotzalco, one in Milpa Alta, two in Cuauhtémoc, and one in Alvaro Obregon. According to data from the *Consejo Nacional de Evaluación de la Política de Desarrollo Social* (CONEVAL), Mexico City has three out of every ten inhabitants in the situation of poverty (about 3 million people) and 4.3% living in extreme poverty, (around 396 thousand people). As for people living in poverty, it is observed that Milpa Alta is the locality with the largest population in this condition (54.7%), followed by Xochimilco (48.2%) and Iztapalapa (43.9%). On the other hand, the municipalities with the lowest amount of poverty are Benito Juárez (7.9%), Miguel Hidalgo (13.5%), and Cuauhtémoc (20.9%).

In a report presented by the *Consejo de Evaluación de la Ciudad de México* on October 20, 2021 (known as EVALUA), a map with the percentage of people living in poverty, according to an Unsatisfied Basic Needs index was introduced. Table 8.1 shows the percentage of people living in poverty in each of the selected localities. It is worth noting that poverty and inequality in Mexico City coexist, as percentages vary widely.

Table 8.1 Percentage of people living in poverty, measured by Unsatisfied Basic Needs Index 2020

Locality	Percentage (%)
Alvaro Obregón	57.70
Azcapotzalco	49.90
Benito Juárez	29.10
Coyoacán	45.60
Cuauhtémoc	48.30
Gustavo A. Madero	56.50
Iztacalco	53.30
Iztapalapa	68.50
Milpa Alta	80.00
Venustiano Carranza	58.50

Source made by the authors

We believed that following a descriptive approach was the best methodological manner to explain in a better way what we were looking for in this chapter. The goal of descriptive research is to describe a phenomenon and its characteristics. This research is more concerned with what rather than how or why something has happened. Therefore, observation and survey tools are often used to gather data (Gall et al. 2007). The questionnaires were sent to 20 civil society organizations, and we received a response from 19.

The key research question is to determine the role inequality played during the pandemic as part of the technological learning process faced by civil society organizations in Mexico City.

As it was mentioned above, to answer this research question, we developed a ten-question questionnaire divided into two groups: those issues dealing with the authorities of Mexico City (one to five) and those aspects dealing with the organizations and the technological learning process experienced during the pandemic (six to ten). Most of the questions were established in a "yes or no" fashion, followed by a space to elaborate more on the answer. Also, in some questions, the answers were provided in a multiple-choice format; Derived from the main research question, there are others, such as knowing whether there was a plan to reduce inequalities or knowing more about the main technological tools that civil society organizations used in this specific period.

8.7.2 Collection of Structured Data: The Questionnaire

See Table 8.2.

8.8 Survey Results

The first question stated, during the pandemic, did you perceive an action plan or policy (implemented by the Government of Mexico City) dedicated to the knowledge/use/training of any technological tool to be used by Civil Society Organizations? The answer, unsurprisingly, was "no" from 18 out of the 19 organizations interviewed. Let us analyze the projects carried out by the *Agencia Digital de Innovación Pública* (ADIP) office in charge of implementing technological innovation tools in the Mexico City context. We will realize that several efforts were made, but most focused on COVID-19. In March 2020, a website for citizen consultation on the actions undertaken by the city government to combat the pandemic was activated, as well as an explanation of the kind of support provided to vulnerable populations. Other actions included websites dedicated to Microcredits Fund for Social Development (FONDESO), unemployment insurance, and a technological tool to provide immediate medical attention to people with COVID-19 symptoms who were in Mexico City without leaving home. Through another development, it was possible

Table 8.2 Survey questions applied for the research

On the Mexico City Government (1–5)	On Civil Society Organizations (6–10)
1. During the pandemic, did you perceive an action plan or policy (implemented by the Government of Mexico City) dedicated to the knowledge/use/training of any technological tool used by Civil Society Organizations? (a) Yes (b) No	6. Which technological tool (of the following options) was the one you used the most during the pandemic? (a) WhatsApp (b) Facebook (c) Zoom (d) Twitter (e) Microsoft offices package (PowerPoint, Word, Excel, etc.) (f) Other
1a. If the previous answer was Yes, could you tell us which? ____________	6a. The tool you marked in question 2, did you already use before the pandemic, or did you learn to use it during it? (a) I already used it (b) I learned how to use it
1b. If the answer to question 1 was No, could you tell us briefly what you would have liked to have done? ________	6b. If you learned to use it, would you say it was an easy or difficult learning process? (a) Easy (b) Difficult
2. During the pandemic, did you perceive that connectivity access was equitable for everyone in Mexico City? (a) Yes (b) No	7. Do you think the pandemic helped improve your technical knowledge of technological tools? (a) Yes (b) No
2a. If the previous answer was Yes, why? Briefly explain________________________	8. Without a pandemic, do you think your organization would have used the technological tools you used during it? (a) Yes (b) No
2b. If the answer to question 3 was No, why? Briefly explain _________	9. Do you consider that the technological tools you used during the pandemic have enough attributes to carry out your activities? (a) Yes (b) No
3. From your point of view, is there a digital gap between different social and economic sectors of Mexico City? (a) Yes (b) No	10. If necessary, would you consider developing your technological tools? (a) Yes (b) No 10a. If answer 10 was yes, what kind of tool(s) would you consider developing? Explain briefly___________________________

(continued)

Table 8.2 (continued)

On the Mexico City Government (1–5)	On Civil Society Organizations (6–10)
3a. If the previous answer was yes, who should be responsible for "closing" this gap? (a) The Government of Mexico City (b) Private companies (c) Civil Society Organizations (d) The general population (e) Other	
4. Do you consider that the Government of Mexico City has done enough to improve the connectivity of the different population sectors? (a) Yes (b) No	
4a. If your answer to question 5 was No, what do you propose to do? ___________	
5. Do you consider that the Government of Mexico City has done enough to reduce the existing digital gap among the population? (a) Yes (b) No	
5a. If your answer to question 6 was No, what should I do? _____________	

Source made by the authors

to organize the delivery of medical kits, pantries, and pantry vouchers to the homes of suspicious persons detected through automated screening (see www.adip.cdmx.gob.mx). As it can be perceived, all these actions were keen on the public of Mexico City. However, none included a systematic plan dedicated to the knowledge/use/training of any technological tool used by Civil Society Organizations.

8.8.1 Beyond Access and Infrastructure

The organization which responded that there was a plan regarding the first question referred to two aspects. Installing infrastructure in the city to access free Internet; in February 2020, there was an increase in the bandwidth of Mexico City's free Wi-Fi network, since the 13,694 Wi-Fi points increased their speed to 100 Mbps. Furthermore, in November 2021, Mexico City was awarded the Guinness Record for the most "connected" city globally, with 21,500 free Wi-Fi points (see *El País* 11.11.2021). The second point mentioned was the digitalization and facilitation of many procedures and payments that could be made through the Internet. Procedures carried out with older adults were not necessary to move with them. In this respect, their significant efforts were to facilitate the paperwork related to permits or licenses provided by the Mexico City government (see www.adip.cdmx.gob.mx). An important aspect to consider is that this civil society organization is situated in a locality where more than 50% of its population lives in poverty, according to the unsatisfied basic needs index 2020.

The answers were quite diverse regarding what the organizations would have liked to be done.

- An online meeting on how to manage the COVID-19 situation, which to manage in order to promote the health of vulnerable populations.
- Online training in technological tools.
- Institutional strengthening tools in the pandemic or post-pandemic context.
- That all hospitals gave care and that doctors were prepared for care. The government gave the necessary supplies for care, had a care protocol policy, and prevention work that had medicare and medicines, and not only gave paracetamol to the population.
- That we would have communicated with the government and other organizations. That training would have been given, about digital tools, to work from home. Financial support for the organizations so they could continue working and their expenses were not suspended.
- That there was Internet in the most vulnerable localities.
- Have offered a series of online options, access to platforms, and virtual courses.
- Facilitate access to technological and training resources for the general population.
- Do not cut the funds that the government granted for health issues, research, or the implementation of related projects in the field.
- Training on the use of information and communication technologies.
- That digital platforms were made available for free, in addition to implementing training courses on them, at the same time that it would have been handy recommendations on time dosage or time management in digital tools since the platforms were initiated gave the feeling of freedom. However, in many cases, the use of digital platforms extended working hours.
- That organizations were more protected with resources. They abandoned us completely.
- More government support in technology, courses, training, and donations of technological tools.
- By working with society, support programs for civil associations would have been carried out.
- That workshops on the new normality were given to train civil society organizations.
- Give correct information about what could be done.
- Train and provide equipment.
- Many students in vulnerable areas dropped out of school because they did not have the technological resources to take online classes because of the pandemic. Specifically, children in preschool and primary school did not have a computer. Therefore, some took classes with their parent's cell phones, which generated

high Internet costs that forced children to leave school. I would have liked to have at least offered options for using the Internet. Maybe at class times or in marginalized areas. I think that is the least they could have supported the student body with.

We believe that all the answers provided become a relevant source of information for those actions that the Mexico City government could take shortly. The first four proposals deal with the COVID-19 context and all the actions that could have been taken. There is a strong feeling in the population that something else could have been done and that a more significant number of lives could have been saved concrete actions would have been taken. These answers show that civil society organizations seemed to be greatly concerned about health issues and how to improve the health system in the post-COVID-19 era. In fact, socio-emotional strategies and psychological support are highly demanded in schools.

8.8.2 Austerity

Another group of answers is related to training, communication, and not leaving the organizations without funding. In previous local administrations, civil society organizations played a crucial role in dealing with particular problems in Mexico City. However, since the arrival of the current administration, there has not been a possibility to work with local authorities, budgets have been cut, adducing austerity, and organizations have been forced to look for resources elsewhere. As almost all of them expressed, a lack of communication has translated into not having the ideal environment to perform their activities.

8.8.3 Learning in Times of Emergency

A third theme identified has to do with education. It can be inferred that increasing or improving the education sector in Mexico City is a point that the participants highlighted. Especially during the pandemic, having a better-educated population could have made a big difference between the different population sectors.

Another point deals with the accessibility to technological tools that accompany education in our times; it is not only about having an excellent educative system but also about providing adequate conditions for students to support their learning process.

8.8.4 Unequal Access to Connectivity

The second question was, did you perceive connectivity access was equitable for everyone in Mexico City during the pandemic? Perhaps this would translate into better connectivity after Mexico City was awarded a Guinness Record for having the most significant number of free Wi-Fi posts globally. However, this was not the case since the answer was an overwhelming "no," with 17 (89%) out of 19 organizations claiming this. Only two of them perceived that access to connectivity was equitable for the population of Mexico City.

The answers as to why it was not equitable access to connectivity during the pandemic were quite varied, as we will present next.

- It was selective by bands and megabytes.
- It depends on private providers.
- Internet access remains costly for many vulnerable sectors.
- There were areas where the Internet was only for those who could hire it. Where there are accessible Wi-Fi networks, the signal is not good.
- There were remote places that did not have access to the Internet.
- Not everyone has the Internet and computers.
- Not all people had access to devices to access video call platforms, not all people had safe spaces to carry out activities within their homes, and not all people had a good Internet connection.
- Not everyone knows the management of these applications or platforms; they do not have the equipment or adequate Internet connectivity.
- Because most people with access to the Internet do so through a payment to private companies. During the pandemic, the lack of economic activity made it difficult to access it.
- Not all homes and organizations had the equipment for computing and Internet connectivity, and not all could access them.
- First is that people can access the Internet, then it is to pay to access a quality service that is fast in downloading, and the cost is not affordable for all people.
- Not everyone has the resources to have the Internet, at least on a cell phone, or not even a cell phone is counted.
- Not all have the same economic resources to access these technological tools.
- Not all people have computers and the Internet.
- When people lost their jobs, they stopped paying for their phones and were left uncommunicated without being able to ask for help.
- The Internet signal does not reach the entire national territory, and when it arrives, it is of inferior quality.

The responses provided by the participants are pretty influenced by an economic aspect, which has to do with the perception that the better service you hire, the better connectivity you could have. However, some of them mentioned that access to the Internet could be difficult in some parts of the city. It was interesting to see that one of them mentioned the free Wi-Fi service provided by the government and claimed that it did not work as expected. There is also an issue related to the capabilities for managing certain types of equipment, an issue that recounts the education factor previously mentioned. Here we can state that the definition of the first scope of the digital inequalities identified by Helsper (2021) *access* is definitely present in the responses. Since the approach establishes that "digital inequalities started by describing differences between people who did and did not have access to desktop computers and the internet... at a more macro level, poor digital infrastructure in lower-income neighborhoods and rural areas was another cause of digital exclusion" (Helsper 2021, p. 29).

One of the most publicized initiatives made by the Mexico City government is the one known as PILARES, defined as (*Puntos de Innovación, Libertad, Arte, Educación y Saberes*, translated as Points of Innovation, Freedom, Art, Education, and Knowledge). According to the website https://pilares.cdmx.gob.mx currently, there are 281 PILARES spread all over Mexico City. In these buildings, a wide variety of training workshops and skills are provided to the population. Each has the equipment and a stable Wi-Fi connection to carry on any technological activity. These places are part of a program where you can finish elementary, middle, and high school, and even a bachelor's degree, with the help of tutors. You can use the computers available at each PILAR; everything is free. People could also join a cyber school as part of the activities organized in these places. We believe it is essential to allude to PILARES since it is aimed at reducing inequalities and accessibility within the population of Mexico City; however, it was not mentioned by any of the organizations.

8.8.5 Digital Inequality: "Mind the Gap"

The third question dealt with the issue of whether the organizations surveyed thought there was a digital gap between different social and economic sectors of Mexico City. The result was relatively straightforward since 18 out of 19 organizations believed there was such a gap. Based on expecting a result of such proportions, a follow-up question was established, so we asked them: who should be responsible for "closing" this gap? We provided five options: the government of Mexico City, private companies, civil society organizations, the people, and other.

The responsibility of the Mexico City government was vastly supported by 47% of the participants, the runner-up was "other" with 26%, and the people occupied the third position with 11%; organizations of the civil society had the same percentage of those who did not respond, with 5%. Regarding the answers given as "other," the participants highlighted four options: a combination of government and business, the government, private companies, and civil society organizations, the three sectors of society have the responsibility to close the "digital gap," but the government sector must design public policies to achieve it and a model of new governance needs to be introduced, it is a work among all to bring technology to vulnerable people.

Despite the organizations identifying the three sectors commonly involved in governance processes, it can be inferred that the leadership to reduce the technological gap is practically given to the Mexico City government. It can be inferred that despite the apparent "divorce" between civil society organizations and the Mexico City government, they still believe that the public administration shall lead the process to reduce the existing social and economic gaps between sectors of the population in Mexico City. It can be established that the second scope identified by Helsper dealing with (skills and use) was found in this question. Helsper establishes that "digital literacy is a broad, multidimensional, and dynamic concept. In an ever-changing world that is increasingly digital, the continuous upgrading of our skills becomes necessary. This implies a renovated way of learning. Process matters. To stay 'equally included' we need to commit to lifelong learning that allows us to undertake (or reject) the next new activity in ways that benefit us" (Helsper 2021, p. 31). Hence, if there is no concern from those actors who would have under their responsibility the reduction of the digital gap, then inequalities will continue to prevail. "Mind the gap."

8.8.6 Further Actions to Be Connected and Learn

The following question dealt with one of the most critical aspects of learning processes using technological tools in cities: connectivity. To the question, do you think the Government of Mexico City has done enough to improve the connectivity of the different population sectors? Participants clearly stated that there is still a lot to be made. 14 out of 19 organizations believed that the government of Mexico City had not done enough, and they thought that some of the following initiatives could help:

- Free Internet as a human right to the entire population.
- Improve the Mexico City network.
- More free hotspots in popular low-income areas.
- That we have a higher educational level in our population and much illiteracy. The government is not interested in having people with education because it would demand their rights.
- That companies, government and CSOs work together.
- Approve more budget to make communities farther away from the Internet.
- Offer plans to purchase equipment and training.
- Joint work between government, private initiatives, and organized civil society to shorten technological gaps that are being unable.
- Grant free interconnection to the most unprotected social extract houses and computer equipment.
- Support ourselves to stay close and communicate, even if we do not have government support.

- Give computer classes at all educational levels and verify that they have the necessary equipment.
- That there is connectivity in public places such as some squares, government buildings, and in cases of emergency at certain times.
- Create "school Internet packages" whose password is given to students.

Despite being awarded the Guinness Record for the city (in the whole world) with the most significant number of free Wi-Fi posts, there seems to be a general perception within the civil society organizations that things are not working as expected. There are mentions of improving the conditions of the network, the number of hotspots, or providing the Internet as a fundamental human right. There is also the idea of joining efforts between the private, public, and civil society sectors. It also appeared, yet again, the need to focus on the educative sector, with such ideas as providing Internet packages, giving a more significant number of lectures, and providing the necessary equipment for every pupil.

8.8.7 The Need for a Long-Term Project

The last question, from the five questions dedicated to the government of Mexico City, touched on a critical topic. The question goes as follows: Do you consider that the Government of Mexico City has done enough to reduce the existing digital gap among the population? Almost all participants agreed that it had not been made enough to reduce the digital gap in the context of Mexico City since 17 out of 19 participants said "no." Again, the organizations provided several options; next, we introduced what they proposed.

- Develop community engineering.
- Provide more training.
- More free hotspots in popular low-income areas.
- Providing better education for the entire population.
- Working together (civil society, government, and private sector).
- Request more budget for technological tools.
- Access to information, dissemination of options.
- Ensuring the continuity of programs beyond a six-year plan.
- Generate public policies for free access. Internet throughout Mexico City.
- Deliver computer equipment and free Internet to the most unprotected population.
- Regulate the cost and quality of Internet access so that it is affordable for the majority of the population in the city.
- Close that digital gap. Find a way to connect within and in all hard-to-reach places.
- Give more excellent resources and access to the most vulnerable people.
- Train all educational levels and provide equipment.
- Free Internet access in public places.
- Eliminate poverty.

- Generate agreements or agreements with companies to provide technological tools for shared or private use at affordable prices.

The digital gap is an aspect that is present in many cities globally. However, it is not presented in such a fashion as the one taking place in Mexico City. Mexico City, as it has been previously cited, is a city that functions at different speeds, you could sense the experience of living in the first world, and then after a couple of blocks, you would be in a deprived part of town. We can interpret that poverty reduction is the core of the solution to reduce the gap. In a document presented in February 2022, the Evaluation Council of Mexico City (EVALUA) presented the poverty information by NBI (unsatisfied basic needs) for the period 2015 and 2020 for Mexico City, based on the Intercensal Survey 2015 and the Population and Housing Census 2020, published by the National Institute of Statistics and Geography (INEGI). The summary of the results establishes that between 2015 and 2020, the population in poverty by NBI in Mexico City fell from 60.9% to 57.4%, while the percentage of the population in extreme poverty decreased from 19.5% to 16.8%. In the same period, a decrease in deficiencies was observed in almost all dimensions of NBI. This was due to a significant increase in Internet access, an expansion in social security and health service coverage, and greater availability of durable goods in households. Between 2015 and 2020, 15 of the 16 localities of Mexico City showed a decrease in the percentage of the population in poverty by NBI. The four municipalities that presented the most significant reduction were: Milpa Alta (from 86.6% to 80%), Cuauhtémoc (from 53.7% to 47.2%), La Magdalena Contreras (from 69.2% to 63.3%), and Cuajimalpa de Morelos (from 58.8% to 53.2%).

Despite the poverty reduction shown in the data, among the participants of this project, there is a perception that the government has not done enough to decrease the digital gap in Mexico City. Long-term planning strategies covering the whole of Mexico City, facilitating access to everyone, and increasing the budget for technological matters, were all answers provided by participants.

8.8.8 *WhatsApp and Zoom as Means of Learning*

The sixth question (and the first dedicated to civil society organizations) concerned the technological tool the 19 participants used the most during the pandemic. WhatsApp and Zoom were entirely utilized, with 42% each, while Facebook was used by 16% of the respondents. This trend matches quite well with the data detected by the Internet Association of Mexico. Another exciting piece of information is that 17 out of 19 organizations used the technological tool before the pandemic. For the two who did not use it for one of them, it was easy to learn, and for the others, it was more complicated.

8.8.9 Technological Tools and Trust

The seventh question referred to how the pandemic accelerated the learning process of technological tools within many organizations. Do you think the pandemic helped improve your technical knowledge of technological tools? 89% (17 out of 19) claimed this was the case. This answer was not a surprise since we have seen in recent times that there has been a significant increase in how people interact and communicate. It can be argued that the pandemic only accelerated processes that otherwise would have taken longer. Perhaps one of the aspects on which we will have to focus shortly to gain more effective use of technological tools is the aspect of trust, as some recent research (see Pansera et al. 2022) shows that this aspect could deprive people of using technology.

8.8.10 Location and Technology

The following question, number 8, established, without a pandemic, do you think your organization would have used the technological tools you used during it? Here the answers were more even. 7 out of 19 (37%) were optimistic that they would still have used the technological tools without a pandemic. The exciting part of these answers is the location of the organizations that responded. Since they operate in diverse parts of the city, it cannot be inferred that the location was an issue that could have affected the use or learning process of technological tools by civil society organizations. The 12 remaining participants (63%) said "no," which is interesting since they were also from diverse areas of Mexico City. Hence, it appears that location became more of an incidental aspect when utilizing technological tools and cognitive processes.

8.8.11 Tech-Based Teaching

The ninth question dealt with the features of the technological tools used by the civil society organizations under research. Do you consider that the technological tools you used during the pandemic have enough attributes to carry out your activities? 13 out of 19 (68%) responded that "yes," the technological tools they used satisfied their needs. Perhaps this was because most of the activities during the pandemic consisted of respecting the no-contact policies established in most countries. Also, all the tools mentioned by the organizations facilitate live interaction and sending documents; furthermore, they are user-friendly. The negative response from the six organizations that replied this way was entirely based on the nature of their work, meaning that they require a more specialized technological tool.

8.8.12 Is Creativity Emerging?

The last question was about being innovative and creating your technological tools. If necessary, would you consider developing your technological tools? The results were interesting since 12 (63%) responded with "yes." Again, there was a mix of locations here, so we can deduce that setting does not matter to developing technological tools. So, even if you face substantial inequalities compared to organizations from affluent neighborhoods, this will not deter you from creating your technical developments.

We introduce the following list of tools organizations would consider developing:

- Effective communication platforms in health and HIV.
- Our media, such as complimentary newspapers and community radio.
- Websites with helpful content for students, without the need to download applications, and that can work without the need for the Internet.
- A platform that could help me train, manage, and link with other CSOs with missions like my A.C.
- Maps to identify social needs by area.
- An app in which complaints can be made about human rights violations, and the complaint immediately reaches the competent and involved instances.
- Software programs or create software that meets our needs and procures resources.
- More training for the optimal mastery of the technological tools we already use.
- App to train my associates in Zoom, WhatsApp, and Instagram management.
- We are currently about to launch a virtual toy library with games for children to learn by playing from a cell phone.

As it can be perceived, these responses were given based on the themes that the organizations deal with on their daily agendas. Hence, we received ideas for platforms dealing with health issues, websites to communicate, student learning, creating maps about human rights violations, technical training, and even games for children. Similarly, to what happened with previous questions, the issue of the location of the organization did not seem to matter to have the will to create a technological tool.

In short, digital inequalities need to be addressed by promoting different public actions, which are beyond access to sophisticated tools and high technology. Gaps can be bridged as new ways of governance are set up and larger learning opportunities for all are created. Poverty and inequality are together blocking creativity and human flourishing. Better and non-traditional ways of using technology are needed in one of the most fascinating and vibrant cities of the world: Mexico City. A summary of topics and main aspects interviewees provide will be shown in Table 8.3 next.

Table 8.3 Summary of topics and main aspects provided by interviewees

Topic	Main aspects highlighted
Beyond access and infrastructure	Installing free Wi-Fi Infrastructure must consider ways to linking such aspects as public health and education to the utilization of technology
Austerity	This aspect has left many civil society organizations without any funding access; it has been translated into the search for financial resources abroad
Learning in times of emergency	It is not only about having an excellent educative system, but also about providing adequate conditions for students to support their learning process
Unequal access to connectivity	The service remains private-oriented and tariff-based on megabytes; it is costly for many vulnerable sectors
Digital inequality: "Mind the gap"	It can be inferred that despite the apparent "divorce" between civil society organizations and the Mexico City government, they still believe that the public administration shall lead the process to reduce the existing social and economic gaps between sectors of the population in Mexico City
Further actions to be connected and learn	Improving the conditions of the network, the number of hotspots, or providing the Internet as a fundamental human right
The need for a long-term project	Long-term planning strategies covering the whole of Mexico City, facilitating access to everyone, and increasing the budget for technological matters
WhatsApp and Zoom as means of learning	These technologies were used before the pandemic and proved to be very convenient. They are easy to learn
Technological tools and trust	The pandemic accelerated trust in the use of technological tools between civil society organizations in the Mexico City context
Location and technology	It cannot be inferred that the location was an issue that could have affected the use or learning process of technological tools by civil society organizations
Tech-based teaching	All the tools mentioned by the organizations facilitate live interaction and sending documents; furthermore, they are user-friendly
Is creativity emerging?	We received ideas for platforms dealing with health issues, websites to communicate, student learning, creating maps about human rights violations, technical training, and even games for children

Source made by the authors
For detailed information about these findings, see Sect. 8.8; similar research developments in the Mexican context can also be found in Pansera et al. (2022). For a general discussion about digital inequalities that can be related some of the findings made here see Helsper (2021)

8.9 Conclusions

There is a perception that the government of Mexico City has not done enough to reduce the economic and social inequalities prevailing. Despite the data provided, in the sense of a reduction in some of the poverty indicators detected by the institutions measuring them, the survey demonstrated that the feeling within these organizations is that many things still need to be done. We can say that organizations identified three core themes over which the inequalities related to technological aspects could be achieved. The first deals with providing better connectivity, but not only via Wi-Fi access. The organizations also mentioned the relevance of adequate equipment and structured long-term planning. Secondly, establishing a plan to improve the capabilities of the general public and civil society organizations, in particular, was not put into practice during the pandemic, let alone in the learning process. Thirdly, particular emphasis was given to the educative sector in creating "technological planning" for all the students in Mexico City.

At the beginning of the research, we thought that the part of the city where the organizations performed their daily activities could have been directly linked to the access or the operation of the technological tools they used. However, the territory did not seem to be a factor in stopping organizations from using determined programs. Instead, the complications arrived from the accessibility to a good network with adequate speed to carry on their agendas. Another element to consider is that most organizations used technological tools before the pandemic. Hence this challenging period only accelerated the learning process that some were experiencing.

This chapter demonstrates that there is still plenty of research regarding accessibility and technological capabilities in the Mexico City context. Furthermore, analyzing the relationship that the government of Mexico City could have with civil society organizations for possible future collaborations remains uncertain. Moreover, the issue of poverty and inequalities in Mexico City still has many angles for future analysis.

References

Agencia Digital de Innovación Pública del Gobierno de la Ciudad de México (2020) Sección COVID-19 dentro del portal de datos abiertos de la Ciudad de México para fortalecer la transparencia y rendición de cuentas en la respuesta a la pandemia. https://adip.cdmx.gob.mx/comunicacion/nota/presentan-seccion-covid-19-dentro-del-portal-de-datos-abiertos-de-la-ciudad-de-mexico-para-fortalecer-la-transparencia-y-rendicion-de-cuentas-en-la-respuesta-la-pandemia

Ali MA, Alam K, Taylor B (2020) Do social exclusion and remoteness explain the digital divide in Australia? Evidence from a panel data estimation approach. Econ Innov New Technol 29(6):643–659. https://doi.org/10.1080/10438599.2019.1664708

Asociación de Internet (2021) MX 17° Estudio sobre los Hábitos de los Usuarios de Internet en México. https://www.asociaciondeinternet.mx/estudios/habitos-de-internet

Asociación de Internet (2022) MX 18° Estudio sobre los Hábitos de Personas Usuarias de Internet en México. https://www.asociaciondeinternet.mx/estudios/habitos-de-internet

Bourdieu P (1986) The forms of capital. In: Richards JC (ed) Handbook of theory and research for sociology of education. Greenwood Press, New York, pp 241–258

Consejo Nacional de Evaluación de la Política de Desarrollo Social CONEVAL (2020) https://www.coneval.org.mx/Medicion/MP/Paginas/Pobreza_2020.aspx

Consejo de Evaluación de la Ciudad de México EVALUA (2021) Evaluación de la Política Social de las Alcaldías de la Ciudad de México (2018–2020), Consejo de Evaluación de la Ciudad de México. https://www.evalua.cdmx.gob.mx/storage/app/media/2021/estadistica/presentacion datos/presentacion-pobreza-2016-2020.pdf

Gall MD, Gall JP, Borg WR (2007) Educational research: an introduction, 8th edn. Pearson, Boston

García-Díaz A (2021) Construyendo escuelas resilientes: La educación flexible, la integración y la COVID-19. Rev Española Educ Comp 38:211–227

Garduño M (2021) Covid-19 lleva a más personas a las redes sociales; registro crece 13%. FORBES

Hale T, Webster S, Petherick A, Phillips T, Kira B (2020) Oxford COVID-19 government response tracker. Coronavirus Government Response Tracker

Helsper E (2021) The digital disconnect, the social causes and consequences of digital inequalities. SAGE Publications, London

Hoehn-Velasco L, Silverio-Murillo A, BalmoridelaMiyar J, Penglase J (2022) The impact of the COVID-19 recession on Mexican households: evidence from employment and time use for men, women, and children. Rev Econ Household 20:763–797. https://doi.org/10.1007/s11150-022-09600-2

Jaramillo M (2021) The pandemic against the poor: Mexico City and COVID-19. https://blogs.lse.ac.uk/latamcaribbean/2021/03/02/the-pandemic-against-the-poor-mexico-city-and-covid-19/. Accessed 2 Mar 2021

Lustig N, Martinez-Pabon V, Sanz F, Younger SD (2020) The impact of COVID-19 lockdowns and expanded social assistance on inequality, poverty, and mobility in Argentina, Brazil, Colombia and Mexico. COVID Econ Vetted Real-Time Pap 46(1):32–67

Lloyd M (2020) Desigualdades educativas y la brecha digital en tiempos de COVID-19 en Educación y Pandemia, Una Visión Académica. Universidad Nacional Autónoma de México Instituto de Investigaciones sobre la Universidad y la Educación

Pansera M, Marsh A, Owen R, De Alba J, Flores A (2022) Exploring citizen participation in innovative city development in Mexico City: an institutional logics approach. Organ Stud 44(10):1679–1701

Robinson L, Cotten S, Schulz J, Williams A, Hale T (eds) (2015) Communication and information technologies annual: digital distinctions and inequalities. Emerald Group Publishing, Bradford

Secretaría de Educación Pública (2020) Agenda Digital Educativa. http://infosen.senado.gob.mx/sgsp/gaceta/64/2020-02-051/assets/documento/Agenda_Digital_Educacion/pdf

Sen A (1997) Human capital and human capability. World Dev 25(12):1959–1961. https://doi.org/10.1016/s0305-750x(97)10014-6

Sen A (1999) Development as freedom. Oxford University Press, Oxford

UNESCO (2021) Global education cooperation mechanism education 2030. https://www.sdg4education2030.org/the-goal

Chapter 9
Advantage and Paradox—Inequalities for Children in Urban Settings

Sebastian Palmas, Jan Beise, Thomas George, and Danzhen You

Abstract This chapter examines the impact of urban living on the wellbeing of children in terms of health, access to services, and education. While cities offer some advantages, such as access to resources and economic opportunities, they also present significant challenges, including high levels of inequality and limited access to basic services. In particular, the chapter highlights the dire conditions faced by children living in slums. Additionally, the review discusses the persistence of inequalities in cities in Sub-Saharan Africa and the urgent need for evidence-based interventions to address these issues. Finally, the review discusses UNICEF's approach and priorities for improving the lives of children in urban settings through targeted programming.

Keywords Slums · Exclusion · Deprivation · Inequalities · Sub-Saharan Africa

The views expressed in this report are those of the author and do not necessarily reflect the views of the United Nations Children's Fund (UNICEF).

S. Palmas (✉) · D. You
Division of Data, Analytics, Planning and Monitoring, UNICEF, New York, NY, USA
e-mail: spalmas@unicef.org

D. You
e-mail: dyou@unicef.org

T. George
Programme Group, UNICEF, New York, NY, USA
e-mail: tgeorge@unicef.org

J. Beise
Middle East and North Africa Regional Office (MENARO), UNICEF, Amman, Jordan
e-mail: jbeise@unicef.org

G. H. Tonon (ed.), *Urban Inequalities*, The Urban Book Series,
https://doi.org/10.1007/978-3-031-59746-6_9

9.1 Cities Will Be Home to an Increasing Share of the Global Child Population

Cities are growing. UN-Habitat projects that the proportion of the world population in urban areas will increase from 56.2% in 2022 to 68.4% by 2050 (UN-Habitat 2022b). This growth will mainly come from less developed regions, projected to have an average annual urban population growth of 2.09%, from 3375 million people in 2020 to 5556 in 2050. Africa, currently the least urbanised region and with the highest urban growth rate (3.17% between 2015 and 2055), will become a primarily urban continent between 2030 and 2035 (UN-Habitat 2022b). Asia, with the second highest urban growth rate (1.41%), will continue to have most of the population of the world's cities.

In 2020, it was estimated that 1.18 billion children lived in urban areas, which accounted for approximately 54% of the world's total number of children (UNICEF 2022). Given that projections suggest that by 2050 nearly 70% of the world's population will live in urban areas (UN-Habitat 2022b), the proportion of children living in cities will inevitably rise in the coming decades.

It is essential to prioritise the examination of inequalities in children's wellbeing in urban areas to promote sustainable development for the future. As the number of children living in cities rises, ensuring their human rights and access to safe water, healthcare and education will be essential for promoting social and economic progress. Neglecting the needs of urban children can lead to long-term negative consequences for individuals, communities, and society. Therefore, it is vital to prioritise investment in child health and education to ensure a prosperous future for all.

9.2 Inequality in Opportunities for Urban Children

Debates surrounding inequality often revolve around two perspectives: inequality of opportunities and inequalities of outcomes (see UN DESA 2015; Paes de Barros et al. 2008). Inequality of opportunities refers to external circumstances which put them at a disadvantage compared to other children in urban areas, focusing on the differences of access to basic needs regardless of the circumstances of their birth: their sex, race, geographic location or other social, economic and cultural factors. On the other hand, the second perspective examines the inequality of outcomes in various material dimensions of human wellbeing, such as educational attainment and health status.

This chapter focuses on the inequalities of opportunities affecting urban children. This refers to external circumstances that put them at a disadvantage compared to other children in urban areas, such as access to water, sanitation, education, or healthcare services. It is essential to understand the inequalities in these opportunities because they can have a devastating impact on the development of children, starting early in their lives. By examining how opportunities are unequally distributed, we can identify the policies, practices, and institutions that perpetuate inequality and take action to address them. From the opportunity perspective, children are more likely to develop their full potential and lead fulfilling lives once multidimensional equality is secured.

9.3 An Urban Advantage for Children Despite Challenges?

Children living in cities face several risks to their health and wellbeing. Some of these risks include exposure to inadequate housing and living conditions, lack of access to safe drinking water and sanitation facilities, increased risk of infectious diseases, inadequate nutrition and food insecurity, limited access and unaffordability to quality education and healthcare, and increased exposure to violence and crime. Cities also present an environmental challenge for children, exposing them to air pollution, noise pollution, and other environmental contaminants. These risks can have both short-term and long-term impacts on a child's physical, mental, and social development and can contribute to persistent inequalities among children in urban areas.

Despite these risks, multiple evidence shows the potential advantages of cities. On average, children living in urban areas benefit from the "urban advantage": the perceived benefits and opportunities of living in cities. Urban areas have better access to health, education, water, and sanitation facilities. On average, urban households earn higher incomes and benefit from improved infrastructure and communication networks. Cities are often melting pots of different cultures, exposing children to new ideas, customs, and beliefs. Cities often have more social activities and events, such as museums, concerts, and festivals, which can enrich children's experiences. Some examples of urban advantages across different themes are:

Access to essential services:

- Coverage of safe drinking water in 2020 was higher in urban areas (86%) than in rural areas (60%) (WHO and UNICEF 2021).
- Safely managed sanitation services coverage was higher in urban areas (62%) than in rural areas (44%) (WHO and UNICEF 2021).
- Basic hygiene services, such as hand-washing facilities with soap and water, are significantly more common in urban areas than in rural areas in most countries (Quispe-Coica and Pérez-Foguet 2022).

Health:

- Urban areas tend to have better healthcare infrastructure, facilities, and resources than rural areas, resulting in better access to healthcare services for urban populations, including children (Ezeh et al. 2017).

- International Labour Organization (ILO) reports that while 56% of the global rural population lacks access to essential healthcare services, only 22% of the urban population is not covered (Scheil-Adlung 2015).
- Data from Demographic Health Surveys (DHS) and Multiple Indicators Cluster Surveys (MICS) show that average access to skilled birth attendance in low-income and middle-income countries was 80% for urban areas and 72.8% for rural areas (Joseph et al. 2018).
- Stunting rates are lower in urban areas than in rural areas (Black et al. 2013; UNICEF 2019).
- Using DHS from 20 Sub-Saharan countries, Gollin et al. (2021) estimated that stunting rates are significantly lower in urban areas in 11 out of 20 countries and anaemia and dietary inadequacy are worse in rural areas for all countries except Zimbabwe and Tanzania.

Education:

- Children in rural areas are often far away from their schools (Bashir et al. 2018), potentially affecting their access and learning performance.
- The likelihood of finding qualified teachers is higher in highly developed urban areas as opposed to rural areas (Luschei and Chudgar 2015).
- The average attendance in early child education in 61 low- and middle-income countries was 37%, with a gap of 16% points between urban and rural areas and a 34% point gap between the richest and poorest quintiles. Children living in urban areas were more likely to attend (47%) than those in rural areas (31%) (UNESCO 2021).
- Data from the Programme for International Student Assessment (PISA) 2015 show that, on average, across OECD countries, students in city schools score 31 score points higher in science than students in rural schools, which is roughly equivalent to one year of schooling (Echazarra and Radinger 2019).
- Household expenditure for education is markedly higher in urban than rural areas (UNESCO 2021).

Housing:

- Gollin et al. (2021) conclude that housing quality is higher in denser areas than in rural areas in 20 sub-Saharan countries.

9.4 The Benefits of the Urban Advantage for Children Are not Uniformly Distributed

However, studies have shown that not all urban children have the same advantages and that there are significant inequalities in child wellbeing within cities (Adams 2018; Shutters et al. 2022). The inequality is significantly evident in the poorest households, which often live in informal settlements with inadequate housing and are located in neighbourhoods that are less well serviced and farther away from the economic centre. Moreover, they have less ability to cope with shocks and stressors such as natural hazards, which can be worsened by climate change or economic turbulence.

Access to basic sanitation services within cities varies greatly between neighbourhoods or communities. This can result in significant inequalities in access to basic sanitation services between the richest and poorest urban populations (WHO and UNICEF 2021). In some cases, children living in urban slums or informal settlements may have minimal access to basic sanitation services, such as toilets. Residents may be forced to rely on inadequate, unsafe, shared or public facilities, or even resort to open defecation. These inequalities can have serious health consequences, particularly for children, who are more vulnerable to diseases transmitted through poor sanitation and hygiene practices.

In many cases, slum residents rely on informal and often unsafe water sources such as unprotected wells, surface water, and vendor-sold water of questionable quality. In India, slum households are likelier to have their primary water source away from the premises (Satapathy 2014).

According to UNICEF (2018a, b), significant inequalities exist between the richest and poorest children in primarily low- and middle-income countries, as evidenced by data from 10 indicators. In half of the countries analysed, children in the poorest quintile are at least twice as likely to die before their fifth birthday than their richest urban peers (Fig. 9.1). Stunting is even more prevalent among the poorest urban quintile, with at least twice the rate compared to peers in the richest quintile in approximately 8 out of 10 countries analysed and four times higher in 2 out of 10 countries. It is worth noting that for stunting and under-five mortality, the relative inequality is higher in urban areas than in rural areas, opposite to the pattern of the other indicators.

Save the Children (2015) reported that under-5 mortality inequality is worsening in too many cities. In almost half of the countries they studied, urban survival gaps have grown, and it finds that in most countries, the poorest urban children are at least twice as likely to die as the richest urban children before they reach their fifth birthday.

UN-Habitat (2022b) highlights the inequalities in vaccination rates for measles between slum and non-slum urban children. In Haiti, for example, only 18.2% of slum children were vaccinated in 2017, compared to 79.1% of non-slum urban children. Additionally, the report found that although access to skilled birth attendance for women in slums increased from 40.3 to 79.3%, it remained significantly lower than the access for non-slum women, who had a skilled birth attendance rate of 92.5% in 2017.

According to UNESCO (2021), education expenditure inequalities between poor and rich households are striking in Ghana, Guatemala, El Salvador, and Pakistan. Inequalities in attendance based on wealth were even more pronounced than urban–rural differences. Specifically, 58% of the richest children attended early childhood education, while only 24% of the poorest children attended this educational level (UNESCO 2021), meaning that the average share of the richest children attending early childhood education was more than twice that of the poorest children.

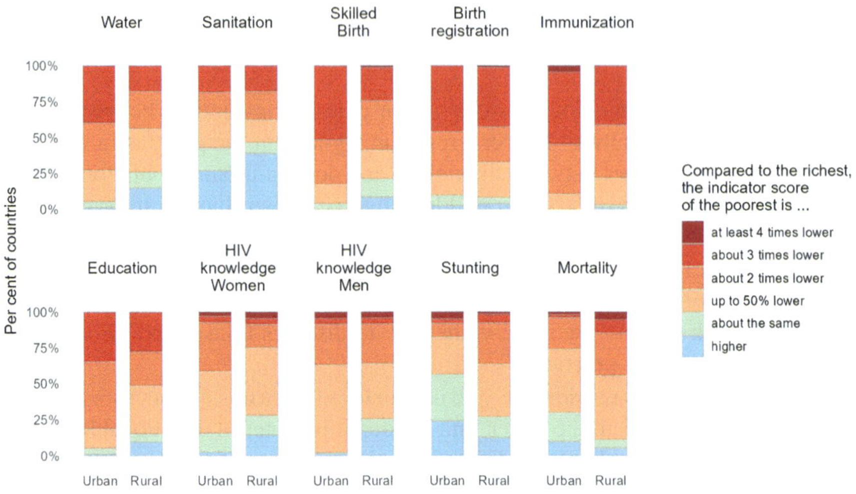

Fig. 9.1 Share of countries and areas with relative wealth inequality gaps (ratio richest to poorest quintiles) by urban and rural (2011–2016). *Source* UNICEF (2018a, b)

9.5 Most Marginalised Urban Children Fare Worse Than Rural Peers—The Urban Paradox

Intra-urban inequality can be so large that many of the most disadvantaged children in urban areas fare worse than children in rural areas in some countries, as shown by UNICEF (2018a, b) (see Fig. 9.2). For example, in approximately 60% of the countries with data, the percentage of children in the poorest urban quintile with access to basic sanitation is lower than that of all rural children. The same is true for DTP3 immunisation coverage in approximately half the countries. For about 40% of countries, the birth registration coverage and primary education completion rates for the urban lowest wealth quintile are lower than children in the rural areas. Similarly, in 44% of the countries, the poorest urban children have a higher stunting prevalence than children in rural areas and their under-five mortality rate is 38% higher.

Even compared to the poorest quintile in rural areas, children in the poorest urban quintile fare worse in quite several countries (see Fig. 9.2). In almost 30% of the countries analysed, the poorest urban children have lower DTP3 immunisation coverage; in 20%, birth registration coverage is lower than for children in the poorest rural quintile. For under-five mortality rate, this share of countries is 24%—this amounts to roughly 4.3 million children among the urban poorest who face a higher risk of dying before their fifth birthday than their rural peers. Similarly, in 18% of countries, the poorest urban children are less likely to complete their primary education than the poorest children in rural areas. In both cases, these numbers refer only to the countries studied in this study—the actual global number will be much higher.

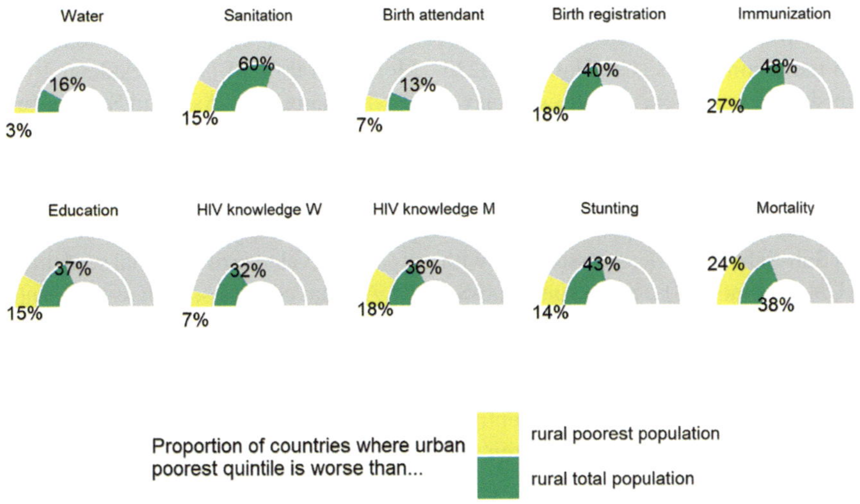

Fig. 9.2 Share of countries in which the poorest urban quintile fares worse than the rural population on average or worse than the poorest rural quintile (2011–2016). *Source* UNICEF (2018a, b)

A recent study of 99 low and middle-income countries estimated that of the 14,030,486 children under one year unvaccinated for DTP1, more than 28% live in urban and peri-urban areas (Wigley et al. 2022). However, they also found that the proportion of unvaccinated children under one living in urban areas was higher in countries with larger urban populations, such as India and China. This large population of unvaccinated children living in urban areas demonstrates that access to vaccines is still an urban issue.

This is the "urban paradox": although urban residents enjoy better access to services and opportunities on average, a substantial part of the urban population, including children, is left behind. It is crucial to understand the urban paradox and the extent to which the urban advantage masks it.

9.6 The Urban Paradox Impacts Impoverished Households the Most

The urban paradox is expected to have a greater impact on the most impoverished households, which typically reside in slums or informal settlements. Within cities, slums are marked by severe deprivation and exclusion, often located on the most hazardous urban land, fewer public services, and overcrowding. Slums are characterised by deprivations such as lack of access to improved water, lack of access to improved sanitation, lack of sufficient living area, lack of housing durability, and lack of security of tenure.

While there has been a reduction in the proportion of the world's population living in slums, from 28% in 2000 to 24% in 2018, there are still an estimated 1 billion people in these conditions (UN-Habitat 2022b). More than 90% of urban residents living in slums live in poor countries (Bolay 2020), the majority in Asia and Sub-Saharan Africa.

In Sub-Saharan Africa, a significant proportion of the urban population, specifically 56%, reside in settlements that are characterised as informal. This percentage is notably higher compared to other developing regions, suggesting that urbanisation in Sub-Saharan Africa is associated with a greater prevalence of informal settlements (UN-Habitat 2020a, b). In some sub-Saharan African countries, such as the Central African Republic and Sudan, over 90% of the urban population resides in slums and informal settlements.

The number of people living in urban slum households worldwide ranges passed 1 billion in 2018 and stands currently at almost 1.1 billion. It is estimated that up to half—or 500 million—are children under 18 (UN-Habitat 2022a). As urbanisation continues to increase rapidly in less developed countries, and with the projected increase in the share of the world's population living in cities, it is crucial to pay attention towards children living in slums.

9.7 Urban Slums: Examining Inequalities in Access to Education, Healthcare, and Basic Services

9.7.1 Access to Improved Water Sources and Sanitation Facilities

Inadequate water availability is a persistent problem in slums due to seasonal changes, intermittent supply, limited operating hours, and frequent malfunctioning of public water sources (Price et al. 2021). With limited space and access to basic amenities such as toilets and bathing facilities, slum dwellers often struggle with maintaining proper hygiene and sanitation, which can lead to the spread of disease (Debela et al. 2018; Otsuka et al. 2019). For example, a survey in a slum in Kampala found that only 16.5% of households had hand-washing facilities and 65.6% had functional sanitation facilities (Ssemugabo et al. 2021). The lack of clean water and sanitation facilities in slums also creates a faecally contaminated Environment, leading to diarrhoea-related mortality (Fry et al. 2002; Adane et al. 2017).

9.7.2 Health

Access to healthcare is a major concern for residents living in slums. Studies have shown that people living in slums face various barriers to accessing healthcare services, including financial constraints, perceived long wait times, inadequate facilities, and a shortage of healthcare providers (Mishra et al. 2019; Sudhinaraset et al. 2016). Regarding vaccination, it has been shown that zero-dose children are more likely to be living in slums than in formal urban areas (Mehmood et al. 2022; Utazi et al. 2022).

Numerous studies have highlighted the poor child survival rates in slum areas compared to non-slum areas. A study conducted in 45 low- and middle-income countries by Kyu et al. (2013) found that infant mortality rates were 4.6% in slum neighbourhoods and 3.0% in non-slum communities on average. The higher mortality rates have been associated with preventable causes such as poor neonatal care (Fry et al. 2002), neonatal tetanus (Awasthi and Pande 1998), diarrhoea, and respiratory infections (Vaid et al. 2007).

Generally, children living in urban slums and informal settlements fare poorly across nutritional indicators. A 2018 comprehensive analysis of nutrition indicators conducted by UNICEF (2018a) found that stunting prevalence among the poorest urban children was at least four times higher in one out of five countries than among children from the wealthiest quintile. The report further showed that in 44% of the countries, the most impoverished urban children had a higher stunting prevalence than children in rural areas (UNICEF 2018a).

As for maternal care, women in urban slums are less likely to access skilled birth attendants during childbirth and less likely to receive antenatal care than women in non-urban areas. Evidence from India shows that, in general, mothers living in urban slums are less likely to have assisted delivery from a skilled birth attendant than those from non-slum areas (Bayou et al. 2016).

a. Overcrowding

   Dwellings in slums are typically characterised by cramped living conditions, with multiple individuals often sharing small, inadequate living spaces, leading to overcrowding. Overcrowding in slums can have several negative consequences on the health and wellbeing of children. It can increase the spread of communicable diseases such as tuberculosis (Ogbudebe et al. 2015), measles (Huey et al. 2020), and COVID-19 (Malani et al. 2021) due to the proximity of people and the limited access to healthcare. Moreover, overcrowding can expose children to social issues such as crime, violence, and social exclusion due to a lack of space and resources (Makinde et al. 2016).

b. Structural quality/durability of dwellings

   Many informal settlements are located on steep slopes, flood plains, or other dangerous areas that may be particularly vulnerable to extreme weather and climate change (Revi et al. 2014). In addition, slums are often built with weak materials and close together, making them particularly susceptible to earthquakes and flooding (De Risi et al. 2013).

c. Education

   The availability of NGO schools is closely linked to enrolment rates in slums due to the lack of public schools (World Bank 2013). In 2014, the likelihood of children attending school in Dhaka's slums was two-and-a-half times lower than the national average (UNICEF 2014). While the gross enrolment rate in slums was 91%, it was lower than the national average of 101%, and the net enrolment rate was 62% compared to the national average of 77% (World Bank 2013). According to a 2019 study, children residing in urban slums are significantly represented in the out-of-school population in Bangladesh, with both gross and net attendance rates for children aged 6–10 being considerably lower than the national average (Quattri and Watkins 2019).

9.8 The Need for Increased Urban Programming

The extent of these relative wealth inequalities illustrates the dangers of generalising children's wellbeing in urban areas and the necessity to first understand the patterns of poverty and exclusion in cities, including the effects of population changes—in particular for Sub-Saharan Africa. Without the development of evidence-based programming specifically targeting children living in poverty, the gaps in inequity may widen, leaving more and more urban children behind.

Promoting equitable children's health, education, and the reduction of climate change vulnerabilities in cities—including slums—should now be an imperative across all regions. Integrated approaches to improve slum conditions (e.g., Corburn and Sverdlik 2017) recognise that improving slums is not only about improving physical infrastructure but also addressing social, economic, and environmental issues. This integrated approach to improving children's conditions is a vision that should be adopted not only for slum upgrading but also for overall urban development.

9.9 Cities in Sub-Saharan Africa Will Face Challenges as Consequence of the Combined Effects of Demographic Dynamics, Economic Development, and Climate Change

9.9.1 Population Growth

It is important to focus on Sub-Saharan Africa because it is experiencing some of the fastest rates of urbanisation in the world and its high poverty rates. Without addressing these challenges, the urban inequality will continue to widen, with serious consequences for the children's wellbeing and the development in the region.

Currently estimated at 1.2 billion people, the population of the region will reach 1.4 in 2030 and 2.1 in 2050 (UN DESA Population Division 2022), almost doubling in 38 years. Countries in Sub-Saharan Africa are expected to contribute more than half of the global population increase anticipated through 2050. The populations of the Democratic Republic of the Congo (DRC) and the United Republic of Tanzania are expected to have some of the highest growth rates, around 2–3% per year between now and 2050. By 2050, Nigeria will become the fourth most populous country (with 375 million people in the world), DRC the eighth (215 million people) and Ethiopia the ninth (with 213 million people).

9.9.2 Natural Growth and Youthful Population

The highest birth rates in the world drive the population growth of Sub-Saharan Africa. According to the United Nations Population Division (2022), the fertility rate in Sub-Saharan Africa was 4.7 births per woman in 2020, significantly higher than the global average of 2.3 births per woman. Higher fertility rates were found to be greater in low-income nations than in middle-income countries (Tesfa et al. 2023).

The high fertility rates in the region, as well as improvements in child survival rates over the past few decades, have resulted in a relatively young population compared to other regions. According to the United Nations, in 2020, approximately 60% of the region's population was under 25, and the median age was 19.4 years. According to the UN DESA Population Division (2023), the Sub-Saharan African population aged 0–19 years is expected to increase from 614 million in 2022 to 712 million in 2030, 817 million in 2040, and 902 million in 2050. This age group represented about 53% of the total population in 2022, making Sub-Saharan Africa the youngest region in the world.

9.10 Urban Trends

Africa, currently the least urbanised region (42%) but with the highest urban growth rate (in average 3.4% between 2020 and 2030), will become a primarily urban continent sometime around 2035 (UN-Habitat 2022b).

The rapid growth of the cities area in Sub-Saharan Africa is attributed to peri-urbanisation, mostly of un-serviced land in a mostly informal process primarily driven by low-income households (UN-Habitat 2020a, b).

Cities in Sub-Saharan Africa have the highest incidence of urban poverty globally, with about 23% of the urban population living below the international poverty line and 29% experiencing multidimensional poverty (UN-Habitat 2022b). Since 2016, South Sudan, Central African Republic, Democratic Republic of the Congo, and Eritrea have seen more than 20% of their urban population living in extreme poverty. These high rates are projected to remain so in the next decade.

Urbanisation in Africa and in Sub-Saharan Africa in particular is occurring at a lower national income level than it happened in other regions (UNICEF 2018a, b). At times, between 1970 and 2000, urbanisation happened even without any increase in GDP (UN DESA 2013). And urbanisation without shared economic growth will create conditions in which a substantial part of the population will not be able to harvest an urban advantage.

9.11 Secondary Cities

Much of the urban growth in Sub-Saharan Africa has occurred in secondary towns and cities. Secondary cities are urban areas that are smaller than primary cities (e.g., global cities or megacities) but are still important centres of economic, political, and social activity within a country or region. Secondary cities could be a subnational capital, transport hub, or large manufacturing centre or may have developed from a process of decentralisation of the main city (Roberts 2014). They may have populations ranging from hundred thousand to several million. By 2015, 46.9% of the urban population in Sub-Saharan Africa was living in towns and cities with less than 300,000 persons.

Despite evidence indicating that secondary cities and towns are the primary drivers of urban growth in Sub-Saharan Africa, urban development and governance interventions focus primarily on primary and megacities, assuming that development will trickle down to secondary towns and cities. Big cities in Sub-Saharan Africa receive the focus of investments in infrastructural development, which attracts businesses, skilled workers and educational institutions. This approach has increased physical and socioeconomic development inequalities between primary and secondary towns, leading to socio-spatial inequalities and multiple deprivations (Roberts 2014).

UNICEF and UN-Habitat (2020) observed the deprivations of health, water and sanitation, housing and income and employment in secondary towns in Sub-Saharan Africa. According to the data, primary cities have a higher average of piped water to homes and plots compared to secondary cities, leading to water scarcity in the latter. Compared to primary cities, secondary cities generally have inadequate toilet facilities, such as pit latrines without slabs and open defecation areas.

Multiple deprivations in secondary cities of Sub-Saharan Africa are related to substandard housing and living environments. According to UNICEF and UN-Habitat (2020), housing conditions in secondary cities are worse than those in primary cities. The contributing factors to the poor housing conditions in secondary cities include inadequate planning by the authorities, unclear land tenure, and insufficient financial resources of the urban authorities.

9.12 Rural-to-Urban Migration

Internal migration, including rural-to-urban migration, is a significant component of the urbanisation process in Sub-Saharan Africa. Despite being less visible than international migration, it is recognised as a core driver of the region's dynamics. DHS data from Sub-Saharan countries consistently show a net rural-to-urban migration trend (Gollin et al. 2021). In fact, internal migration can sometimes involve far more significant numbers of people than international migration and may have greater poverty reduction potential through remittances (UNDP 2009). Moreover, a literature review found that quasi-experimental studies suggest that migrating households have an equal or higher income after migration (Selod and Shilpi 2021).

Although rural–urban migration is not without challenges, studies have found that it can lead to better living conditions for migrants than households in rural areas. Gollin et al. (2021) argue that rural–urban migration in Sub-Saharan Africa is associated with improvements in income, access to services, and living conditions.

9.13 Urban Spread and Infrastructure

Rapid urban growth can strain existing infrastructure, resulting in inadequate access to basic services like water, sanitation, and healthcare. This is particularly true in cities of developing countries where urbanisation has been rapid and unplanned (Angel et al. 2016). Consequently, slums or informal settlements emerge with limited access to basic services. Therefore, infrastructure development must keep pace with urban growth to ensure that urban residents can access basic services.

9.14 Effect of Climate Change on Urban Settings

Climate change is also a factor to consider. The effects of climate change are not uniform across regions and cities, with the poorest neighbourhoods having lower resilience to their impact. Not only that, but climate change is also estimated to increase the number of people living in extreme poverty.

As temperatures rise due to climate change, cities will likely experience more frequent and intense heat waves, which can have serious health consequences for urban residents (Tong et al. 2021). Heat waves can lead to various health issues, from dehydration and heat exhaustion to more severe conditions like heatstroke, which can sometimes be fatal (WHO 2023). Vulnerable populations, such as children, are particularly at risk. High temperatures can also exacerbate air pollution, which can have negative health impacts, including respiratory problems (Piracha and Chaudhary 2022).

As a result of climate change, rising sea levels pose a major threat to coastal cities worldwide. This threat is especially acute for coastal cities in Sub-Saharan Africa, where several large urban areas such as Lagos, Abidjan, Luanda, Dar es Salaam, and Dakar are projected to undergo significant growth in the coming decades. However, it is not just the large cities that are at risk; small coastal towns across the continent are also expected to experience growth due to the high rates of population growth and urbanisation in the coastal zone (Neumann et al. 2015).

9.15 Refugees in Urban Settings

Although estimating the exact number of refugees living in urban areas can be challenging, according to UNHCR, approximately 60% of the world's refugees reside in urban areas globally (UNHCR 2022b). In addition, an estimated 48% of internally displaced people (IDP) were found to be living in urban areas based on available location data for 2021, although the actual number is likely to be higher (UNHCR 2022a).

Uganda, already recognised as one of the countries hosting the highest number of refugees, has a significant refugee population in urban areas such as Arua and Yumbe (UNHCR 2023). However, it has been reported that the overwhelming demand has hindered access to education for refugees in urban areas such as Arua (Easton-Calabria et al. 2022). Additionally, refugee children living in urban areas face language barriers and financial constraints that impede their access to quality education. The study also revealed that refugees in the area encounter challenges when trying to access the already burdened health system.

9.16 UNICEF's Approach and Priorities for Children in Cities

Given the current rapid urbanisation trend and the specific challenges faced by children in urban areas, UNICEF advocates developing programming approaches to address three interconnected drivers of change:

- Equity and inclusion (the core vision of change): Increasingly, the urban poor live in underserved and inaccessible areas, which further deepen their exclusion, marginalisation, and vulnerability to economic and environmental shocks. The greatest inequalities appear in cities in low- and middle-income countries, where the urban poor are concentrated in informal areas with limited access to basic services, employment, and infrastructure and have unequal social standing and influence on decision-making. Informal and excluded areas are often home to marginalised urban populations, migrants, refugees, and internally displaced people who face additional barriers to inclusion.
- Urbanisation, planning, and the environment (the spatial organisation of change): Rapid urban expansion is underway in low- and middle-income countries. The fastest rates of growth are in sub-Saharan Africa and South Asia. Most developing countries are unprepared for rapid urban expansion, particularly in small and medium-sized cities, where urban land expansion often outpaces the governance capacity of cities and towns. The lack of city-scale planning, regard for equity, inclusion, and the environment in market-driven urban projects or municipal investment in public space and safe mobility makes the provision of public goods and services more expensive and contributes to environmental degradation. These

conditions increase risks and hazards for children in urban areas—particularly the most vulnerable.

- Adaptive urban governance systems (the management and implementation of change): Within the complex system of a city, adaptive urban governance is a key driver and manager of urban change. Cities are characterised by governance structures that include national, regional, and municipal levels across sectors, often with overlapping jurisdictions. This complexity makes it more difficult to deliver assistance while offering more partnership opportunities. Capacities—technical, human, and financial—can vary widely between cities. Low or limited capacity can prevent urban governments from reaching every child, especially those most vulnerable to poverty and exclusion. With a big push for decentralisation in countries worldwide, local governments, including in urban contexts, are increasingly accountable for delivering services to children and families.

Based on an integrated analysis of the drivers of change and lessons learnt from the past, UNICEF identified the following priority areas of work for children in urban settings, including humanitarian and fragile contexts (see UNICEF 2018b, 2022; UN CEB 2019).

A. Evidence and Data

A critical first step towards reducing urban inequality is identifying the patterns of poverty. UNICEF will build on its strength and experience in data to strengthen the collection and analysis of intra-urban data for monitoring the situation of children in an urban setting and to highlight urban inequalities.

B. Local planning, financing, and budgeting

UNICEF's approach for subnational and local levels addresses urban complexity by adopting and adapting strategies for programming in different territorial contexts, such as smaller cities and towns and large cities. UNICEF supports the development of systems, capacities, and resources for evidence-based local plans and budgets that equitably respond to the priorities of children and their families, focussing on slums and informal settlements.

C. Community engagement

The voice and participation of the poor are often lacking in the urban decision-making process, with children's issues likely to be further excluded due to their absence from political processes. Even when children are included, their participation is often tokenistic and has little lasting impact. This becomes even more complex in slums and informal settlements as governments sometimes do not recognise them officially. UNICEF will work to establish effective and inclusive child participatory mechanisms in cities, with special attention to equal participation of girls and boys and children from disadvantaged groups, especially children living in slums and informal settlements.

D. Quality and equitable services for children

For middle-income countries, UNICEF often focuses on reducing urban marginalisation, reaching children and families who have been excluded and underserved and reducing barriers to access, or improving equity in spending and service quality across urban areas. However, for lower-income countries,

UNICEF will explore ways to introduce integrated models for coordinating quality services (including social protection) with local governments at the urban level. UNICEF will also explore best practices for managing and responding to crises/shocks in urban areas.

E. Child-responsive urban planning

Child-responsive national planning standards—UNICEF will work with partners to ensure that the standards set for land use and built space are key to ensuring children's wellbeing, such as housing, infrastructure for child-related services, safe and accessible public spaces, child and environmentally friendly spaces, transportation, and support interventions for road safety for children.

Slums and informal settlements should be consistently and systematically included in all urban plans and strategies. This includes adequate budget allocations that enable access to quality services for all children, adolescents and their caregivers, especially the most marginalised.

9.17 Importance of Addressing Inequities in Urban Inequality: A Call for Commitment to Equity and Social Justice

As this chapter shows, urbanisation can exacerbate inequality, particularly for children, by concentrating wealth and resources in certain areas while neglecting others. As cities grow and develop, services such as quality education, healthcare, and safe living environments become more concentrated in some neighbourhoods, usually those better connected and closer to the economic centre consisting of higher-income households, leaving children in low-income areas with limited access to these services. Thus, addressing urban inequality is crucial for ensuring that all children have access to the services and opportunities necessary for a healthy and stimulated development allowing them to reach their full potential.

Urban inequality concerning children can take many forms, including inequalities in education, healthcare or housing access. Due to this complexity, addressing urban inequality requires a comprehensive approach considering the unique circumstances and needs of different communities, especially the poorest (UNICEF 2012, 2018b). Solutions must be tailored to address the multidimensionality of inequality within each community and must be developed with the involvement of affected community members. This may require policy changes, community-led initiatives, and targeted investments in underserved areas.

While this chapter focuses on inequalities, it is crucial to emphasise the importance of addressing inequities. Inequity refers to unjust or unfair differences that result from systemic or structural factors, such as discrimination or historical disadvantage. Addressing urban inequality requires a commitment to equity and social justice and a willingness to challenge the underlying unjust and discriminatory factors that perpetuate inequalities, like those presented in this chapter. Further data on urban inequity for children is essential to improving their lives in urban areas and developing targeted interventions to address systemic barriers and ensure equity and justice for all.

References

Adams EA (2018) Intra-urban inequalities in water access among households in Malawi's informal settlements: toward pro-poor urban water policies in Africa. Environ Dev 26:34–42. https://doi.org/10.1016/j.envdev.2018.03.004

Adane M et al (2017) Sanitation facilities, hygienic conditions, and prevalence of acute diarrhea among under-five children in slums of Addis Ababa, Ethiopia: baseline survey of a longitudinal study. PLoS ONE 12(8):e0182783. https://doi.org/10.1371/journal.pone.0182783

Angel S et al (2016) Atlas of urban expansion, 2016 edn. Vol 1: Areas and densities. New York University, UN-Habitat, Lincoln Institute of Land Policy, New York, Nairobi, Cambridge

Awasthi S, Pande VK (1998) Cause-specific mortality in under fives in the urban slums of Lucknow, North India. J Trop Pediatr 44(6):358–361. https://doi.org/10.1093/tropej/44.6.358

Bashir S et al (2018) Facing forward: schooling for learning in Africa. World Bank, Washington. http://hdl.handle.net/10986/29377

Bayou YT et al (2016) The adequacy of antenatal care services among slum residents in Addis Ababa, Ethiopia. BMC Pregnancy Childbirth 16:142. https://doi.org/10.1186/s12884-016-0930-z

Black RE et al (2013) Maternal and child undernutrition and overweight in low-income and middle-income countries. Lancet 382(9890):427–451. https://doi.org/10.1016/s0140-6736(13)60937-x

Bolay JC (2020) Urban planning against poverty: how to think and do better in cities in the global south. Springer, Berlin. https://doi.org/10.1007/978-3-030-28419-0

Corburn J, Sverdlik A (2017) Slum upgrading and health equity. Int J Environ Res Public Health 14(4):342. https://doi.org/10.3390/ijerph14040342

Debela TH et al (2018) Fecal contamination of soil and water in sub-Saharan Africa cities: the case of Addis Ababa, Ethiopia. Ecohydrol Hydrobiol 18(2):225–230. https://doi.org/10.1016/j.ecohyd.2017.10.003

De Risi R et al (2013) Flood risk assessment for informal settlements. Nat Hazards 69:1003–1032. https://doi.org/10.1007/s11069-013-0749-0

Easton-Calabria E et al (2022) Urban refugees and IDPs in secondary cities: case studies of crisis migration, urbanisation, and governance. Cities Alliance/UNOPS, Brussels. https://www.citiesalliance.org/resources/publications/cities-alliance-knowledge/urban-refugees-and-idps-secondary-cities

Echazarra A, Radinger T (2019) Learning in rural schools: insights from PISA, TALIS and the literature. OECD education working paper no. 196. https://doi.org/10.1787/19939019

Ezeh A et al (2017) The history, geography, and sociology of slums and the health problems of people who live in slums. Lancet 389(10068):547–558. https://doi.org/10.1016/S0140-6736(16)31650-6

Fry S et al (2002) Health of children living in urban slums in Asia and the near east: review of existing literature and data. Environmental Health Project, U.S. Agency for International Development. https://pdf.usaid.gov/pdf_docs/PNACQ101.pdf

Gollin D et al (2021) Do urban wage premia reflect lower amenities? Evidence from Africa. J Urban Econ 121:103301. https://doi.org/10.1016/j.jue.2020.103301

Huey S et al (2020) Nutritional status and measles antibody titer in children living in urban slums of Mumbai. Curr Dev Nutr 4(2):1524. https://doi.org/10.1093/cdn/nzaa068_009

Joseph G et al (2018) Socioeconomic inequalities in access to skilled birth attendance among urban and rural women in low-income and middle-income countries. BMJ Glob Health 3:e000898. https://doi.org/10.1136/bmjgh-2018-000898

Kyu HH et al (2013) Association of urban slum residency with infant mortality and child stunting in low- and middle-income countries. BioMed Res Int 604974. https://doi.org/10.1155/2013/604974

Luschei TF, Chudgar A (2015) Evolution of policies on teacher deployment to disadvantaged areas. UNESCO, Paris. https://unesdoc.unesco.org/ark:/48223/pf0000232401

Makinde O et al (2016) Overcrowding as a risk factor for domestic violence and antisocial behaviour among adolescents in Ejigbo, Lagos, Nigeria. Glob Mental Health 3:E16. https://doi.org/10.1017/gmh.2016.10

Malani A et al (2021) Seroprevalence of SARS-CoV-2 in slums versus non-slums in Mumbai, India. Lancet Glob Health 9(2):e110–e111. https://doi.org/10.1016/S2214-109X(20)30467-8

Mehmood M et al (2022) Prevalence, geographical distribution and factors associated with pentavalent vaccine zero dose status among children in Sindh, Pakistan: analysis of data from the 2017 and 2018 birth cohorts enrolled in the provincial electronic immunisation registry. BMJ Open 12:e058985. https://doi.org/10.1136/bmjopen-2021-058985

Mishra K et al (2019) A study on the health seeking behavior among caregivers of under-five children in an urban slum of Bhubaneswar, Odisha. J Fam Med Prim Care 8(2):498–503. https://doi.org/10.4103/jfmpc.jfmpc_437_18

Neumann B et al (2015) Future coastal population growth and exposure to sea-level rise and coastal flooding—a global assessment. PLoS ONE 10(3):e0118571. https://doi.org/10.1371/journal.pone.0118571

Ogbudebe CL et al (2015) Reaching the underserved: active tuberculosis case finding in urban slums in southeastern Nigeria. Int J Mycobacteriol 4(1):18–24. https://doi.org/10.1016/j.ijmyco.2014.12.007

Otsuka Y et al (2019) Comprehensive assessment of handwashing and faecal contamination among elementary school children in an urban slum of Indonesia. Tropical Med Int Health 24(8):954–961. https://doi.org/10.1111/tmi.13279

Paes de Barros R et al (2008) Measuring inequality of opportunities for children. World Bank policy research working paper 4659. World Bank, Washington

Piracha A, Chaudhary MT (2022) Urban air pollution, urban heat island and human health: a review of the literature. Sustainability 14(15):9234. https://doi.org/10.3390/su14159234

Price HD et al (2021) Daily changes in household water access and quality in urban slums undermine global safe water monitoring programmes. Int J Hyg Environ Health 231:113632. https://doi.org/10.1016/j.ijheh.2020.113632

Quattri MA, Watkins K (2019) Child labour and education—a survey of slum settlements in Dhaka (Bangladesh). World Dev Perspect 13:50–66. https://doi.org/10.1016/j.wdp.2019.02.005

Quispe-Coica A, Pérez-Foguet A (2022) A new measure of hygiene inequality applied to urban-rural comparison. Int J Hyg Environ Health 239:113876. https://doi.org/10.1016/j.ijheh.2021.113876

Revi A et al (2014) Urban areas. In: Field CB et al (eds) Climate change 2014: impacts, adaptation, and vulnerability. Part A: Global and sectoral aspects. Contribution of working group II to the fifth assessment report of the intergovernmental panel on climate change. Cambridge University Press, Cambridge. https://www.ipcc.ch/report/ar5/wg2/

Roberts BH (2014) Managing systems of secondary cities. Cities Alliance/UNOPS, Brussels. https://www.citiesalliance.org/resources/publications/global-knowledge/managing-systems-secondary-cities

Satapathy BK (2014) Safe drinking water in slums: from water coverage to water quality. Econ Polit Wkly 49(24):50–55. https://www.jstor.org/stable/24480195

Save the Children (2015) State of the World's Mothers 2015: the urban disadvantage. https://resourcecentre.savethechildren.net/document/state-worlds-mothers-2015-urban-disadvantage/

Scheil-Adlung X (2015) Global evidence on inequities in rural health protection: new data on rural deficits in health coverage for 174 countries. ILO, extension of social security series no 47. https://www.ilo.org/secsoc/information-resources/publications-and-tools/Workingpapers/WCMS_383890/lang--en/index.htm

Selod H, Shilpi F (2021) Rural-urban migration in developing countries: lessons from the literature. Reg Sci Urban Econ 91:103713. https://doi.org/10.1016/j.regsciurbeco.2021.103713
Shutters ST et al (2022) Urbanisation favors high wage earners. Urban Sustain 2(6). https://doi.org/10.1038/s42949-022-00049-x
Ssemugabo C et al (2021) Characteristics of sanitation and hygiene facilities in a slum community in Kampala, Uganda. Int Health 13(1):13–21. https://doi.org/10.1093/inthealth/ihaa011
Sudhinaraset M et al (2016) Decision-making for delivery location and quality of care among slum-dwellers: a qualitative study in Uttar Pradesh, India. BMC Pregnancy Childbirth 16:148. https://doi.org/10.1186/s12884-016-0942-8
Tesfa D et al (2023) The pooled estimate of the total fertility rate in sub-Saharan Africa using recent (2010–2018) Demographic and Health Survey data. Front Public Health 10:1053302. https://doi.org/10.3389/fpubh.2022.1053302
Tong S et al (2021) Urban heat: an increasing threat to global health. BMJ 375:n2467. https://doi.org/10.1136/bmj.n2467
UN CEB (2019) UN system-wide strategy on sustainable urban development. CEB/2019/4/Add.4
UN DESA (2013) World economic and social survey 2013, sustainable development challenges
UN DESA (2015) Concepts of inequality. Development issues no. 1
UN DESA Population Division (2022) World population prospects 2022, online edn
UN DESA Population Division (2023) Data portal. Available at: https://population.un.org/dataportal. Accessed on 1/3/2023
UNDP (2009) Human development report 2009: overcoming barriers—human mobility and development. UNDP, New York
UNESCO (2021) Global education monitoring report 2021/2: non-state actors in education: who chooses? Who loses? UNESCO, Paris
UN-Habitat (2020a) World cities report 2020
UN-Habitat (2020b) UN-Habitat Sub-Saharan Africa atlas
UN-Habitat (2022a) Urban indicator database
UN-Habitat (2022b) World cities report 2022
UNHCR (2022a) Global trends: forced displacement in 2021
UNHCR (2022b) World cities day: cities lead the way in protecting forcibly displaced against impact of COVID-19. Available at https://www.unhcr.org/uk/news/press/2020/10/5f9c01704/world-cities-day-cities-lead-way-protecting-forcibly-displaced-against.html. Accessed on 1/3/2023
UNHCR (2023) Uganda comprehensive refugee response portal. Available at https://data.unhcr.org/en/country/uga. Accessed on 1/3/2023
UNICEF (2012) The state of the world's children 2012: children in an urban world. UNICEF, New York
UNICEF (2014) All children in school by 2015: global initiative on out-of-school children; South Asia regional study covering Bangladesh, India, Pakistan and Sri Lanka. UNICEF, Kathmandu
UNICEF (2018a) Advantage or paradox? The challenge for children and young people of growing up urban
UNICEF (2018b) Shaping urbanisation for children: a handbook on child-responsive urban planning
UNICEF (2019) The state of the world's children 2019: children, food and nutrition
UNICEF (2022) Strategic note on UNICEF's work for children in urban settings, 2nd edn
UNICEF, UN-Habitat (2020) Analysis of multiple deprivations in secondary cities in Sub-Saharan Africa
Utazi CE et al (2022) Assessing the characteristics of un- and under-vaccinated children in low- and middle-income countries: a multi-level cross-sectional study. PLOS Glob Publ Health 2(4):e0000244. https://doi.org/10.1371/journal.pgph.0000244
Vaid A et al (2007) Infant mortality in an urban slum. Indian J Pediatr 74(5):449–545. https://doi.org/10.1007/s12098-007-0075-1
WHO (2023) Heat and health. Available at: https://www.who.int/news-room/fact-sheets/detail/climate-change-heat-and-health

WHO, UNICEF (2021) Progress on household drinking water, sanitation, and hygiene 2000–2020: five years into the SDGs. WHO, UNICEF, Geneva

Wigley A et al (2022) Estimates of the number and distribution of zero-dose and under-immunised children across remote-rural, urban, and conflict-affected settings in low and middle-income countries. PLOS Glob Public Health 2(10):e0001126. https://doi.org/10.1371/journal.pgph.0001126

World Bank (2013) Bangladesh education sector review: seeding fertile ground—education that works for Bangladesh. World Bank, Dhaka. http://hdl.handle.net/10986/17853

Chapter 10
Children and Adolescents in Urban Settings: Inequality in Nigerian Cities

Enrique Delamonica, Alberto Minujin, and Placid Ugoagwu

Abstract The unequal development that characterizes the fast urbanization in Africa, and very especially in Nigeria, is having serious impact in their inhabitants. This is particularly relevant to children. Millions of them see the possibilities of survival and development of their potential as member of the society compromised. However, there is practically no reliable information to quantify the situation of urban children, the extent of their deprivations, and to assess the impact of urban inequality in their lives. The objective of this chapter is to analyze, from a variety of perspectives, the impact of urban deprivation and inequalities on children and adolescents in the two largest cities of Nigeria: Kano and Lagos. The objective is to present a model to analyze intra-urban disparities as well as to establish a baseline regarding the situation in the two largest metropolitan areas of Nigeria. After the introduction and a brief background section, the methodology used to assess the living conditions of children in metropolitan Kano and Lagos is introduced. In the following section, results showing the impact of urban inequalities on children are described for indicators reflecting outcomes covering the life cycle of children and adolescents. All of these analyses are based on the micro-data of the 2016 Multiple Indicator Cluster Survey. A brief concluding section closes the paper with challenges and limitations of the analysis and suggestions for further analysis.

Keywords Children · Adolescents · Adequate living conditions · Slums · Child outcomes

E. Delamonica (✉)
Division of Data, Analytics, Planning & Monitoring, UNICEF Headquarters, New York, USA
e-mail: edelamonica@unicef.org

A. Minujin
The New School and Executive Director Equity for Children, New York, USA
e-mail: minujina@newschool.edu

P. Ugoagwu
Jos University Teaching Hospital, Rikkos, Nigeria

G. H. Tonon (ed.), *Urban Inequalities*, The Urban Book Series,
https://doi.org/10.1007/978-3-031-59746-6_10

10.1 Introduction

The unequal development that characterizes the fast urbanization in Africa, and very specially Nigeria, is having serious impact in their inhabitants. This is particularly relevant to children. Millions of them see the possibilities of survival and development of their potential as members of the society compromised. However, there is practically no reliable information to quantify the situation of urban children, the extent of their deprivations, and to assess the impact of urban inequality in their lives.

It is quite common in economics and public health to use "inequality" to describe an observed quantitative difference among groups and use "inequity" for differences which are: (a) avoidable and (b) unfair (Evans et al. 2001; Vandemoortele 2012). However, working within sociological frameworks, Therborn (2006) makes conceptually the same distinction but calling them "difference" and "inequality," respectively, (and provides a categorization of various types of inequalities). The word "inequality" is purposefully used throughout the chapter in this latter, sociological meaning.

The objective of this chapter is to analyze, from a variety of perspectives, the impact of urban deprivation and inequalities on children and adolescents in the two largest cities of Nigeria: Kano and Lagos (located in the homonymous states) for two reasons. One is to provide the information that could be used as a baseline for further studies and updates of the intra-urban inequalities suffered by children in these two cities. And the second one is to show, via this example, a simple but powerful and practical way to assess intra-urban inequalities among children.

The purpose is to unpack the living conditions of children growing up in these cities and to analyze the difference between those who live in worse and better-off conditions. The underlying objective is to present evidence and set up standards that help to advocate for urban planning and policies that work on reducing inequalities.

Thus, concretely, in this paper, we attempt to answer the following questions, to show the model to analyze intra-urban inequalities as well as to establish a baseline regarding the situation in the two largest metropolitan areas of Nigeria:

- How many children live in unacceptable housing conditions, and how many in inadequate and in acceptable conditions?
- What is the situation of children in the different conditions?
- What are the gaps or disparities between children in unacceptable and in acceptable housing conditions regarding selected life cycle variables?

The paper is organized in the following way:

After this introduction, a brief background section focuses on why analyzing urban areas is quite relevant and urgent as well as some of the characteristics of cities in Nigeria. In the third section of the paper, the methodology used to assess the living conditions of children in metropolitan Kano and Lagos is presented. In this section, following and relying on the latest literature and quantitative analyses (UNICEF 2016; UN Habitat 2016), three groups of children are identified based on housing materials, overcrowding, and access to basic amenities. This approach allows us to understand where the poorest of the poor children live. These are the first-ever, city-specific disaggregated measurements of children's living conditions across a range of life cycle indicators following internationally accepted measurement tools and criteria in Africa. All of these analyses are based on the micro-data of the most recent Multiple Indicator Cluster Survey.

In the fourth section, results showing the impact of urban inequalities on children are presented. Understanding this impact is important to develop policies and programs to ensure equitable and universal access to basic social services, which would effectively reach the poorest and most marginalized children. In this section, indicators reflecting access to basic social services (e.g., birth registration or primary schools) and well-being outcomes (e.g., nutrition or violence) are analyzed. Moreover, the outcomes cover the life cycle of the child and adolescent (e.g., primary and secondary school attendance). Thus, they provide evidence about some of the different challenges (and opportunities) faced by the poorest adolescents, those in acceptable circumstances, and those from the better-off households. This evidence illustrates the unequal pathways determining transitions into adulthood for adolescents.

A brief concluding section closes the paper. Challenges and limitations of the analysis are presented along with suggestions for further analysis.

10.2 Background on Urban Inequalities

10.2.1 Global Trends

An identifying marker of the new century is the accelerated process of urbanization especially in the developing world. The UN estimates that by 2050, 66% of the world's population would be urban. The world already passed the threshold of the 50% of urban population that was the standard used to consider an urbanized world. It is important to stress that the fastest process of urbanization is happening in Africa and Asia (Fig. 10.1).

Urbanization provides opportunities for many and can even improve the possibility of achieving inclusion and better social and economic opportunities for all. Yet city dwellers can also face negative outcomes such as overcrowded living, housing issues, unemployment, loss of social and community networks, and other social problems such as crime and violence. If urbanization is not properly managed and planned, these trends can overburden urban systems. Cities and metropolitan areas would suffer long-term stresses and would be constrained when confronting multiple disruptive shocks.

Living conditions of the urban population are far from equitable. For instance, some recent studies show that Gini coefficient is usually higher in urban areas than

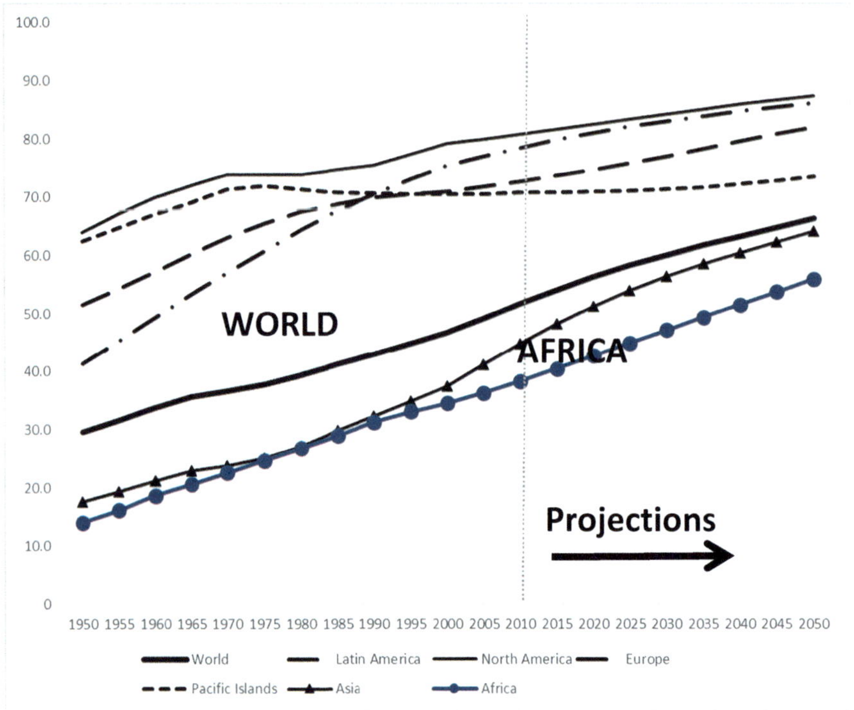

Fig. 10.1 Past and projected urbanization trends. *Source* Authors' calculations based on UN population division data

the countries as a whole (UN Habitat 2012). The available evidence seems to indicate that material deprivations, as well as income poverty, disproportionately affect children and adolescents in urban areas. Although it is accurate to state that children living in urban areas are, on average, better off than their rural peers, it is also true that millions of children in urban areas struggle with poverty and have no access to the "urban advantage" (Bartlett 2008; Montgomey 2009; Satterthwaite and Bartlett 2002). In fact, urban averages make invisible the daily struggles to survive of millions of families in urban settings (UNICEF 2019). This situation has clear negative implications for the effective realization of child rights and the possibilities children enjoy to develop their potential and contribute to society as full citizens.

Over the last few years, the number of studies focusing on urban inequalities has significantly increased. These studies often include issues related to children and comparisons between countries and among regions (UN Habitat 2011; UNICEF 2016). However, in spite of pioneering studies (e.g., Mugisha 2005, 2006), African urban children are rarely included or prominently portrayed in these analyses.

International studies on urban issues generally emphasize how the living conditions in slums especially related to infrastructure and access to services must be prioritized, as they constitute a determining factor in urban poverty, and have a direct impact on the well-being of individuals (e.g., UN Habitat 2003, 2011; Cohen 2011). Another topic frequently analyzed is related to urban segregation aiming to understand the links between society and urban space, or more precisely, determining the different ways in which people live depending on socio-economic status (e.g., de Pablos Ramírez and Arbucias 2010; Veiga and Rivoir 2001; Villaça 2011).

These approaches have highlighted the fact that differences in living conditions among children and adolescents are a result of several factors (among them, economic inequality, lack of public policies, indifference, discrimination, and lack of long and medium-term strategic urban planning). There are also valuable several studies about children in urban settings focusing on single dimensions or issues such as child labor or crime and violence (Rizzini 2009a, b). Some of these mostly qualitative studies provide a general overview of the situation under discussion but do not evaluate the factors behind intra-urban inequality.

In spite of the growing concerns related to child poverty and of the analyses on quality of life and well-being of children in developing countries, quantitative information is scarce. Although studies present information on the differences between urban and rural populations (Gordon et al. 2003), there is still much to learn about intra-urban inequalities.

The Sustainable Development Goals (SDG) recognize the critical importance of promoting equity. This is mentioned in an explicit way with the adoption of an SDG 10 on inequality and the fact that the 17 goals and 169 associated targets that the world's governments have committed to achieving are universal, linked by a pledge "that no one will be left behind … and we will endeavor to reach the furthest behind first" represent a significant step forward. Moreover, achieving most of the goals by 2030 depends heavily on what will happen in urban areas. The New Urban Agenda, adopted at the Habitat III Conference in Quito, Ecuador, and still guiding international efforts, also stresses the need to include equality in the center of the discussion.

Children and adolescents are not only one of most affected by urban inequalities but also, crucially, they are an important asset that societies have as agents of social change and peace. Therefore, following the recommendations provided in Equity for Children (2015), it is essential to incorporate the improvement of children's quality of life in the urban agenda. This process implies the need to recognize the particularities of growing up in that specific context, not as an isolated aspect, but as a crucial component for urban planning. Children's knowledge and experience as urban citizens need to be incorporated in the discussion; consequently, it implies to include them as key stakeholders.

10.2.2 Children in Urban Areas in Nigeria

An estimated 85 million Nigerians live in urban settlements—about half the total population and about double the 42.8 million estimated to live in urban areas in 2000 (World Bank 2014). Nigeria's urban population is still growing rapidly, at an average of 4.8% a year from 2000 to 2018. The urbanization level rose from 35 to

Table 10.1 Basic child indicators in Nigeria (2016)

	Nigeria	Urban	Rural	Girls	Boys
Birth registration	46.9	69.5	37.0	46.5	47.3
Stunting	22.8	10.3	23.2	18.9	20.6
Pre-primary school attendance	35.6	56.3	26.4	35.4	35.7
ITN	49.1	42.9	51.8	49.9	48.3
NAR Primary	60.9	80.0	52.5	59.2	62.6
NAR Secondary	46.9	66.4	37.7	46.2	47.4

Source Authors' calculations based on MICS 2016

47% during this period, higher than the average urban population share of 37% in Sub-Saharan Africa. Cities with a population of more than 300,000 doubled, from 21 to 42 (World Bank 2016). By 2020, urban population was expected to rise to 55% of the total (World Bank 2016).

It is estimated that 66% of the urban population live in slums (UN Habitat 2014). In Lagos, up to 50% of the population lives in informal settlements in some 200 different locations across the city (Gandy 2006).

However, the analysis of the situation of children in urban areas in Nigeria is hindered by the lack of data and evidence on the extent of urban poverty, intra-urban inequalities and urban-specific vulnerabilities (UNICEF 2016). Urbanization in Nigeria is not planned or managed. It is not surprising, then, that it is unequal, exclusionary, and fragmented. Thus, as it is described below, many children are deprived of access to the most basic services, which puts their future development and safety at stake.

The urban advantage is lost to many city dwellers, as urban areas in Nigeria tend to have low access to piped water, sewerage, fixed line telephones and electricity and cities typically lack drainage, flood protection, or proper vehicle access and roads, especially in slums and informal settlements (UN Habitat 2014). All this combined with a lack of barriers to access to health services heavily affects children's wellbeing. Admittedly, their situation is often better than in rural areas. Nevertheless, as it is seen in Table 10.1, there are still substantial constraints, in terms of malnutrition and secondary schooling.

10.3 Methodology

The living conditions are a crucial aspect that sets the framework, limitations, and opportunities for children's survival, development and social and economic realization. The living space and environment, the public services available, and the social and community relations are central aspects of everyday life of any member of the family and particularly relevant for children.

Based on the indicators included in Nigeria MICS 2016, we defined three dimensions to classify households: (1) Overcrowding; (2) quality of housing construction; and (3) characteristics of the household related to available amenities, services, and environmental conditions. For each of these dimensions, a few indicators are used. For each of these indicators, thresholds of no, moderate, and severe deprivation are used. Combining the presence of deprivation in each indicator, moderate and severe deprivation in each dimension is estimated. Combining the three dimensions households can be classified according to their overall housing conditions.

The indicators used in this research are similar to those used by many of the studies on multidimensional poverty, in particular child poverty (Gordon et al. 2003; Delamonica 2021). They are also aligned with the definitions used in the UN Habitat research on slums (UN Habitat 2002, 2022).

The data used for this study are from the 2016 of Multiple Indicator Cluster Survey Data for Nigeria. As the scope and the interest of this study is to explore the intra-urban inequality of children living in different housing conditions in Nigeria in the two largest urban cities in the country (Kano and Lagos), the data for all individuals of ages younger than 18 years were extracted from the House List (HL) dataset, children dataset, women dataset and household dataset in MICS 2016. As Kano and Lagos are populous and important states, they have been oversampled in the survey, which allows for further disaggregation than in the other states. This has allowed us to pick up just the metropolitan area of the cities of Kano and Lagos. We consider the housing conditions for all those households with at least one child 0–17 years old.

10.3.1 Overcrowding

Persons per room were used as the indicator of overcrowding. Following the recommendations from UN Habitat in the metadata for the SDG target on adequate, safe, and affordable housing (indicator 11.1.1), households with less than three persons per room were considered as not overcrowded, while households with three or four persons per room were considered as moderately deprived (i.e., moderately overcrowded). Households with five or more persons per room were classified as severely deprived or severely overcrowded (Table 10.2). Based on these criteria, in Kano, 22.8% of the child population lives in overcrowded households while 36.5% of the children live in overcrowded conditions in Lagos.

10.3.2 Housing Construction Quality

Housing quality was measured based on three indicators namely: main material for floor, roof, and wall. The elements to determine deprivation are listed, for each indicator, in Tables 10.3, 10.4, 10.5, and 10.6.

Table 10.2 Percentage of children in overcrowded households

	Kano	Lagos
Not overcrowded	40.4	26.2
Moderate overcrowded	36.8	37.3
Severely overcrowded	22.8	36.5
Total	100.00	100.0

Source Authors' calculations based on MICS 2016

Table 10.3 Elements to determine deprivation of floor material

Households were considered not deprived if the floor was primarily made of one of the following materials:	The following floor materials were used to classify a household as deprived:
Vinyl carpet	Earth/Sand
Ceramic tiles	Dung
Cement	Wood planks
Rug (wall to wall)	Palm/Bamboo
Vinyl tiles	
Parquet or polished wood	

Source Authors' classification of items in the MICS 2016 questionnaire

Table 10.4 Elements to determine deprivation of roof material

Households were considered as not deprived if the roof was made of:	The following roof materials were used to classify a household as deprived:
Metal	Thatch/Palm leaf/Sod
Tin	Rustic mat
Zinc	Palm/Bamboo
Iron sheets	Wood planks
Wood	Cardboard/Plastic sheeting
Calamine	No roof at all
Cement fiber	
Ceramic tiles	
Cement and roofing shingles	

Source Authors' classification of items in the MICS 2016 questionnaire

Households deprived in at least two of the three elements comprising housing quality (floor, roof, and wall material) were considered severely deprived. Households suffering "only" one of the deprivations were classified as moderately deprived. Households with no deprivation in any of the three elements were classified as not deprived. Thus, three out of four children in Kano and almost all the ones in Lagos live in non-deprived households in terms of the quality of constructions materials.

Table 10.5 Elements to determine deprivation of wall material

Households were considered as not deprived if the walls were made of:	The following wall materials were used to classify a household as deprived:
Cement	Cane/Palm/Trunks/Thatch
Stone with lime or cement	Dirt/Earth
Bricks, cement blocks	Bamboo with mud
Covered adobe	Stone with mud
Wood planks and shingles	Uncovered adobe/Mudbrick
	Plywood
	No walls

Source Authors' classification of items in the MICS 2016 questionnaire

Table 10.6 Percentage of children in households deprived in terms of housing construction quality

	Kano	Lagos
Basic indicators		
Deprivation in terms of roof	7.0	1.3
Floor	14.3	0.8
Wall	3.3	0.4
Aggregate summary of housing construction quality		
Not deprived	78.4	97.9
Moderate	18.8	1.7
Severe	2.8	0.4

Source Authors' calculations based on MICS 2016

10.3.3 Housing Characteristics

Four indicators were used to classify households according to their amenities and environmental conditions: water (main source of drinking water), time of roundtrip to fetch water (in minutes), sanitation (Toilet facilities), and type of fuel used for cooking. Tables 10.7, 10.8, and 10.9 illustrate the elements used to consider households as deprived (or not) in each of these indicators. In addition, if it takes more than 30 mins (roundtrip), the household is regarded as deprived.

Households deprived in two or three of these four elements were considered severely deprived (e.g., close to half the households in Kano), while households with "only" one deprivation were considered moderately deprived (e.g., about two-thirds of the households in Lagos). Households with no deprivations in any indicator were classified as not deprived. About a fifth and a third of the children in Kano and Lagos, respectively, were estimated to live in households not deprived in terms of these housing characteristics (Table 10.10).

Table 10.7 Elements to determine deprivation of safe drinking water

Households were classified as not deprived if water supply comes from:	The following sources of drinking water were used to classify a household as deprived:
Piped into dwelling	Unprotected well
Piped into compound, yard or plot	Protected spring
Piped to neighbor	Unprotected spring
Public tap/Standpipe	Rainwater collection
Tube well	Tanker-truck
Borehole and protected well	Cart with small tank/Drum
	Stream, dam, lake, pond, canal, irrigation channel
	Bottled water
	Sachet (pure) water

Source Authors' classification of items in the MICS 2016 questionnaire

Table 10.8 Elements to determine deprivation of sanitation

Households were classified as not deprived if sanitation consists of:	The following sources of sanitation were used to classify a household as deprived:
Flush to piped sewer system	Flush to somewhere else
Flush to septic tank	Flush to unknown place/Not sure/DK where
Flush to pit (latrine)	Pit latrine without slab/Open pit
Ventilated Improved Pit latrine (VIP)	Composting toilet
Pit latrine with slab	Bucket
	Hanging toilet
	Hanging latrine
	No facility, bush, field

Source Authors' classification of items in the MICS 2016 questionnaire

Table 10.9 Elements to determine deprivation in cooking fuel

Households were classified as not deprived if the main fuel for cooking was:	The following sources of cooking fuel were used to classify a household as deprived:
Electricity	Coal/Lignite
Liquefied petroleum gas (LPG)/Cylinder	Wood
Biogas	Straw/Shrubs/Grass
Kerosene	Animal dung
	Agricultural crop residue

Source Authors' classification of items in the MICS 2016 questionnaire

Table 10.10 Percentage of children in households deprived in terms housing characteristics

	Kano	Lagos
Basic indicators		
Water source	43.5	66.1
Water time	15.7	6.2
Sanitation	38.9	6.8
Cooking energy	40.1	1.1
Aggregate summary of housing characteristics		
Not deprived	22.0	29.7
Moderate	37.5	64.2
Severe	40.5	6.1

Source Authors' calculations based on MICS 2016

Table 10.11 Percentage of children living an acceptable, inadequate, and unacceptable housing condition

Overall housing condition	Kano	Lagos
Acceptable	18.2	17.2
Inadequate	26.6	55.3
Unacceptable	55.2	27.5
Total	100.0	100.0

Source Authors' calculations based on MICS 2016

10.3.4 Measurement of Overall Housing Condition

Once moderate and severe deprivation is determined in each of the three dimensions (overcrowding, construction quality, and housing characteristics), households can be classified in three groups. These groups consist of those households in acceptable living conditions, living in inadequate housing conditions, or suffering from unacceptable housing conditions.

Households without any deprivation in the three dimensions were classified as "acceptable" (Table 10.11). Households with any one deprivation (i.e., in anyone, but only one, of the three dimensions) were classified as "inadequate" while those suffering from two or more deprivations were considered as living in "unacceptable" conditions. Very few children live in acceptable housing conditions, less than one in five in both Kano and Lagos.[1] While most households can be considered to provide unacceptable housing conditions in Kano (55.2 of children live there), the situation in Lagos is marginally better with the majority of households suffering "only" moderate deprivations and, thus, providing inadequate housing conditions to their children (55.3 of the children live there).

[1] This is not surprising. As it was mentioned above it is estimated that two-thirds of the Lagos population live in slums.

Table 10.12 Percentage of children whose birth was registered in metropolitan Kano and Lagos according to housing conditions

	Nation-wide	State-wide	Urban areas (state)	Metropolitan areas (state)	Acceptable	Inadequate	Unacceptable
Kano	46.9	35.3	66.3	69.0	85.6	79.3	58.6
Lagos	46.9	82.3	83.0	84.5	100.0	89.7	83.3

Source Authors' calculations based on MICS 2016

10.4 Results

In this section, results showing the impact of urban inequalities on children are presented. Indicators related to the life cycle of the child (from birth registration or secondary school attendance) are used to analyze this impact. These indicators cover a range of basic social services and well-being outcomes. In all cases, inequalities across households in acceptable, inadequate, and unacceptable conditions are highlighted.

In terms of the percentage of children (aged five years old or younger) whose birth has been registered, it can be observed that the situation in urban Kano and Lagos is better than the national average (Table 10.12). Moreover, in Kano, children living in acceptable housing conditions are registered at birth at a rate 50% higher than children living in unacceptable conditions.[2]

A different situation attains in terms of the percentage of children (under five years of age) whose height is below the international norm (stunting[3]). Urban areas in Kano are just around the national average (Table 10.13). Children in households with unacceptable living conditions suffer higher levels of stunting than the rest of the children in Kano metropolitan area. However, state-wide, in Lagos, the percentage is about a tenth of the national average. Given the low rates (i.e., small number of cases), it is possible that the observed differences in children living in acceptable, inadequate, or unacceptable living conditions in metropolitan Lagos may not be statistically significant.

Although sleeping under mosquito nets is not widely popular in Nigeria (just about 50% nation-wide for children under 5), in Kano, the rate of utilization is relatively high (over 80%). The rate of utilization is even higher among children in urban areas and specifically among children living in acceptable housing conditions in metropolitan Kano. The gradient in metropolitan Lagos seems counterintuitive. One possibility is that given the better-off households may not "feel" the need for nets as they take comfort in their relatively well-conditioned house which may often include mosquito mesh in windows (Table 10.14).

[2] Unless otherwise indicated, when referring to significant differences, confidence intervals have been taken into account.

[3] i.e. their height is two standard deviations below the international norm for their age.

Table 10.13 Percentage of children suffering from stunting in metropolitan Kano and Lagos according to housing conditions

	Nation-wide	State-wide	Urban areas (state)	Metropolitan areas (state)	Acceptable	Inadequate	Unacceptable
Kano	22.8	22.6	22.0	18.9	16.5	16.3	20.9
Lagos	22.8	2.7	2.1	2.2	0.5	0.5	2.5

Source Authors' calculations based on MICS 2016

Table 10.14 Percentage of children using insecticide-treated mosquito nets (ITN) in metropolitan Kano and Lagos according to housing conditions

	Nation-wide	State-wide	Urban areas (state)	Metropolitan areas (state)	Acceptable	Inadequate	Unacceptable
Kano	49.1	82.7	84.5	87.2	96.00	84.42	86.9
Lagos	49.1	39.1	38.2	71.8	64.0	68.8	89.4

Source Authors' calculations based on MICS 2016

In Kano metropolitan area, the percentage of children under five years of age who are attending pre-primary school/early child development centers is about the same as the national average (Table 10.15). However, there are stark differences among children from households where living conditions are acceptable or unacceptable—the former more than doubling the latter. In Lagos Metropolitan area, the difference is not as high, given the high average attendance rate.

Children in both metropolitan Kano and metropolitan Lagos enjoy high levels of both primary and secondary schooling. In particular, children of primary school age in households providing acceptable living conditions have almost 100% attendance rates. While in Kano (State-wide), the average net attendance rate is below the national average, urban areas are higher than the national average, and in metropolitan Kano, the rate is almost 100%, with an intuitive gradient (Table 10.16).

In terms of secondary schooling, the gradient in Kano seems less intuitive. However, the differences are small and within the margin of error. This means no

Table 10.15 Percentage of children attending pre-primary school education/ECD in metropolitan Kano and Lagos according to housing conditions

	Nation-wide	State-wide	Urban areas (state)	Metropolitan areas (state)	Acceptable	Inadequate	Unacceptable
Kano	35.6	18.1	29.9	32.5	47.4	46.4	18.2
Lagos	35.6	86.5	86.5	85.6	100.0	94.1	86.3

Source Authors' calculations based on MICS 2016

Table 10.16 Percentage of children of primary school age attending primary school (net attendance rate) in metropolitan Kano and Lagos according to housing conditions

State	Nation-wide	State-wide	Urban areas (state)	Metropolitan areas (state)	Acceptable	Inadequate	Unacceptable
Kano	60.9	54.8	72.9	96.2	99.0	93.7	96.2
Lagos	60.9	95.7	95.9	99.4	100.0	98.7	99.4

Source Authors' calculations based on MICS 2016

major differences in net secondary school attendance for children in different households are observed. In metropolitan Lagos, the better-off children of secondary school age do seem to have an advantage over the other children. However, those living in inadequate or unacceptable housing conditions, seem to attend secondary schooling at the same rate (Table 10.17).

In Table 10.18, the results are summarized by simultaneously contrasting the relative and absolute gaps for each of the indicators analyzed along the life cycle of children and adolescents. Relative gaps are calculated as the ratio between the indicator for children in households with acceptable housing conditions and those in unacceptable housing conditions. Thus, a relative gap larger than one indicates the advantage of children in households with acceptable housing conditions compared to children in the other group. It is expressed as "how many more times" the outcome is likely among the better-off children. A relative gap of 1 shows parity between the two groups while a relative gap of 2.5 means children in households enjoying acceptable housing conditions are two and half times more likely to have a positive result than children in unacceptable housing conditions. The absolute gap is the difference between the same groups and is expressed in percentage points. Equity is indicated by values close to 0.

It can be observed that the differences (both in terms of absolute and relative gaps) in primary and secondary education in both metropolitan areas are relatively small. This indicates that universal education is close to a reality in those areas—as opposed to all urban areas or state-wide (which indicates the lagging situation in rural areas).

However, in "newer" services such as birth registration and early child development/pre-primary schooling, the gaps are wider. This is the case in both metropolitan areas.

Table 10.17 Percentage of children of secondary school age attending secondary school (net attendance rate) in metropolitan Kano and Lagos according to housing conditions

	Nation-wide	State-wide	Urban areas (state)	Metropolitan areas (state)	Acceptable	Inadequate	Unacceptable
Kano	46.9	42.3	67.0	88.7	88.8	87.1	89.5
Lagos	46.9	81.6	85.3	85.2	100.0	80.0	85.5

Source Authors' calculations based on MICS 2016

Table 10.18 Summary of absolute and relative intra-urban gaps

	Kano		Lagos	
	Absolute gap	Relative gap	Absolute gap	Relative gap
Birth registration	27.0	1.5	16.7	1.2
Stunting[a]	4.4	1.27	2.5	5.0
Insecticide-treated nets	9.1	1.1	– 1.4	0.8
ECD/Pre-primary school NAR	29.2	2.6	13.7	1.2
NAR Primary	2.8	1.0	0.6	1.0
NAR Secondary	–	1.0	14.5	1.2

Source Authors' calculations based on MICS 2016
[a]*Nota Bene* As stunting is a bad outcome the relative gaps show stunting is more widespread among children in unacceptable housing conditions

As stunting is a bad outcome, the ratio has been inversed to show positive results. In other words, the absolute gaps of 4.4 and 2.5 (for Kano and Lagos, respectively, please see above that for Lagos the difference may not be statistically significant) indicate the advantage of children in households enjoying acceptable housing conditions. The counterintuitive results for the percentage of children sleeping under insecticide-treated mosquito nets in Lagos have been addressed in the previous section.

10.5 Conclusions

Four conclusions can be offered, based on these results.

First, in terms of the baseline of the situation of urban children, only about one in five children live in acceptable conditions in Kano and Lagos. However, this similarity disappears when considering unacceptable conditions. Roughly twice as many children live in unacceptable conditions in Kano (55%) compared to Lagos (27%). Evidently, this situation is reversed for children living in inadequate conditions with baseline figures of 27% (Kano) and 55% (Lagos).

Secondly, there are clear and often very large differences in favor of children from households with acceptable housing conditions, both in Lagos and Kano. This is the case for most indicators along the cycle which were explored in this chapter (for instance birth registration and per-primary education). This is not a surprising result as it conforms to evidence from other similar studies in other regions.

Thirdly, for several indicators (in particular traditional schooling), the gaps are not as large as they could have been expected to be. This may be the result of the relatively widespread access to schooling in the metropolitan areas or the need to further disaggregation.

Fourthly, there seems to be a large difference between the metropolitan areas and other urban areas in the state. This is particularly striking in Kano State.

Besides these empirical results, the chapter intended to illustrate the application of a practical way to assess intra-urban inequalities. From this perspective, three further steps in the analysis could be envisioned for additional research. One of them could be to add more indicators. In particular, emphasis should be on indicators more specifically relevant for adolescents and not associated with the provision of services. These could include violence and HIV/AIDS knowledge. The second expansion of the analysis could involve refining the classification and disaggregation of households. It is possible that three groups are not sufficient and further disaggregation may be needed (as long as the sample size allows it). This would lead to sensitivity analysis (i.e., slightly altering thresholds of deprivation to assess if the classification of households is reliable and consistent), which is the third suggestion for future work.

References

Bartlett S (2008) Climate change and urban children: impacts and implications for adaptation in low-and middle-income countries. Environ Urbanization 20(2):501–519. http://pubs.iied.org/10556IIED.html

Cohen M (2011) Growth and recovery in a time of default: lessons from the role of the urban sector in Argentina (No. 2011/10). WIDER working paper. http://www.wider.unu.edu/publications/working-papers/2011/en_GB/wp2011-010/

de Pablos Ramírez JC, Arbucias JS (2010) Vida urbana: entre la desigualdad social y los espacios del habitar. Anduli Rev Andaluza Cien Soc 9:119–142

Delamonica E (2021) Child rights, poverty, and well-being: measurement debates and empirical advances. Popul Rev 60(1):141–170

Equity for Children (2015) Approaches to equity. http://equityforchildren.org/2017/07/approaches-to-equity-executive-summary

Evans T, Whitehead M, Bhuiya A, Diderichsen F, Wirth M (2001) Introduction. In: Evans T, Whitehead M, Bhuiya A, Diderichsen F, Wirth M (eds) Challenging inequities in health: from ethics to action. Oxford University Press

Gandy M (2006) Planning, anti-planning, and the infrastructure crisis facing metropolitan Lagos. In: Murray M, Myers G (eds) Cities in contemporary Africa. Palgrave Macmillan, New York, pp 247–264

Gordon D, Nandy S, Pantazis C, Townsend P, Pemberton SA (2003) Child poverty in the developing world. Policy Press

Montgomey M (2009) Urban poverty and health in developing countries. Popul Bull 64(2):24–38

Mugisha F (2005) Urban-rural differentials in child labor: how effective is the MICS in informing policies that promote the human rights of children in urban areas? An example of Kenya. In: Minujin A, Delamonica E, Komarecki M (2005) Human rights and social policies for children and women: the multiple indicator cluster survey (MICS) in practice. New School University, New York

Mugisha F (2006) School enrollment among urban non-slum, slum and rural children in Kenya: is the urban advantage eroding? Int J Educ Dev 26(5):471–482

Rizzini I (2009a) População childil e juvenil: direitos humanos, pobreza e desigualdades. In: Direitos humanos e questão social na América Latina. Gramma, Rio de Janeiro, pp 81–88

Rizzini I (2009b) População Childil e Juvenil: Direitos Humanos, Pobreza e Desigualdades. In: Freire SDM (ed) Direitos humanos e questão social na América Latina. Gramma, Rio de Janeiro

Satterthwaite D, Bartlett S (2002) Poverty and exclusion among urban children. Innocenti Digest, Florence
Therborn G (2006) Meaning, mechanisms, patterns, and forces: an introduction. In: Therborn G (ed) Inequalities of the world. Verso, London, pp 1–58
UN Habitat (2002) Cities without slums, first session, Nairobi
UN Habitat (2003) Global report on human settlements 2003: the challenge of slums. UN Habitat, Nairobi
UN Habitat (2011) State of the world's cities 2010/2011. Bridging the urban divide. UN Habitat, Nairobi
UN Habitat (2012) Estado de las ciudades de América Latina y el Caribe 2012. Rumbo a una nueva transición urbana. UN Habitat, Río de Janeiro
UN Habitat (2014) State of African Cities 2014: Re-imagining sustainable urban transitions. UN Habitat, Nairobi
UN Habitat (2016) World Cities report: urbanization and development
UN Habitat (2022) Global urban monitoring framework a guide for urban monitoring of SDGs and NUA and other urban-related thematic or local, national and global frameworks. UN Habitat, Nairobi
UNICEF (2016) Global urban strategic framework: mapping annex 1. Country reports
UNICEF (2019) Advantage or paradox? The challenge for children and young people of growing up urban. UNICEF, New York
Vandemoortele J (2012) Equity begins with children. In: Minujin A, Nandy S (eds) Global child poverty and well-being: measurement, concepts, policy and action. Policy Press, Bristol
Veiga D, Rivoir AL (2001) Desigualidades sociales y segregación en Montevideo. Universidad de la República, Facultad de Ciencias Sociales, Departamento de Sociología, Universidad de la República, Montevideo
Villaça F (2011) São Paulo: segregação urbana e desigualdade. In: Estudos avançados 25(71), San Pablo
World Bank (2014) World development indicators. World Bank, Washington, DC
World Bank (2016) From oil to cities: Nigeria's next transformation. Directions in development. World Bank, Washington, DC. https://doi.org/10.1596/978-1-4648-0792-3

Chapter 11
Unequal Childhoods, Socio-economic Status, and Children's Subjective and Psychological Well-Being in South Africa: A Population-Based Study

Shazly Savahl and Sabirah Adams

Abstract Social and economic inequality and inequity are key social concerns in contemporary societies. Empirical research has consistently demonstrated that children are disproportionately encumbered by these systemic challenges. Despite international policies, such as the United Nations (UN) 2030 Agenda on Sustainable Development and the UN Convention on the Rights of the Child, continental policy such as Africa's Agenda for Children 2040, and local policy such as the National Plan of Action for Children (2019–2024), children living in various geographical regions in South Africa are faced with numerous inequalities in their daily lives. Despite the South African government has commenced with several initiatives to lighten the burden of social inequality and deprivation for children and society at large, fundamental factors in relation to poverty, access to primary healthcare services, safety, education, and demarcated safe natural spaces for children still plague the majority of children.

Keywords Inequality · Children · Well-being · Local policy · Poverty · South Africa

11.1 Background

Social and economic inequality and inequity are key social concerns in contemporary societies. Empirical research has consistently demonstrated that children are disproportionately encumbered by these systemic challenges. Despite international policies, such as the United Nations (UN) 2030 Agenda on Sustainable Development

S. Savahl (✉)
Department of Psychology, University of the Western Cape, Cape Town, South Africa
e-mail: ssavahl@uwc.ac.za

S. Adams
Centre for Higher Education Development, University of Cape Town, Cape Town, South Africa

G. H. Tonon (ed.), *Urban Inequalities*, The Urban Book Series,
https://doi.org/10.1007/978-3-031-59746-6_11

and the UN Convention on the Rights of the Child (UNCRC), continental policy such as Africa's Agenda for Children 2040, and local policy such as the National Plan of Action for Children (2019–2024), children living in various geographical regions in South Africa are faced with numerous inequalities in their daily lives. Regardless of the South African government implementing various efforts to alleviate socio-economic inequality, poverty, and deprivation for children and society as a whole, core issues regarding poverty, access to primary healthcare services, safety, education, and designated safe areas continue to affect the large percentage of children (Adams and Savahl 2022).

Within the framework of emerging economies, socio-economic status (SES) has become a pivotal determinant. SES is a multifaceted construct that encompasses individuals' economic resources, social standing, and access to opportunities within a given society (Adler and Stewart 2007). It reflects the relative privilege or disadvantage experienced by individuals or groups within a society; it influences access to resources, opportunities, and social determinants of health. The conceptualisation of SES is located in social stratification theory, which suggests that societies are structured hierarchically based on differential access to resources and power (Marmot and Wilkinson 2006). Despite substantial SES variations among countries, it stands as a prominent metric for measuring the well-being of children (Bradley and Corwyn 2002). Coleman (1988) introduced a comprehensive framework for understanding socio-economic status (SES), delineating it into three components. First, financial capital refers to material resources and encompasses the economic aspects of SES. This component includes aspects such as income, wealth, and access to financial resources, reflecting the economic well-being of individuals and families.

Second, human capital encompasses non-material resources, with education being a central component. Human capital represents the skills, knowledge, and capabilities possessed by individuals, which contribute significantly to their socio-economic outcomes. Education, in particular, is a key determinant of human capital, influencing individuals' earning potential, occupational attainment, and social mobility.

Finally, social capital represents resources acquired through social relationships and networks. This component encapsulates the social connections, norms, and trust within communities, which facilitate access to opportunities, support systems, and collective resources. Social capital plays a crucial role in shaping individuals' access to social support, job opportunities, and other resources critical for socio-economic advancement (Krieger et al. 1997).

Empirical evidence indicates a substantial correlation between SES and diverse facets of well-being among children and young people, encompassing public health outcomes, educational achievements, psycho-social development, and resource accessibility (Bradley and Corwyn 2002). Socio-economic factors such as household income levels, and recent income declines have been associated with reduced levels of subjective well-being (SWB) among children (Bradshaw and Richardson 2009; Bradshaw et al. 2011; Rees et al. 2011). Studies consistently indicate that SES correlates with children's SWB, albeit the nature and intensity of this correlation may differ based on distinct socio-economic indicators and cultural settings (Bradshaw et al. 2013). Children from higher socio-economic backgrounds commonly exhibit elevated levels of life satisfaction and positive affect relative to those from lower socio-economic backgrounds (Gilman and Huebner 2003). Moreover, higher SES tends to coincide with reduced levels of negative affect and emotional distress among children (Rees and Main 2015).

An essential measure of social inequality within a country is the Gini coefficient, which serves as the international standard for gauging the distribution of income and wealth. It is expressed as a ratio ranging from 0 to 1, with "0" signifying equal income distribution among all individuals, and "1" indicating that all income is concentrated in a single individual.

Despite the foundational principle of equality enshrined in the South African Constitution, inequality persists. The latest World Bank Report (2022) has again ranked South Africa as the most unequal country in the world; "race" was identified as a determining factor as 10% of the population own 80% of the country's wealth. This can be traced to the history of social oppression owing to colonisation, and subsequently apartheid. This is evident in intergenerational poverty and inequality, as among the most pronounced globally specifically as it relates to wealth and income. South Africa illustrates a significant disparity between the privileged and disadvantaged in terms of both wealth and income distribution. However, we note that "the burdens of these multiple overlapping layers of inequality are often endured by children who necessitate care and supervision from adults for both safety and basic tenets of their well-being" (Hall et al. 2012, p. 24).

A comprehensive exploration of the impact of socio-economic status (SES) and social disparities on well-being necessitates recognition of the oppressive historical context of apartheid. Among apartheid's most profound legacies is the pervasive social inequality experienced across diverse segments of the population. This has precipitated the division of communities and neighbourhoods into either privileged, high-SES enclaves or disadvantaged, low-SES areas.

High-status communities are characterised by substantial wealth, advanced educational accomplishments, notable employment rates, and low levels of violence. Conversely, disadvantaged neighbourhoods exhibit lower educational achievements and income and increased instances of substance abuse, unemployment, crime, and violence. Amidst these extremes, a growing demographic is marked by a moderate socio-economic status (SES). This segment largely comprises individuals who have experienced upward mobility since the onset of democracy in 1994.

Frequently denoted as the burgeoning middle class, this group encompasses formerly marginalised individuals who, through education, perseverance, and other favourable psychological attributes, have achieved a significant degree of financial stability.

Regrettably, this ongoing state of underdevelopment has endured, marked by the transgenerational transmission of trauma, enduring poverty, and high levels of inequity (Savahl et al. 2019). A report released in 2011 jointly by UNICEF and the Human Rights Commission delineated racial disparities within South Africa. The findings revealed that compared to a white child, a black child exhibits the following disparities: one and a half times less probability of accessing early child development programmes; twice as unlikely to access sufficient sanitation and water; twice as improbable to finish secondary education; twelve times more prone to hunger; and eighteen times more susceptible to residing in poverty.

In 1994, the newly formed South African government implemented a series of legislations aimed at improving the conditions for children (Savahl et al. 2015). However, despite these legislative advancements, the actual living conditions premised on the objective indicators of health and well-being of children are challenging.

As a result, there has been a shift towards researching children's subjective perspectives, experiences, and evaluations of their well-being. The term "subjective" implies relying on self-reported reflections and evaluations. To grasp the subjective elements effectively, it's essential to directly interact with children, augmenting objective indicators across diverse scenarios, environments, and cultural paradigms.

SWB encompasses a multifaceted and expansive spectrum of experiences, perceptions, reflections, and evaluations individuals attribute to their lives (Diener 1984, 2000). This concept comprises cognitive aspects, such as overall and domain-specific life satisfaction, as well as affective dimensions, including positive (PA) and negative affect (NA) (Diener et al. 1999). In the context of children's SWB, our focus lies in understanding their evaluations of their overall life satisfaction, satisfaction with various life domains, and emotional experiences (Savahl et al. 2021). Eudaimonic well-being, often conceptualised as psychological well-being, delves into the degree to which individuals are fully engaged in society. It encompasses elements such as life purpose, autonomy, meaning in life, and self-realisation (Nahkur and Casas 2021; Rees et al. 2020; Ryff and Singer 2008).

11.1.1 The Current Study

South Africa's urban population has steadily increased from 2010 (62.22%) to 2020 (67.35%). Given that the large percentage of children live in urban areas, it is important to understand the interplay between urbanisation and inequities and its impact on children's well-being and development. Key to this understanding is the role of the SES of the child's family, which can have a significant impact on their development and well-being. In the current chapter, we use data that includes a national sample of children, to measure children's SWB and PWB across three levels of SES. Our research is driven by the overarching objective of investigating children's SWB, encompassing cognitive life satisfaction, domain-specific cognitive life satisfaction, and positive and negative affect, as well as PWB (eudaimonic well-being), across three socio-economic status (SES) levels (lower, lower-middle, and middle). Our study will conclude by offering recommendations for policy aimed at addressing disparities and enhancing children's overall quality of life.

11.2 Method

11.2.1 Research Design

The study used secondary data sourced from the third wave of the South African Children's Worlds Study (see www.isciweb.org). The Children's Worlds Study is a global survey currently in its fourth iteration. The survey aims to assess children's subjective perceptions and evaluations of their lives and well-being and specific factors that influence their well-being, across different contexts and countries (Rees et al. 2020). The third wave of the study in South Africa employed a cross-sectional survey design and used a nationally representative stratified random sample of children aged 10 and 12 years old.

11.2.2 The Dataset: Sampling and Participants

The South African Children's Worlds study was conducted across the nine provincial regions in South Africa. The sampling frame was school-based and included children registered at primary schools across all nine provinces from two age groups (10 and 12 years old). In South Africa, children in these age groups are generally in Grades 4 and 6. The stratification for the larger study was based on Grade (4 or 6), geographical location (urban or rural), and provincial region. For this study, we limit our analysis to the children in the urban locations. For the sampling protocol, we used a 99% confidence interval and 1.5% margin of error. The final cleaned dataset consisted of 3614 participants (boys = 45.3%, girls = 54.7%) between the ages of 10 to 12 years (Mage = 11.08, $SD = 0.997$).

11.2.3 Instrumentation

The instruments comprised several standardised scales. For each wave of the survey, countries adapted and translated the instrument to the specific country context using the standards as outlined by the International Test Commission (2017). In South Africa, this process comprised the backward translation method, cognitive testing with children from the local contexts, and the consideration of an expert translation committee. We translated the questionnaires into seven languages (English, Afrikaans, isiZulu, isiXhosa, Sesotho, Setswana, Tshivenda, and Sitsonga. The scales used in the current study are presented below:

11.2.3.1 The Children's Worlds Subjective Well-Being Scale (CW-SWBS)

We developed the CW-SWBS from Huebner's Students' Life Satisfaction Scale (Huebner 1991). This amended scale consisted of six items, three taken from the original SLSS, and three new items distilled from the qualitative research. The scale response options were on an 11-point 0–10-point unipolar agreement scale with end-labelled verbal anchors of "Not all agree" (0) to "Totally agree" (10). The CW-SWBS has shown acceptable internal consistency (0.93) and an appropriate fit structure (Savahl et al. 2021).

11.2.3.2 The Children's Worlds Domain-Based Subjective Well-Being Scale

The Children's Worlds: Domain-Based Subjective Well-Being Scale (CW-DBSWBS) is based on the Personal Well-Being Index—School Children (Cummins and Lau 2005) and the Brief Multidimensional Students' Life Satisfaction Scale (Seligson et al. 2003). The scale comprises 11 items that assess children's domain-based cognitive SWB, with each item representing a particular life domain. The scale comprised a mix of concretely worded and abstractly worded items. Response options were on an 11-point unipolar satisfaction scale ranging from "Not at all" (0) to "Totally/Completely satisfied" (10). The CW-DBSWBS has shown acceptable internal consistency (0.87) and an appropriate fit structure (Savahl et al. 2021).

11.2.3.3 The Children's Worlds Positive and Negative Affect Scale

The Children's Worlds Positive and Negative Affect Scale (CW-PANAS) measures state-level affective well-being. We constructed the scale using six items drawn from Russell's Model of Core Affect. The scale comprised six items assessing three positive (happy, calm, and full of energy) and three negative (sad, stressed, and bored) affective states. For both PA and NA, these items reflect a pleasant-unpleasant, activated-deactivated, and neutral affect (see Rees 2019). Response options were on an 11-point unipolar intensity scale ranging from "Not at all" (0) to "Extremely" (10). Savahl et al. (2021) report internal consistency scores of 0.60 for the PA sub-scale, and 0.68 for the NA sub-scale.

11.2.3.4 The Children's Worlds: Psychological Well-Being Scale

The Children's Worlds-Psychological Well-being Scale (CW-PWBS) was constructed from the psychological well-being scales of Ryff (1989, 1995), Ryan and Deci (2000), and Keyes (2006) and adapted for use with children (see Rees et al.

2020). The scale comprises six items that reflect each of the dimensions of psychological well-being (self-acceptance, positive relationships, environmental mastery, personal growth, autonomy, and purpose/meaning in life). Response options were on an end-labelled 11-point unipolar agreement scale ranging from "Not at all agree" (0) to "Totally agree" (10). Rees et al. (2020) report an internal score of 0.86 for the CW-PWBS.

11.2.3.5 Single Item on Overall Life Satisfaction

We included an item assessing Overall Life Satisfaction (OLS) (Cummins and Lau 2005) on an end-labelled 0–10 scale using the following wording: "How 'satisfied' are you with your life as a whole?".

11.2.4 Data Analysis

A central co-ordinating committee managed the data, oversaw the quality, and conducted the cleaning and depuration of the final dataset. This process included the identification and exclusion of cases with a high proportion of missing data and the exclusion of cases with systematic response patterns. We weighted the final dataset based on the proportion of children per province in South Africa. We conducted the data analysis in two phases. In the first phase, we used confirmatory factor analysis (CFA), in AMOS 28, to test the fit structure and viability of the scales used to measure the constructs. The results of the CFA are reported in Savahl et al. (2023). Here we used a "separate components" approach as our goal was not to contribute to the discussion on the viability of various structural configurations. We used the full information maximum likelihood technique to manage missing data. To address non-normality of data, we used the maximum likelihood estimation method with bootstrap coefficients (500 samples). Following recommendations by Kline (2011), we used the Comparative Fit Index (CFI), Root Mean Squared Error of Approximation (RMSEA), and Standardised Root Mean Residual (SRMR) as fit indices to determine model fit. Scores higher than 0.950 for the CFI and scores below 0.05 for the RMSEA and SRMR were accepted as indicators of a good fit (Casas 2017).

In this phase, we also conducted multigroup confirmatory factor analysis (MGCFA) to test the measurement invariance of the various latent constructs across the nine provinces, age, gender, geographical context, and SES. Measurement invariance is a pre-requisite for meaningful group comparisons (Millsap and Olivera-Aguilar 2012) and is the most widely used technique to test measurement invariance across groups (Steenkamp and Baumgartner 1998). The MGCFA is presented in this chapter. It comprised three sequential steps, characterised by the application of incremental constraints, wherein we assessed configural, metric, and scalar measurement invariance. We accepted the tenability of each subsequent constrained model if it met the criteria of reduced fit of < 0.01 on the CFI (Cheung and Rensvold 2002), < 0.015 on the SRMR and RMSEA (Chen 2007).

In the second phase, we generated descriptive statistics (mean scores and standard deviations for each of the SWB subscales and the psychological well-being scale) using the Stata 14 (Stata Corporation LLC) programme, taking into consideration complex survey design effects such as sample weighting and clustering. We further conducted a comparative means analysis and presented the statistics across the three levels of SES.

11.2.5 Procedure and Ethics

We obtained ethics clearance for the study from the Institutional Review Boards of the participating universities, and the nine provincial Education Departments. We held an information session with the participants from the selected schools, where we discussed the aims and objectives of the study and the nature of their involvement. During this session, we also advised the participants on their rights and the ethics principles of informed consent, confidentiality, the right to withdraw, privacy, and the use of the data. We followed an active consent process, wherein consent was requested from parents/guardians and children.

11.3 Results

11.3.1 Confirmatory Factor Analysis

The reliability coefficients, skewness, and kurtosis scores are provided in Savahl et al. (2023). The results of the confirmatory factor analysis indicated satisfactory fit for all the measurement scales (see Table 11.1 for fit statistics), with each item demonstrating significant loadings on latent constructs and meeting the predetermined threshold criteria for the fit indices. Additionally, all factor loadings were statistically significant and adequately loaded onto the latent constructs.

11.3.2 Measurement Invariance Testing Across SES Groups

We evaluated the measurement invariance of various scales across the three SES groups using multigroup confirmatory factor analysis. Our analysis supported scalar invariance, indicating that each constrained model demonstrated a reduction in fit of less than 0.01 on the Comparative Fit Index (CFI) (Cheung and Rensvold 2002), and

Table 11.1 Fit indices

Model	X^2	*df*	Sig	CFI	RMSEA	SRMR
1. Cognitive SWB	12.732	5	0.026	0.999	0.015 (0.005–0.025)	0.0059
2. Cognitive SWB configural SES	52.144	15	0.00	0.997	019 (0.014–0.025)	0.0073
3. Cognitive SWB metric SES	94.092	23	0.00	0.995	021 (0.017–0.026)	0.0109
4. Cognitive SWB scalar SES	137.580	31	0.00	0.992	021 (0.014–0.028)	0.0113
5. Cognitive domain SWB	618.192	44	0.000	0.959	0.044 (0.041–0.047)	0.0279
6. Cognitive domain SWB configural SES	835.637	132	0.000	0.951	028 (0.026–0.030)	0.0311
7. Cognitive domain SWB metric SES	955.996	152	0.000	0.945	028 (0.026–0.029)	0.0334
8. Cognitive domain SWB scalar SES	1200.719	172	0.000	0.939	029 (0.028–0.031)	0.0332
9. Positive/negative affect	242.971	8	0.000	0.941	0.060 (0.058–0.073)	0.0440
10. Positive/negative affect configural SES	240.031	24	0.000	0.936	040 (0.035–0.044)	0.0472
11. Positive/negative affect metric SES	253.416	32	0.000	0.935	035 (0.031–0.039)	0.0473
12. Positive/negative affect scalar SES	322.929	40	0.000	0.926	035 (0.032–0.039)	0.0473
13. Psychological well-being	22.863	9	0.007	0.998	0.015 (0.007–0.023)	0.0085
14. Psychological well-being configural SES	70.455	27	0.000	0.994	015 (0.011–0.020)	0.0114
15. Psychological well-being metric SES	108.492	37	0.000	0.990	017 (0.013–0.021)	0.0148
16. Psychological well-being scalar SES	128.618	47	0.000	0.989	016 (0.013–0.019)	0.0147

less than 0.015 on the SRMR and RMSEA (Chen 2007). These findings validate the comparability of the scores on the measurement scales across the three levels of SES in terms of correlations, regressions, and mean scores.

Table 11.2 Mean scores

	Low SES		Lower-middle SES		Middle SES	
	M	SD	M	SD	M	SD
Cognitive LS	82.12	1.76	87.72	1.78	88.18	1.98
Domain LS	80.23	1.51	85.81	1.39	86.42	1.57
PA	80.10	2.10	82.47	1.90	87.36	1.89
NA	46.68	2.55	44.67	2.96	40.14	2.10
PWB	80.65	1.86	85.05	1.60	88.25	1.77
OLS	80.80	2.31	88.49	2.27	89.10	1.99

* Mean scores on a 100-point scale

11.3.3 Means Analysis

Mean scores and standard deviations (adjusted to a 100-point scale) are presented in Table 11.2. Across the three SES groups, scores for the two cognitive SWB scales and positive affect exceeded 80 on the transformed 100-point scale. Conversely, negative affect scores were consistently low, ranging between 40 and 47. Context-free and OLS scales yielded the highest scores, all falling within the anticipated range of 70–90. Significant mean differences across SES were observed for all scales. We note that the middle SES scored significantly higher than the low and lower-middle SES for all except the NA, where the low SES scored significantly higher.

11.4 Discussion and Conclusion

In this study, we explored the level of children's subjective and psychological well-being of children in low, lower-middle, and middle socio-economic status groups. Our key findings are as follows:

- The three SES groups met the criteria for scalar measurement invariance—this means the scores of the three groups across the different measurement scales are comparable by correlations, regressions, and means.
- The middle SES scored significantly higher on all the well-being measures; with the exception of negative affect.
- Children from the low-SES group scored significantly lower than the lower-middle and middle SES on all the scales.

These results are not surprising given the well-established relationship between SES and well-being. In the South African context, with its tainted socio-political history, it is likely that children from low-SES families, are those that have been historically marginalised and oppressed. They are likely experiencing different forms of poverty and material deprivation, residing in neighbourhoods with poor infrastructure, attending schools that are poorly resourced, and exposed to various forms of violence. The confluence of these factors would be driving the lower levels of well-being.

The findings from the current study have implications for community-based programmes, services, and policies. We contend that policies and services designed for children must consider their SES to ensure that they are effective and equitable. For example, children from low-SES families require more resources and intensive support to negotiate a range of inordinate circumstances. We recommend that policies and services must be tailored to meet their specific and unique needs. One way to achieve is to ensure that children from this SES group are provided with more intense investment and more extensive support and services. Adams and Savahl (2022) see this as a form of redistributive justice. Key to this is the accessibility to health, education, and social services. For example, healthcare services should be designed and delivered in ways that are accessible to all children. Similarly, quality educational opportunities should be equitable and accessible to all children, regardless of their family's financial position. By promoting and implementing equitable access, policymakers can ensure that all children have an equal opportunity to succeed.

Further to that, we also advise that policymakers implement the provision of additional resources and support to children from low-SES families. Using education and school as an example, support should be afforded to programmes that provide subsidised meals, school supplies, and transportation. The provision of these resources will make a substantive difference in children's success at school. In the same way, programmes that provide access to healthcare and mental health services would ensure that children from low-SES families have access to the care they need to stay healthy and flourish.

In conclusion, SES is an essential factor that must be considered when developing policies and services for children. It is axiomatic that children from low-SES families face inordinate circumstances that negatively impact their development, well-being, and overall quality of life. By placing an emphasis on SES, policymakers can ensure that all children have access to a range of resources and support they need to flourish. Therefore, it is essential to advocate for policies and services that are accessible and equitable, and provide additional and sustainable resources and support to children from low-SES families.

References

Adams S, Savahl S (2022) A social justice perspective on children's well-being: considerations for children's rights in the context of COVID-19. In: Tonon G (ed) Social Justice for Children in the South. Springer

Adler NE, Stewart J (2007) The MacArthur scale of subjective social status. MacArthur Research Network on SES and Health

Bosch A, Roussouw J, Claassens T, du Plessis B (2010) A second look at measuring inequality in South Africa: a modified gini coefficient. School of Development Studies, Working paper (58). University of Kwazulu-Natal
Bradley RH, Corwyn RF (2002) Socioeconomic status and child development. Annu Rev Psychol 53(1):371–399
Bradshaw J, Keung A, Rees G, Goswami H (2011) Children's subjective well-being: international comparative perspectives. Child Youth Serv Rev 33:548–556
Bradshaw J, Martorano B, Natali L, De Neubourg C (2013) Children's subjective well-being in rich countries. Child Indic Res 6(4):619–635
Bradshaw J, Richardson D (2009) An index of child well-being in Europe. Child Indic Res 2(3):319–351
Casas F (2017) Analysing the comparability of 3 multi-item subjective well-being psychometric scales among 15 countries using samples of 10 and 12-Year-Olds. Child Indic Res 10:297–330. https://doi.org/10.1007/s12187-015-9360-0
Chen FF (2007) Sensitivity of goodness of fit indexes to lack of measurement invariance. Struct Equ Modeling 14(3):464–504
Cheung GW, Rensvold RB (2002) Evaluating goodness-of-fit indexes for testing MI. Struct Eqn Modell 9:235–255
Coleman JS (1988) Social capital in the creation of human capital. Am J Sociol 94:S95-120
Cummins RA, Lau ADL (2005) Personal wellbeing index: school children (PWI-SC), 3rd edn. Deakin University
Diener E (1984) Subjective well-being. Psychol Bull 95:542–575. https://doi.org/10.1037/0033-2909.95.3.542
Diener E (2000) Subjective well-being: the science of happiness and a proposal for a national index. Am Psychol 55:34–43
Diener E (2006) Guidelines for national indicators of subjective well-being and ill-being. J Happiness Stud 7:397–404. https://doi.org/10.1007/s10902-006-9000-y
Diener E (2009) Subjective well-being. In: Diener E (ed) The science of well-being: the collected works of Ed Diener. Springer, pp 11–58. https://doi.org/10.1007/978-90-481-2350-6_2
Diener E, Suh EM, Lucas RE, Smith HL (1999) Subjective well-being: three decades of progress. Psychol Bull 125(2):276–302. https://doi.org/10.1037/0033-2909.125.2.276
Gilman R, Huebner S (2003). A review of life satisfaction research with children and adolescents. Sch Psychol Q 18(2):192–205. https://doi.org/10.1521/scpq.18.2.192.21858
Hall K, Woolard I, Lake L, Smith C (eds) (2012) South African child gauge 2012. Children's Institute, University of Cape Town, Cape Town
Huebner ES (1991) Initial development of the students' life satisfaction scale. Sch Psychol Int 12:231–240
International Test Commission (2017) ITC Guidelines for translating and adapting tests, 2nd ed. Retrieved from: www.intestcom.org
Keyes CL (2006) Subjective wellbeing in mental health and human development research worldwide: an introduction. Soc Indic Res 77(1):1–10
Kline RB (2011) Principles and practice of structural equation modelling, 3rd edn. Guilford, New York
Krieger N, Williams DR, Moss HW (1997) Measuring social class in US public health research: concepts, methodologies, and guidelines. Annu Rev Public Health 18:341–378
Marmot M, Wilkinson RG (2006) Social determinants of health, 2nd edn. Oxford University Press, Oxford
Millsap RE, Olivera-Aguilar M (2012) Investigating measurement invariance using confirmatory factor analysis. In: Hoyle RH (ed) Handbook of structural equation modeling. Guilford, pp 380–392
Nahkur O, Casas F (2021) Fit and cross-country comparability of children's worlds psychological well-being scale using 12-year-olds samples. Child Indic Res. https://doi.org/10.1007/s12187-021-09833-0

Rees G (2019) Variations in children's affective subjective well-being at seven years old: an analysis of current and historical factors. Child Indic Res 12:141–160. https://doi.org/10.1007/s12187-017-9516-1

Rees G, Main G (2015) Children's views on their well-being: findings from the Children's Society's 2014 subjective well-being survey. The Children's Society

Rees G, Pople L, Goswami HD (2011) Understanding children's well- being: links between family economic factors and children's subjective well-being—initial findings from wave 2 and wave 3 quarterly surveys. The Children's Society, London. http://www.childrenssociety.org.uk/sites/default/files/tcs/research_docs/Economic%20Factors%20March%202011.pdf

Rees G, Savahl S, Lee BJ, Casas F (eds) (2020) Children's views on their lives and well-being in 35 countries: a report on the Children's Worlds project, 2016–19. Children's Worlds Project (ISCWeB), Germany. https://isciweb.org/wp-content/uploads/2020/07/Childrens-Worlds-Comparative-Report-2020.pdf

Ryan RM, Deci EL (2000) Self-determination theory and the facilitation of intrinsic motivation, social development, and well-being. Annu Rev Psychol 55:68–78

Ryff CD (1989) Happiness is everything, or is it? Explorations on the meaning of psychological wellbeing. J Pers Soc Psychol 57:1069–1081

Ryff CD (1995) Psychological well-being in adult life. Current Directions Psychol Sci 4(4):99–104. http://www.jstor.org/stable/20182342

Ryff CD, Singer BH (2008) Know thyself and become what you are: a eudaimonic approach to psychological well-being. J Happiness Stud 9:13–39

Savahl S, Adams S, Benninger E, Florence M, Jackson K, Mpilo M, Bawa U, Isobell D (2019) Researching children's subjective well-being in South Africa: considerations for method, theory and social policy. In: Eloff I (ed) Quality-of-life in African societies. Springer, pp 407–430. https://doi.org/10.1007/978-3-030-15367-0

Savahl S, Adams S, Hoosen P (2023) The subjective and psychological well-being of children in South Africa: a population-based study. Appl Res Qual Life 18:2315–2347. https://doi.org/10.1007/s11482-023-10187-8

Savahl S, Casas F, Adams S (2015) Validation of the children's hope scale amongst a sample of adolescents in the Western Cape region of South Africa. Child Indic Res 9:701–713

Savahl S, Casas F, Adams S (2021) The structure of children's subjective well-being. Front Psychol. https://doi.org/10.3389/fpsyg.2021.650691

Savahl S, Lee BJ, Casas F (eds) (2022) Children's views on their lives and well-being during Covid-19: a report on the Children's Worlds project, 2020–2022. Children's Worlds Project (ISCWeB), Germany

Seligson JL, Huebner ES, Valois RF (2003) Preliminary validation of the brief multidimensional students life satisfaction scale (BMSLSS). Soc Indic Res 61:121–145. https://doi.org/10.1023/A:1021326822957

Steenkamp JBE, Baumgartner H (1998) Assessing measurement invariance in cross-national consumer research. J Consum Res 25(1):78–90

World Bank (2022) World Development Report 2022: finance for an equitable recovery. Washington, DC: World Bank. https://doi.org/10.1596/978-1-4648-1730-4

Chapter 12
Vital Inequalities in Urban Contexts of Argentina's Great North: Approximations Based on Child Mortality Rate and Health Vulnerability Index (2005–2020)

Fernando Longhi

Abstract Different research has shown that social and economic inequalities have been on the rise since the 1980s—even more so as society becomes more urbanized—both in the most developed countries and emerging ones. For Latin America in general—and for Argentina in particular—inequality is one of the most critical dimensions on the process to achieving the so-called territorial equality, which would imply better living conditions for the population. Even in recent years, where a plethora of policies connected to mitigating inequality have taken place, the inherent characteristics of territorial structures, together with other phenomena (policy biases and vacuums, their form of application, in addition to different international situations, among other factors) prevented the disruption of the regional inequality circle in Argentina. We have conceived a particular tendency to analyze vital inequalities, that is, those socially constructed inequalities that offer human beings unequal life opportunities. The selected area of study is Argentina's Great North, a territory that is already defined as a context of persistent poverty. In this territory, the urban dimension is represented by the nine provincial capitals. The results warn, both under a diachronic and synchronous look, the characters in magnitude, trend, and spatial distribution that vital inequalities acquired in these urban settings in recent time periods.

Keywords Vital inequalities · Urban contexts · Child mortality · Health · Argentina

F. Longhi (✉)
Facultad de Humanidades, Ciencias Sociales y de la Salud (Universidad Nacional de Santiago del Estero), Santiago del Estero, Argentina
e-mail: fernandolonghi@conicet.gov.ar

Instituto Superior de Estudios Sociales, Consejo Nacional de Investigaciones Científicas y Técnicas—Universidad Nacional de Tucumán, San Miguel de Tucumán, Argentina

G. H. Tonon (ed.), *Urban Inequalities*, The Urban Book Series,
https://doi.org/10.1007/978-3-031-59746-6_12

12.1 Introduction

Different research has shown that social and economic inequalities have been on the rise since the 1980s—even more so as society becomes more urbanized—both in the most developed countries and emerging ones (Piketty 2015; Stiglitz 2012; Therborn 2015). In response to this process, the ECLAC (2016: 16) has emphatically stated that equality is a "strategic development horizon." Despite this worrisome increase and the urgent need to address this issue, it is clear that we are living in a society where coexistence with inequality has somehow become naturalized, or what is worse, we have become accustomed to living with it. Inequality, as explained by Therborn (2015), is a distinctive feature of today's world.

For Latin America in general—and for Argentina in particular—inequality is one of the most critical dimensions on the process to achieving the so-called territorial equality, which would definitely imply better living conditions for the population as a whole. This inequality arises as the result of different asymmetries accumulated over time (in terms of infrastructure, basic services, production capabilities, among others), which are reflected at different scales and gives rise to multiple developmental gaps, from the economic to social, environmental, and gender areas (Abeles and Villafañe 2022). However, this is manifest not only in different ways, but it is also replicated on various geographical scales, hence the network of inequalities existing within cities, for example, which constitutes a particularly relevant focus of analysis. Studying inequalities between cities and within cities is a highly relevant line of analysis for the broad field of inequality research.

In our country, the problem of inequality is also a phenomenon that can be traced back to the origins of the nation. Over the years, and during the different phases and stages of the country's political, social, and economic processes, the ideal of equality has been—to a greater or lesser degree—present in the discourse and action of public policies. However, it is evident that the result has been rather poor, since the gap between equal rights and socioterritorial inequalities has not been yet overcome. Research has confirmed that even in recent years, where a plethora of policies connected to mitigating inequality have taken place, the inherent characteristics of territorial structures, together with other phenomena (policy biases and vacuums, their form of application, in addition to different international situations, among other factors) prevented the disruption of the regional inequality circle in Argentina (Calvi 2010; Rofman and Romero 1997; Rapoport 2012; Salvia 2011; Kessler 2014). In this circle, the provinces of Argentina's Great North (NGA) are among the most disadvantaged areas (Bolsi et al. 2006).

Regional differences—in terms of social, political, economic, and spatial features—manifest the different territorialities that have been put into play over time. The persistence of these aspects for territorial construction—as well as the emergence of new ones, oftentimes connected to the globalization process and the most recent transformations of the world economy—have given rise to what is called territorial fragmentation. Nowadays, more than ever, it is possible to verify different territorialities in conflict, not only in urban areas but also throughout the territory. At this point, we do not intend to examine these inequalities and the territorial fragmentation derived from this aspect, as it was already analyzed in Paolasso et al. (2019). Rather, we intend to delve into urban inequalities in an enclave of structural poverty, such as Argentina's Great North.

Following Therborn's proposal (2015), we have conceived a particular tendency to analyze vital inequalities, that is, those socially constructed inequalities that offer human beings unequal life opportunities. This type of inequality is exemplified in Therborn's work based on the evaluation of mortality rates, life expectancy, and health index (life years without serious illness, and other child health indicators, such as weight at birth and malnutrition rates), among other factors. In this proposal, the perspective to observe these vital inequalities is two complementary indicators: on the one hand, child mortality rate—both in neonatal and post-neonatal stages—will be considered for interurban comparative analysis. Child mortality is not only a demographic indicator, but also a socioeconomic indicator that segregates different contexts where life takes place and the risk of death in childhood is present (Longhi 2022). On the other hand, the health vulnerability index (Rosati 2022) will be used to analyze inequalities within cities. This index reveals the degree or extent to which the local population is susceptible to health service access and can be cartographically analyzed.

The selected area of study is Argentina's Great North, a territory that is already defined as a context of persistent poverty. In this territory, the urban dimension represented by the nine provincial capitals (Greater San Miguel de Tucumán, Greater Salta, Santiago del Estero-La Banda, Greater Corrientes, Greater Posadas, Greater San Salvador de Jujuy, Formosa, Greater San Fernando del Valle de Catamarca, and Greater Resistencia) will be considered. These cities gather 42% of the regional population, representing 52% of the urban population of Argentina's Great North and 9.5% of Argentina's urban population.

12.2 Inequality, Living, and Health Conditions

In the last decades, social sciences have made significant advances in the interpretation of what we could call *well-being* or *well-living* in society. This condition is directly related to the concept of living conditions, which refers to the combination of certain levels of satisfaction and lack of different economic, social, and environmental dimensions that are associated with what we usually know as poverty (lack) and quality of life (achievement). This conceptualization of society's well-being helps us analyze the entire population according to its levels of satisfaction but, above all, it is an ideal term if we intend to conduct an analysis of the spatial distribution degrees of this well-being status among the population. At this point, the equality or inequality of said distribution acquires substantial relevance, which is therefore expressed in different contexts where health is cared for and promoted, or the necessary conditions for illness to proliferate are identified.

As we have already mentioned, inequality operates as one of modern society's distinguishing features. To a greater or lesser degree, these societies are configured from different levels of inequality. In pre-industrial or traditional societies, the origin of this inequality was attributed to divine or natural factors. It had to be accepted, as it was a given attribute. However, it was mainly from the new ideas introduced by eighteenth-century philosophers of the Enlightenment that equality and equity in societies have progressively been introduced as ideal objectives to be achieved. Subsequent contributions from other classical authors, such as Smith, Marx, Weber or Parsons, or more current authors such as Giddens or Bourdieu, have contributed to a better interpretation of this problem.

In his study on the origin of inequality among men in 1775, Rousseau distinguished two kinds of inequalities: one physical or natural, and another moral or political, the latter "established, or at least authorized, by the common consent of mankind," consisting of "various privileges, some of which are enjoyed to the detriment of others" (2010: 27).[1] The origin of inequality should be traced, according to Rousseau, in the development of our faculties and in the progress of the human spirit, legitimizing itself through the laws and the institution of property (2010: 99). Ever since, inequality has become a political and moral problem (Therborn 2015). This is the notion of inequality that has prevailed: there is broad consensus that inequality must be understood fundamentally as a sociocultural construction (Di Virgilio et al. 2010; Dubet 2015).

The classical theoretical development of the concept has been mainly aimed at accounting for the socioeconomic differences that exist in each society. These, in turn, have repercussions in the most diverse areas: housing, health, security, education, etc. Traditionally, inequality has been understood as income inequality. However, it does not only include differences in income levels; there are other types of needs that are equally or more important for human well-being, such as emotional, social, collective, political, spiritual needs, among others (Salvia 2011). It is for this reason that we speak of material, symbolic, and existential resources. According to Therborn (2015), who follows the ideas of Sen and Nussbaum, the capabilities approach proposed by these two authors identifies—*prima facie*—three types of inequalities: (1) vital inequalities (differentials in life expectancy, fertility, etc.); (2) existential inequalities (includes the unequal allocation of attributes that constitute a person); (3) resource inequalities (considers differences in terms of material and symbolic resources). Our interest is focused on this first expression of inequality, which basically implies the different opportunities that human beings have according to the context where they are born and is expressed in different indicators related to health and its care. In this analysis, we have selected two complementary indicators as indicators to analyze vital inequality: child mortality rate and health vulnerability index.

Child mortality is an indicator related to health and socioeconomic dimensions, as well as well-being, in each geographic area (Spinelli et al. 2000; Díaz Elejalde et al. 2008; Longhi 2013a, b). Within mortality's demographic studies, there is no other

[1] In current terms, the former refers to what is understood by *difference*, while the latter refers to what is strictly understood by *inequality*.

age that takes—and involves—such transcendental interest as the group of children under one year of age, given the complexity and multiplicity of factors affecting that age range, as well as the consequences it entails. Although current literature suggests considering the first thousand days elapsed not from childbirth, but from conception, the incidence of socioenvironmental determinants continues to be more prevalent in the period from birth to the first year of age (González Hernández et al. 2016). Child mortality rate (IMR) is the internationally defined indicator to measure child mortality in a territory to measure its significance, trend and evolution, among other characteristics, so as to make the necessary comparisons and assessments (Longhi 2013a, b).

In the case of health vulnerability, it is assumed that this condition involves unequal access to health services, both in qualitative and quantitative terms (Rosati 2022). To quantify this vulnerability, different obstacles may arise. One of the most prominent aspects is the need to include different factors to explain this access (or lack of it) to the health system. The main related factors include high levels of poverty and exclusion, being part of ethnic or social minorities, pre-existing diseases, lack of health coverage, among others. In addition, there are other features operating at the environmental and non-individual levels. This analysis was made with a high-resolution spatial approach, including the so-called census block, and enabled the detection of spatial concentration areas or "hot zones," characterized by extreme lack of access to the health system.

It would then be clear that forms of social, legal, and economic organization (with all their associated complexities) in modern society would determine the degree of (in)equality between human beings. In addition, these forms of social organization cannot be understood without considering that they are developed in specific places and at specific times in very different ways, by which different territories are generated.

12.3 Measuring Child Mortality and Health Vulnerability

As it was previously mentioned, child mortality is an indicator of mortality and socioeconomic level in a geographical area. The child mortality rate (IMR) connects child deaths to the total number of live births in each period, and it is expressed per thousand live births. Therefore, it measures the risk of death for children under one year of age. It comprises a complex issue, where the health-disease-death processes are presented as biological phenomena within a social context.

From the perspective of the deceased age, the mortality of children under one year of age can be distinguished between neonatal mortality (deaths occurring during the first 28 days of life) and post-neonatal mortality (deaths occurring after the 28th day of life and before one year of age). The importance of this differentiation has to do with the different causes that determine the death of a child in each of these age groups. The neonatal mortality rate is strongly affected by errors and inaccuracies, mainly due to imprecisions in the evaluation of the stillborn and due to errors and omissions in the declaration and registration of births and deaths. As a result, lower rates to the actual values are usually registered, especially in developing countries. In this first age group, biological and genetic factors take on the greatest role in the etiological profile of mortality.

The post-neonatal mortality rate (PMR) has less underreporting than the NMR, since a child's declaration of the death is more likely to be effectively declared and not omitted or confused in the case of a stillborn. Unlike the previous one, in this age group, the socioeconomic component in deaths acquires a greater importance.

Two great paradigms have defined the study of child mortality: biological and social. Spinelli et al. (2000) state that within the biological or dominant paradigm, research studies connected child mortality to maternal age, birth interval, number of children, birth order, low birth weight, maternal education level, occupation of the head of household, parental level of education, housing conditions, access to health services, etc. Meanwhile, child mortality studies that include the social (non-dominant paradigm) attempt to operationalize different variables, including social class. Within this group are studies on poverty, migratory processes, their relationship with health, and the relationship between crisis and health (Spinelli et al. 2000). Studies on child mortality state that the health-disease process is determined by a set of situations, both biological (inherent to human beings) and social (conferred by the environment). In this way, neonatal and postnatal components might display these decisive features.

Health vulnerability, additionally, is a polysemic concept that refers to those contextual determinants of health which might increase the risk of disease. In this paper, following Rosati et al. (2020) guidelines, health vulnerability consists of the following associated factors:

- Access to health benefits and services provided by the State: in this dimension, the main indicator used was the proximity to State health facilities throughout the country, estimating the walking time from different points to the nearest health center.
- Socioeconomic Status of the Population (SES): to make the SES, information from the 2010 National Population, Households and Housing Census was used, implementing a series of relevant variables—educational level, unsatisfied basic needs indicators, etc.—and was combined.

To make the final sanitary vulnerability index, both variables were combined at the census block level, which allows for its expression and cartographic analysis, detecting nuclei of said vulnerability.

A final methodological reference refers to limitations of approach and scale. When using the child mortality rate in urban agglomerations through a comparative and diachronic approach, the spatial unit in which the vital event is recorded (department) does not correspond directly to the concept of agglomeration, which is somewhat larger. Nevertheless, it represents an important spatial proportion of the agglomeration, so its contribution to the awareness of child mortality levels and trends in that spatial unit is significant. This constraint is not found in the analysis of health vulnerability, since the approach delves into the agglomeration—from a synchronic and cartographic point of view—trying to detect nuclei of greater vulnerability and concentration nuclei. The latter, following the postulates of Tobler's Law, is carried out with spatial statistical tools at the census block level.

12.4 Child Neonatal and Post-neonatal Mortality as Fundamental Inequalities in NGA Capital Cities

In a generalized context, child mortality is on a downward trend at the national, regional, and provincial levels. This is clearly a tendency toward a process, but the pace of decline, the levels reached, and the characteristics of this process have different implications. In the territory under study, we observed that this downward trend has been replicated in the capital cities of Argentine Northwest (NOA) (Fig. 12.1) and Argentine Northeast (NEA) (Fig. 12.2). However, there are some distinctive features that should be mentioned.

The main advances have occurred in the first half of the period, after which the figures have generally stabilized or even increased, as in the case of the city of Santiago del Estero. In this province, there is notable evidence as regards the degree of coverage and quality of its vital statistics. Fernández et al. (2008) and UNICEF (2017) have highlighted this condition and the trends toward improving

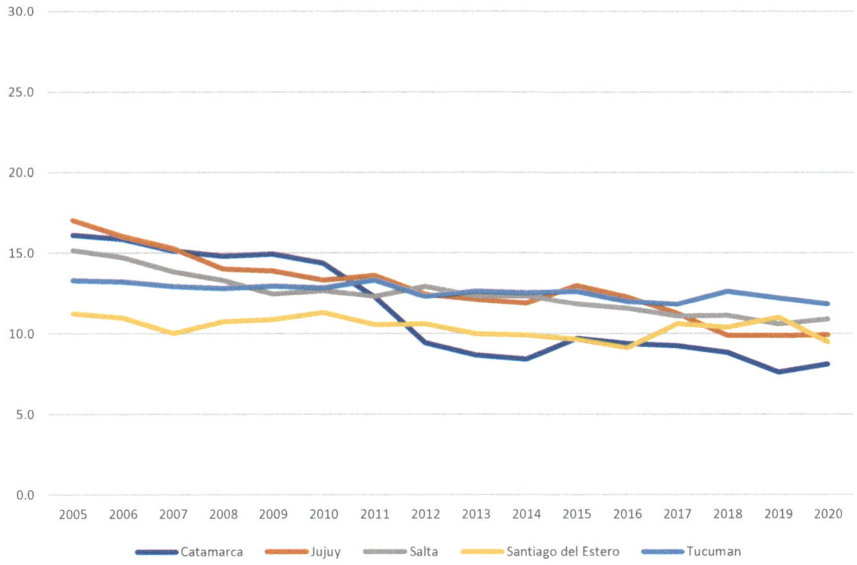

Fig. 12.1 Capital cities of the Argentine Northwest. Evolution of child mortality rates 2005–2020. *Source* National Health Statistics Program 2005–2020

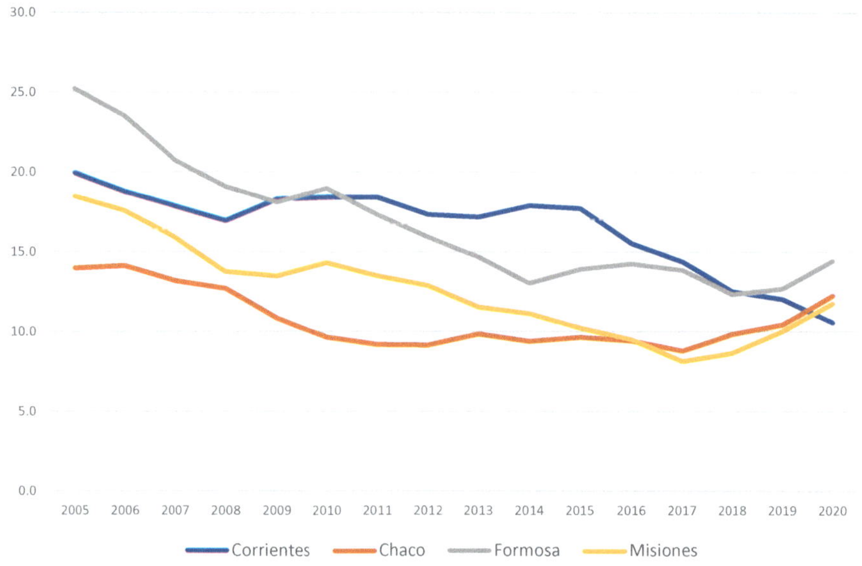

Fig. 12.2 Capital cities of the Argentine Northeast. Evolution of child mortality rates 2005–2020. *Source* National Health Statistics Program 2005–2020

records in recent years, so we consider that the observed increase is more connected to improvements in the quality of records than an increase in child mortality.

The most critical performance was found in the cities of Salta and Tucumán, both in terms of magnitudes and trends. Both have the highest child mortality rates at the end of the period, and the average variation only reaches 10%, which is not very promising, especially when compared to other regional capitals such as Jujuy and Catamarca, where average variation reached 50%. In these two locations, the IMRs in 2005 were the highest in this territory, but in 2020, they were ranked as the cities with the lowest rates. In this process, Jujuy has had a declining performance during the whole period, but Catamarca has experienced an abrupt decline during 2010 and 2015.

In the cities of the Argentine Northeast, in general, higher rates are more prevalent, compared to those recorded in the NOA, as well as a disruptive performance in said trend, fundamentally as of 2017. In fact, this decline was more prominent in the first half of the period; then it was stabilized and even increased in the last years, as in the cases of the cities of Formosa, Resistencia, and Posadas. Formosa despite having managed to reduce its IMR by 72% was the jurisdiction with the highest rates at the beginning and end of the period. Corrientes has achieved an average decrease of 62%, which is largely explained by the events that occurred as of 2015; prior to this date, the IMR performance was stable, while the rest of the cities showed a sustained decrease. Finally, a marked break in the downward trend is found in Resistencia and Posadas, which prevailed until 2017. From then on, health conditions in those locations declined notably, causing an exacerbated increase in the lower IMRs that had been then achieved.

Neonatal mortality rate expresses the mortality that occurs during the first month of life: Most of the causes of death involved in this stretch are endogenous (derived from congenital malformations or weaknesses, including immaturity (prematurity), birth trauma, congenital weakness, and malformations) (Longhi 2013a, b). The study of endogenous child mortality has gained growing interest at present, due to its quantitative importance in view of the decreasing trend in child mortality and, therefore, the greater proportion of endogenous diseases and causes of death. Figure 12.3 shows the recent trend of this indicator in the NOA's capital cities, while Fig. 12.4 shows the trend in the NEA's capital cities. In the first case, we observed substantial progress in the capital cities of Catamarca and Jujuy, as the former experienced a reduction of 54%, while the latter had a 37% decline, which reflects the role played by neonatal death control to curb child mortality in these cities. In fact, back in 2005, Catamarca had the highest numbers in neonatal mortality in this regional context, reaching 16.1 per thousand, but these figures were reduced and, in 2020, it was ranked as the NOA city with the lowest rate (8.1 per thousand), showing effective control over endogenous death causes. Although this paper does not delve into the causes of controlled death causes, or the factors that made such control possible, we believe that greater investment in neonatal health services and gestational control and diagnosis might have been decisive elements to explain this process.

On the contrary, the downward tendency is mitigated and stabilized in the cities of Tucumán, Salta, and Santiago del Estero. Tucumán has the most critical cases.

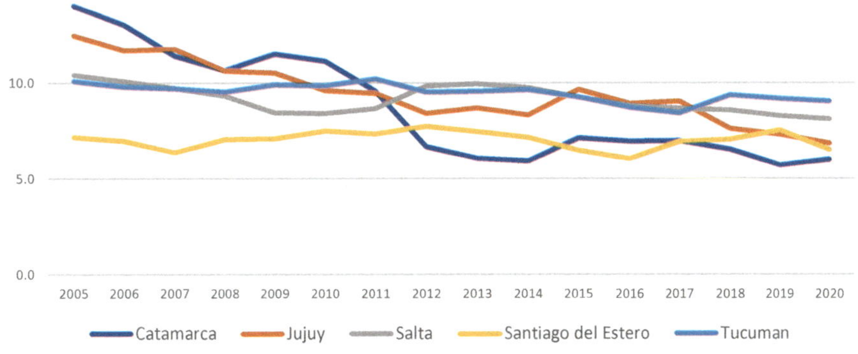

Fig. 12.3 Capital cities of the Argentine Northwest. Evolution of neonatal mortality rates 2005–2020. *Source* National Health Statistics Program 2005–2020

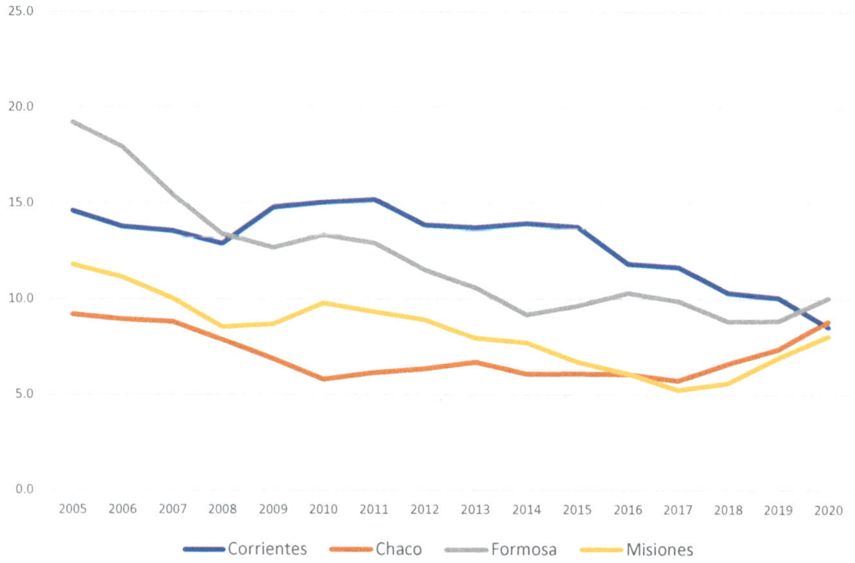

Fig. 12.4 Capital cities of the Argentine Northeast. Evolution of neonatal mortality rates 2005–2020. *Source* National Health Statistics Program 2005–2020

As a matter of fact, the neonatal mortality rate went from 13.3 per thousand in 2005 to 11.9 per thousand in 2020, showing a minimum decrease of 10% in this fifteen-year period. San Miguel de Tucumán has one of the most outstanding maternity services at regional level (Instituto de Maternidad y Ginecología Nuestra Señora de las Mercedes) where more than 50% of provincial births are assisted and is a referent in the control of high-risk obstetric cases in the Argentine Northwest (Mora 2008), with over-demanded and overcrowded capacity levels (Viñas et al. 2002).

In the case of NEA capital cities, there is mainly downward trend; surprisingly, in three of the four cities studied, this index has experienced an increase since 2017 (Fig. 12.4). The exception to this phenomenon is Corrientes, which showed a sustained and marked decrease in neonatal mortality, with an average variation of 41%. However, the greatest decrease occurred in Formosa, which showed significant control over mortality in this age group until 2014, after which there was an abrupt break in this trend. The capitals of Chaco and Misiones also evidenced notable advances in the first half of the period, but thereafter the downward trend halted, the rate stabilized and increased in subsequent years, and the 2020 rate was similar to figures recorded fifteen years earlier. The assessment that can be made about this process is very worrisome, since the increase of this problem as of 2017 shows an even more pronounced trend to the inclination observed in its decline.

In the case of post-neonatal mortality, there is a higher incidence of exogenous conditions that explain child death in this age group, since the phenomenon of exogenous mortality is fundamentally due to the social environment in which death

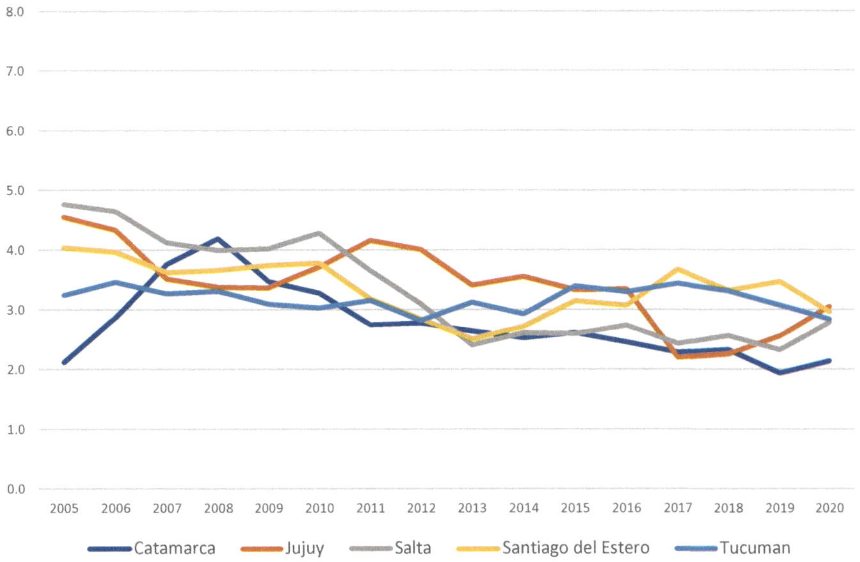

Fig. 12.5 Capital cities Argentine Northwest. Evolution post-neonatal mortality rates 2005–2020. *Source* National Health Statistics Program 2005–2020

occurs, either due to negligence or lack of necessary resources.[2] This manifestation of mortality is more feasible to control than endogenous mortality; developed countries have greatly reduced exogenous mortality and thus have drastically reduced child mortality rates. However, evidence in cities of the NOA and NEA reflects slight decreases in post-neonatal mortality as regards neonatal mortality. Probably, the significant pockets of urban poverty that stand out in these areas are the main barriers that prevent reduction.

Figure 12.5 shows the course of post-neonatal mortality in the NOA capitals. In this case, a stable performance prevails, with major progress in the first half of the period. During this period, the greatest progress was observed in the cities of Salta and Santiago del Estero. However, from 2013 onwards, the gradual downward trend was reversed, showing an increase in post-neonatal mortality in the cities of Santiago del Estero and Tucumán, reaching worrisome figures in both cases. In addition, in fifteen years of study, the city of Tucumán only managed to reduce its post-neonatal mortality by 3%, while the city of Santiago del Estero did so by 7%.

In the case of the NEA provincial capitals (Fig. 12.6), a clear declining performance prevails, compared to the NOA capitals. This trend is most clearly found in the capitals of Corrientes and Misiones, with an average decrease of 22% and 20%, respectively. In the capitals of Chaco and Formosa, the rate decreased until 2012.

[2] The causes of exogenous death include epidemic diseases (measles, whooping cough, epidemic meningitis, tuberculosis, etc.); nervous system diseases (convulsions, non-epidemic meningitis and ear diseases); respiratory system diseases (bronchitis, bronchopneumonia, pleurisy, etc.); and digestive system diseases (mainly childile diarrhea).

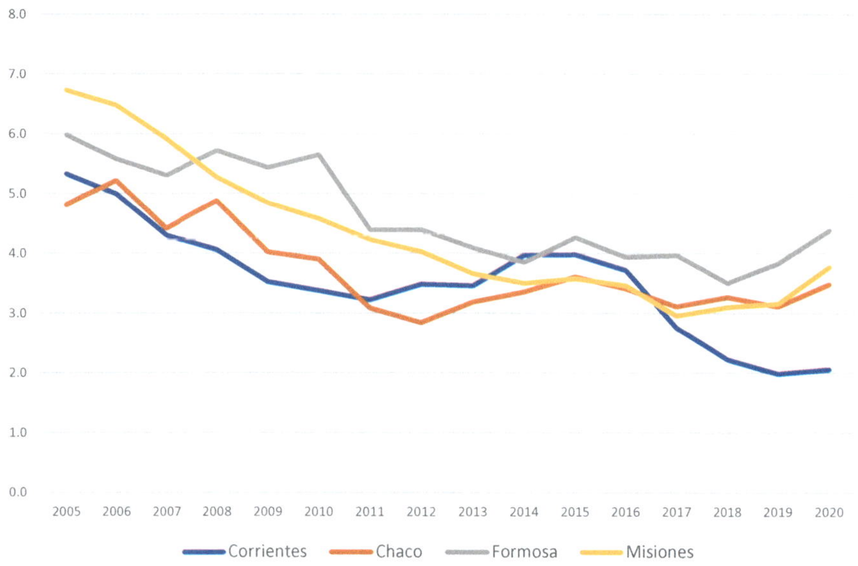

Fig. 12.6 Capital cities Argentine Northeast. Evolution post-neonatal mortality rates 2005–2020. *Source* National Health Statistics Program 2005–2020

After that, a gradual increase in post-neonatal mortality began, triggering a critical situation toward 2020. This critical situation is also experienced by the capital of Misiones; despite the decline previously mentioned, this tendency changed in 2014 and began to increase, ranking closer to the capitals of Chaco and Formosa, which experience a similarly critical situation. It is likely that an important part of child mortality, mainly of an exogenous nature, is related to the communities of aboriginal peoples residing in urban contexts; however, the characteristics of Argentina's vital statistics do not allow us to corroborate this assumption.

Clearly, internal welfare gaps, existing inequalities, barriers to healthcare access, and poverty pockets are serious obstacles to control child death in this age group. It is therefore imperative to detect the spatial concentration of these barriers to identify priority nuclei to implement health policies to control this problem, an aspect that is developed below, under the analysis of health vulnerability.

12.5 Vital Inequalities in Territorial Contexts: Approximations Based on Health Vulnerability Index

To detect areas of extreme health vulnerability, the aforementioned index was applied, using digital mapping tools. In the previous section, we mentioned significant barriers that reversed the downward trends in vital indicators, particularly those of exogenous origin. With the analysis proposed here, we intend to delve into the spatial location of these barriers in each of territorial context studied, to generate useful cartographic tools for the formulation of public policies that seek to address these problems. In this section, given that the information is collected at the census block level, we can extend the analysis to the entire scope of each city, exceeding the departmental limit allowed by vital statistics.

To investigate the pattern of territorial concentration, we applied spatial autocorrelation analysis, which allows us to detect grouping clusters based on individual values and the values of the radii in contact, also known as neighbors. The indicator used is Moran's I Index, which compares the values of each location with the values of neighboring locations. The results of this index vary from − 1 to 1, representing the highest minimum (maximum dispersion) and maximum (maximum concentration) correlations respectively, where zero stands for a totally random spatial pattern. Table 12.1 shows the value of spatial autocorrelation in the main urban agglomerations in the NGA, which allows us to identify maximum spatial concentrations of health vulnerability, which shows a high level of fragmentation in these territories. We observe the higher correlation in the city of Corrientes, and the lower correlation in the city of Jujuy. However, in all cases, there is a pronounced spatial concentration process with quotients higher than 0.6, which enables to identify nuclei with higher health vulnerability.

The following maps (1–9) show these concentration patterns, with the most critical sectors of the city in red and the less critical sectors in blue. The spatial model in these nine provincial capitals is the center-periphery model, where the central areas have higher levels of well-being, while the city margins contain sectors of greatest deprivation, and it is here where vital inequalities, and the very possibility of survival, are mainly present. It is also clear that this periphery is not the same within each city; there are more critical peripheries with respect to one another, and it is also clear that the size of each radius in this cartographic image affects the perception of

Table 12.1 Capital cities of the Argentine Great North. Moran's value, according to the health vulnerability index 2010–2018

City	Spatial autocorrelation quotient
Greater Jujuy	0.634
Greater Santiago del Estero-La Banda	0.684
Greater Catamarca	0.709
Greater Resistencia	0.720
Greater Posadas	0.726
Greater Salta	0.735
Greater Formosa	0.805
Greater Tucumán	0.809
Greater Corrientes	0.844

Source Census 2010 (NSE indicators), hospitals and health centers' official records in 2018 (site addresses) and OpenStreetMap in 2018 (road grid for distance routing)

vulnerability, and this size is closely related to the amount of population contained. This reflection makes us differentiate, on the one hand, that the marginal sectors identified are not necessarily the most populated areas in the city, however; although the affected population is proportionally smaller than the population living in the central areas; the levels of deprivation they suffer express a high level of vulnerability, putting at risk the survival of the populations settled in that area. And this is the main expression of how vital inequalities are expressed in the territory and affect the lives of large sectors of the population, within the general landscape of a territory already defined by its poverty levels.

The following cartography identifies the most critical nuclei of health vulnerability in each capital city of Northern Argentina. This contribution is intended as an input that could be useful for public policies seeking to address inequalities and assist the most socially disadvantaged population (Map 12.1, 12.2, 12.3, 12.4, 12.5, 12.6, 12.7, 12.8 and 12.9).

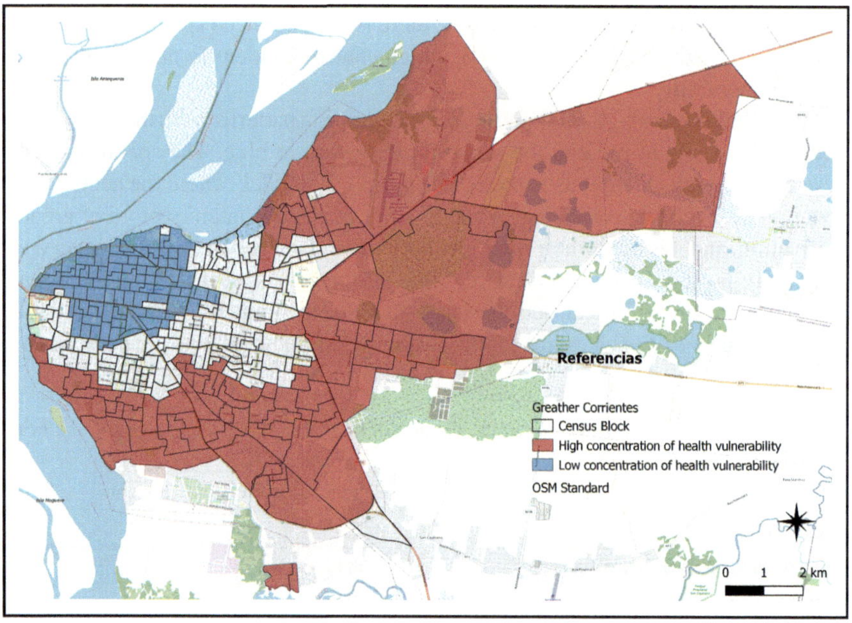

Map 12.1 City of Corrientes. Nuclei of higher and lower concentration of health vulnerability 2010–2018. *Source* Census 2010 (NSE indicators), hospitals and health centers' official records in 2018 (site addresses) and OpenStreetMap in 2018 (road grid for distance routing)

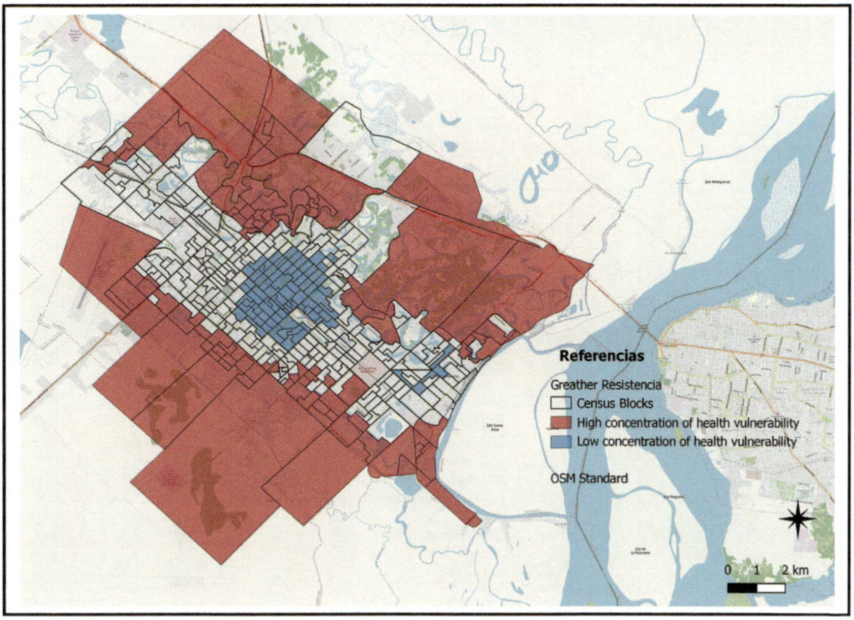

Map 12.2 City of Resistencia (Chaco). Nuclei of higher and lower concentration of health vulnerability 2010–2018. *Source* Census 2010 (NSE indicators), hospitals and health centers' official records in 2018 (site addresses) and OpenStreetMap in 2018 (road grid for distance routing)

12.6 Final Considerations

The proposed approach to vital inequalities provides an appropriate framework for analyzing this population attribute in societies where a significant part of the population falls into categories of poverty or vulnerability. In this case, the territory encompassing the Argentine Great North is a structural enclave of poverty, where great deal of the universe of deficiencies that characterize its population is inserted in urban areas, and where inequality is a leading trait in terms of the use and distribution of resources, clearly following the model that characterizes Latin America in general.

Within this approach, indicators of child mortality and health vulnerability were selected to understand both the temporal evolution and current characteristics of this problem and its spatial distribution. This articulated conjunction between time and territory allows us to glimpse a general panorama of the nuances—in the urban areas analyzed—that reach the vital inequalities and concentration nuclei.

A decreasing trend in child mortality was observed in all the cities analyzed, but the pace of decline was different until the mid-1910s. Ever since, a sustained decline was halted, both in overall child mortality and in neonatal and post-neonatal components, a process that has worsened even as of 2019. Different factors can explain this disruptive trend condition, such as poverty affecting the population, an increase in food prices, the deterioration of education, and urban habitat conditions; however,

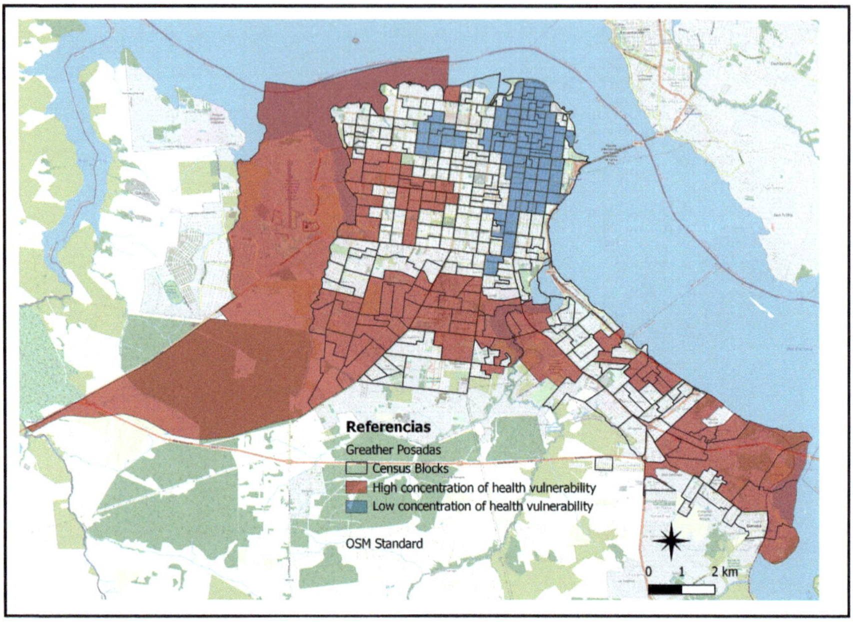

Map 12.3 City of Posadas, nuclei of higher/lower concentration health vulnerability 2010–2018. *Source* Census 2010 (NSE indicators), hospitals and health centers' official records in 2018 (site addresses) and OpenStreetMap in 2018 (road grid for distance routing)

the extension and characteristics assumed by the territorial pattern of distribution for health vulnerability are a key element to explain this process. Indeed, large marginal sectors of NGA cities have extremely critical levels, which could become the main barrier to a downward trend in child mortality.

In this context, it is clear that the main element to be considered for the attenuation of vital inequalities is addressing health vulnerability and considering this concept as a hybrid between socioeconomic conditions and health access. As long as this problem is not addressed in a comprehensive manner, the differences in life expectations and direction based on where individuals are socially born will be systematically reproduced, as it has occurred up to the present.

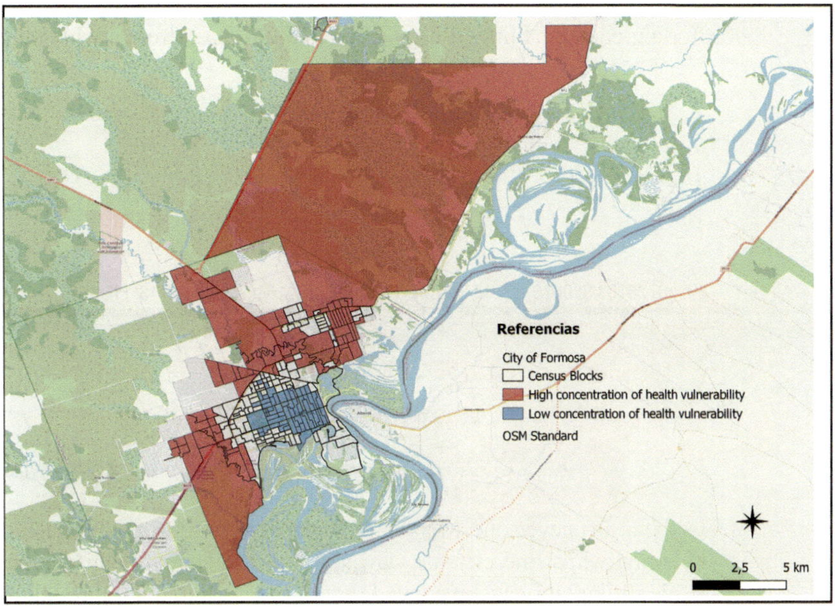

Map 12.4 City of Formosa, nuclei higher/lower concentration of health vulnerability 2010–18. *Source* Census 2010 (NSE indicators), hospitals and health centers' official records in 2018 (site addresses) and OpenStreetMap in 2018 (road grid for distance routing)

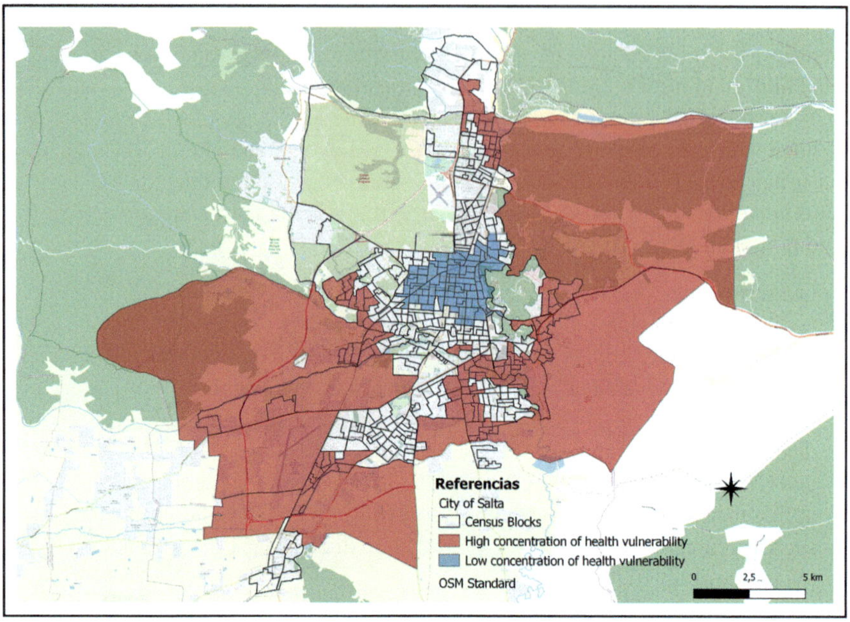

Map 12.5 City of Salta. Nuclei of higher and lower concentration of health vulnerability 2010–2018. *Source* Census 2010 (NSE indicators), hospitals and health centers' official records in 2018 (site addresses) and OpenStreetMap in 2018 (road grid for distance routing)

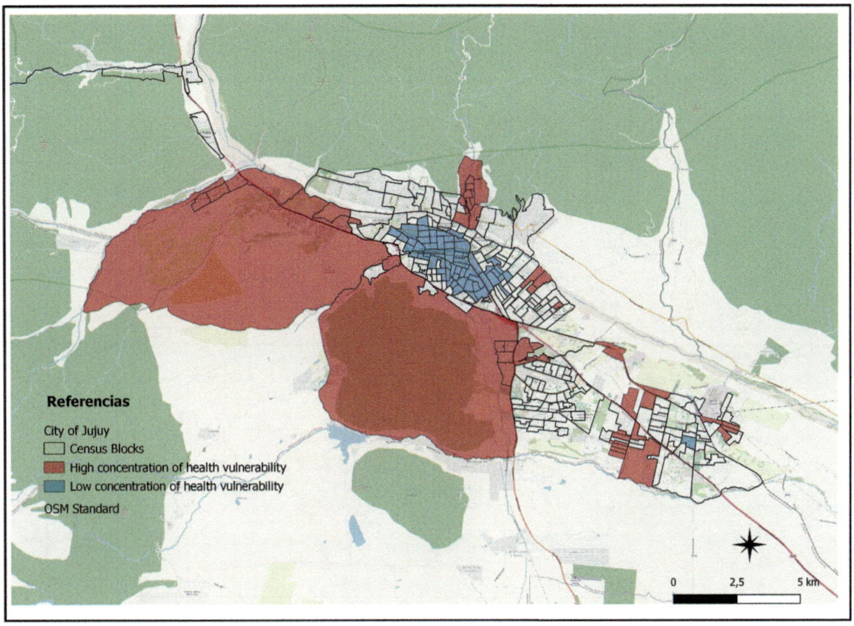

Map 12.6 City of Jujuy. Nuclei of higher and lower concentration of health vulnerability 2010–2018. *Source* Census 2010 (NSE indicators), hospitals and health centers' official records in 2018 (site addresses) and OpenStreetMap in 2018 (road grid for distance routing)

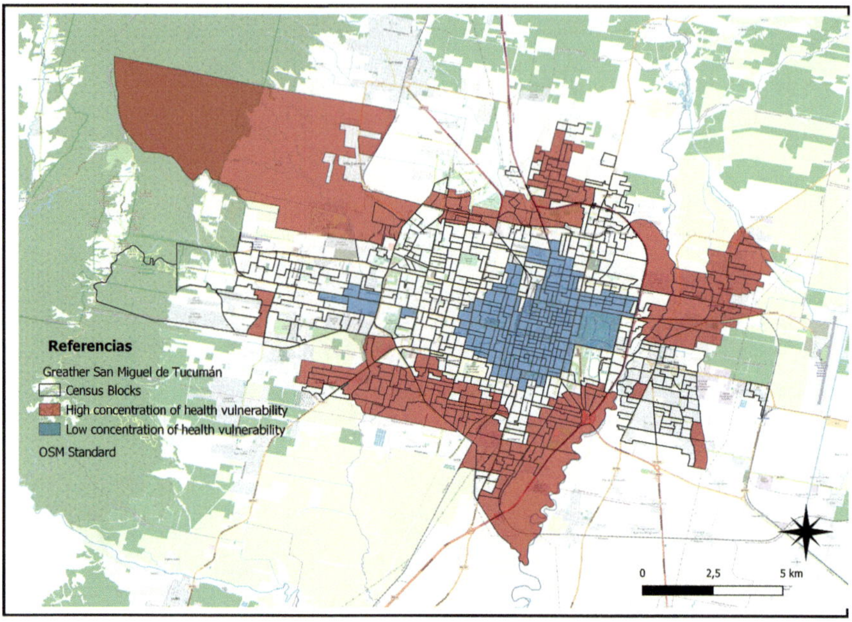

Map 12.7 City San Miguel de Tucumán Nuclei of higher/lower concentration of health vulnerability 2010–2018. *Source* Census 2010 (NSE indicators), hospitals and health centers' official records in 2018 (site addresses) and OpenStreetMap in 2018 (road grid for distance routing)

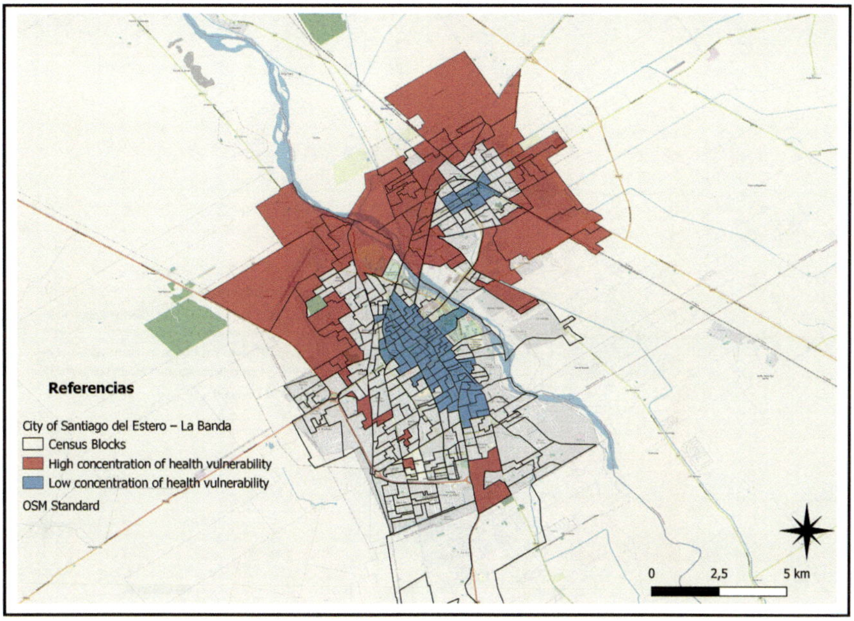

Map 12.8 City Santiago del Estero-La Banda. Nuclei of higher/lower concentration health vulnerability 2010–2018. *Source* Census 2010 (NSE indicators), hospitals and health centers' official records in 2018 (site addresses) and OpenStreetMap in 2018 (road grid for distance routing)

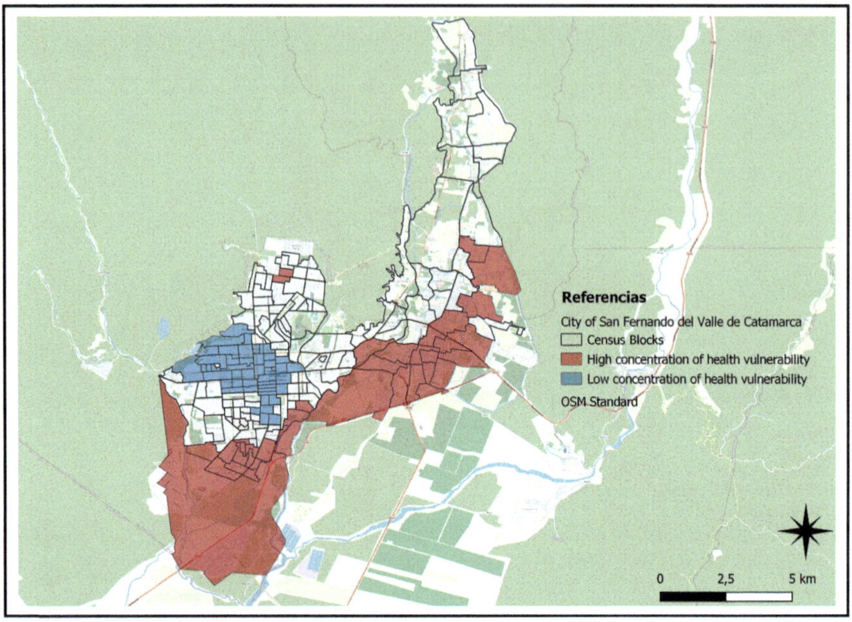

Map 12.9 City of San Fernando del Valle de Catamarca (Catamarca). Nuclei of higher and lower concentration of health vulnerability 2010–2018. *Source* Census 2010 (NSE indicators), hospitals and health centers' official records in 2018 (site addresses) and OpenStreetMap in 2018 (road grid for distance routing)

References

Abeles M, Villafañe S (2022) Asimetrías y desigualdades territoriales en la Argentina: aportes para el debate. Comisión Económica para América Latina y el Caribe (CEPAL)

Bolsi A, Paolasso P, Longhi F (2006) El Norte Grande Argentino entre el progreso y la pobreza. Población Sociedad 12–13:227–266. https://www.redalyc.org/pdf/3869/386939739007.pdf

Calvi G (2010) Las metamorfosis de la desigualdad de ingresos en la Argentina (1974–2005). In: Di Virgilio M, Otero M, Boniolo P (eds) Pobreza y desigualdad en América Latina y el Caribe. CLACSO-CROP, pp 93–139

CEPAL [Comisión Económica para América Latina y el Caribe] (2016) La matriz de la desigualdad social en América Latina. Naciones Unidas. CEPAL

Di Virgilio M, Otero MP (2010) Pobreza y desigualdad en América Latina y el Caribe. CLACSO/CROP

Díaz Elejalde Y, Alonso Uría R (2008) La mortalidad infantil, indicador de excelencia. Revista Cubana de Medicina General Integral 24(2). http://scielo.sld.cu/scielo.php?script=sci_arttext&pid=S0864-21252008000200008&lng=es&tlng=es

Dubet F (2015) ¿Por qué preferimos la desigualdad? (aunque digamos lo contrario). Siglo XXI

Fernández M, Guevel C, Krupitzki H, Marconi E, Massa C (2008) Omisión de registro de nacimientos y muertes infantiles: magnitud, desigualdades y causas. Organización Panamericana de la Salud - Ministerio de Salud

Gonzáles Hernández N, López G, Prado L (2016) Importancia de la nutrición: primeros 1000 días de vida. Acta Pediátrica Hondureña 7(1):597–607. https://www.camjol.info/index.php/PEDIATRICA/article/view/6941/6582

Kessler G (2014) Controversias sobre la desigualdad: Argentina 2003–2013. Fondo de Cultura Económica

Longhi F (2013) Pobreza y mortalidad infantil: Una aproximación teórica al estudio de sus relaciones. Andes 24(2):325–365. https://www.redalyc.org/comocitar.oa?id=12730581010

Longhi F (2022) Mortalidad infantil en Tucumán (Argentina) Brechas estadísticas y cambio secular. Cuadernos Del CENDES 39(109):51–81. http://saber.ucv.ve/ojs/index.php/rev_cc/article/view/24219/144814490552

Longhi F (2013) Pobreza y mortalidad infantil: Una aproximación teórica al estudio de sus relaciones. Andes 24(2). http://www.scielo.org.ar/pdf/andes/v24n2/v24n2a02.pdf

Mora JC (2008) Hipertensión y embarazo: evaluación epidemiológica en el Instituto de Maternidad y Ginecología nuestra señora de las Mercedes, San Miguel de Tucumán. Revista del Hospital Materno Infantil Ramón Sardá 27(1):3–10. https://www.redalyc.org/articulo.oa?id=91227102

Paolasso P, Longhi F, Velázquez G (2019) Desigualdades y fragmentación territorial en la Argentina durante la primera década del siglo XXI. Ediciones Imago Mundi

Piketty T (2015) El capital en el siglo XXI. Fondo de Cultura Económica

Rapoport M (2012) Historia económica, política y social de la argentina, 1880–2003. Emecé

Rofman A, Romero L (1997) Sistema socioeconómico y estructura regional en la Argentina. Amorrortu

Rosati G (2022) La integración de fuentes diversas de información en los estudios territoriales. Reflexiones sobre dos investigaciones. Quid 16: Revista del Área de Estudios Urbanos 17:38–60. https://publicaciones.sociales.uba.ar/index.php/quid16/article/view/7004/pdf_1

Rosati GF, Olego T, Vázquez Brust H (2020) Building a sanitary vulnerability map from open source data in Argentina (2010–2018). Int J Equity Health 19(1):1–16. https://doi.org/10.1186/s12939-020-01292-3

Rousseau J (2010[1775]) El origen de la desigualdad entre los hombres. Ediciones Libertador

Salvia A (2011) Deudas sociales en la Argentina posreformas. Algo más que una pobreza de ingresos. Biblos

Spinelli H, Alarzaqui M, Calvelo L, Arakaki J (2000) Mortalidad infantil, un indicador para la gestión local. Análisis de la mortalidad infantil de la provincia de Buenos Aires en 1998. Organización Panamericana de la Salud

Stiglitz J (2012) El precio de la desigualdad. Taurus

Therborn G (2015) Los campos de exterminio de la desigualdad. Fondo de Cultura Económica

UNICEF (2017). Registro de nacimientos en Argentina. Un estudio sobre la cobertura legal y estadística. UNICEF. https://www.unicef.org/argentina/media/751/file/Registro%20de%20nacimientos.pdf

Viñas M, Graiff O, Luna C (2002) Aportes al diagnóstico de situación perinatal de la provincia de Tucumán. Estudio del Servicio de Neonatología. Instituto de Maternidad y Ginecología Nuestra Señora de las Mercedes. Año 2000. Revista del Hospital Materno Infantil Ramón Sardá 21(4):164–169

Chapter 13
Children in Street Situations in Brazil as Unequal Subject of Rights

Daniel Stoecklin and André Cardozo Sarli

Abstract The UN General Comment 21 on Children in Street Situations (CSS) recalls the rights that are particularly crucial for this population and guides upon the measures and proceedings that should be reinforced to guarantee access to these rights. It therefore evokes some of the factors that favour or impede the translation of the formal liberties contained in the Convention on the Rights of the Child into real freedoms. Observations and interviews in residential care and youth service centres in some of the largest cities of Brazil highlight further discriminations that are mostly invisible. The latter are linked to the institutional order which superimposes its own rationality to the rights-based approach: the composition of the population in these centres reveals that not only CSS but a larger population of children is cared for in terms that cannot fully stem from the General Comment 21 on Children in Street Situations. Hidden frameworks are naturalized in daily interactions within the centres and between the centres and the administrative and political authorities. The translation of social processes into an administrative logic that is eventually what counts as good governance is made through "transactional horizons" centred on an entrepreneurial mode of action.

Keywords Children · Rights · Street · Inequalities · Transactional horizons

13.1 Introduction

Inequalities is a topic that has been consistently framed as an economic issue, and therefore an element to be measured (Waldman 1977; Forbes 2000; Cowell 2011). Approaches from social sciences struggled around issues such as neglecting groups, ignoring contexts and power and focusing too much on differences (Hollander and Howard 2000). They also ascribed to utilitarist or societal-focused approaches or overemphasized the individual level, therefore showing incomplete pictures about inequalities (Banks 2007). Children in streets situations are known to be a symptom of

D. Stoecklin (✉) · A. C. Sarli
University of Geneva, Geneva, Switzerland
e-mail: daniel.stoecklin@unige.ch

G. H. Tonon (ed.), *Urban Inequalities*, The Urban Book Series,
https://doi.org/10.1007/978-3-031-59746-6_13

important inequalities, such as poverty, race, gender, and age-groups. The question of definition and conceptualization, however, do not fall short of issues when the nuances are presented (De Moura 2002). In this chapter, we explore how inequalities also lie in the ways social problems are constructed and represented, and hence how statistics and relying on economic framings highlight and further power imbalances between those who are categorized and the rule-entrepreneurs of the dominant views. For this, we will make use of data emerging from a dedicated fieldwork that took place in Brazil in 2022, in which 32 children with trajectories in the streets and 20 staff of care institutions were interviewed. The interviews made use of the heuristic tool called "Actor's System" (Stoecklin 2017) and the associated theory of "transactional horizons" (Stoecklin 2021). The analysis assesses how the inequalities are shaped and shape the definitions, hence confirming the underpinnings of the structuration theory (Giddens 1984), and suggesting the recursive power of the entrepreneurial mode of action.

13.2 The Numbers of CSS and the Experienced Phenomenon

The numbers of CSS are difficult to assess, notably because the definitions of "street children" have varied over time (Aptekar and Stoecklin 2014). The General Comment on Children in Street Situations (hereafter GC on CSS) was therefore also meant to bring some clarification with a standard definition. It defines CSS as comprising two sub-categories, namely:

> (a) children who depend on the streets to live and/or work, whether alone, with peers or with family; and (b) a wider population of children who have formed strong connections with public spaces and for whom the street plays a vital role in their everyday lives and identities. This wider population includes children who periodically, but not always, live and/or work on the streets and children who do not live or work on the streets but who regularly accompany their peers, siblings or family in the streets. (UNCRC 2017, p. 3).

Yet, the extent to which this definition is operationalized within the 196 State parties (all except the USA) to the UNCRC remains unknown. Our observations in Brazil suggest that practices are not fully compliant with these UN guidelines. The GC on CSS itself is also part of the social construction of the problem, as its definition of CSS is founded in the modern centration on individuals who hence become the units of any social problem that is therefore approached foremost in quantitative terms. Consequently, the bigger the number of people considered in a problematic situation, like street life, the bigger the problem seems to be. The association of the public problem with the social problem reduced to numbers of CSS results in interventions that are targeting people and not processes: as a matter of fact, processes involve many more people than just CSS, but there is a political reluctance to target of these people who are actually also causing the problem of CSS. Consequently, while it is important to assess the numbers of children in street situations to adapt policies, the numbers alone do not tell much the construction of the problem. They tell even less about street life as it is experienced.

This is why it is necessary to understand how the problem is experienced by children in street situations to address it with fine-tuned relevance. This requirement is comprised in the child's right to be heard (Art. 12 CRC) and other civil freedoms, and if implemented properly, including in scientific research, these rights prevent the paradoxical coercion of children implied by purely protectionist policies. Therefore, public policies must find the right balance between their consideration for individual autonomy and responsibility on the one side, and, on the other side, the structural and material conditions that allow some to fulfil the requirements more or better than others. Such attention to local conditions is, however, not favoured by the overarching order instituted by the law. Consequently, legislation is always directed towards abstract individuals, which, as we have seen with "street children", partly misses the logic of action that it was written for.

The focus on processes of marginalization that were documented in numerous monographies seems to be somewhat "lost in translations" (Stoecklin 2017): while the GC on CSS gives a substantive definition of these children (population a and b), the expression "children in street situations" remains "a concept in search of an object" (Lucchini and Stoecklin 2020). Our approach is therefore not centred on individuals but rather on an "object" (the social process) experienced by them. This displaces the focus from substantial characteristics of this population to processual experiences that they are exposed to and that they simultaneously construct. The "object" we study is therefore the "forms" emerging from symbolic interactions that take place in the streets and within institutions sheltering children who are (still) connected to the streets. We shall hence look at children's rights not just as a list of the rights that are most important for CSS, but at the "forms" that children's rights actually take in the street space and in institutions that are supposed to provide these children with minimum standards regarding the respect of their rights. With this focus on processes, we pursue the endeavour of qualitative research:

> The expression 'children in street situations' disentangles children from their situations, stressing that the problem is not inherent to the children but to the situations in which they have to make a living. It obliges to address the situations, and not just the children experiencing them. The plural form (situations) not only emphasises the diversity of these situations, it also stresses the evolutionary dynamics of the streets, as opposed to the static view that implicitly conveys the image of waves of uncontrolled children filling in dead-ends and submerging cities, as is especially portrayed by the Chinese words for 'street children' – *liulang ertong* – 'wandering children' (Stoecklin 2000; Aptekar and Stoecklin 2014). The expression 'children in street situations' therefore entails a change of perspective which looks valid cross-culturally: a widening of the focus. However, this change is difficult for various reasons: political reluctance, a cultural habit of individualized selves, and, last but not least, theoretical blindness. As researchers, our responsibility is to work on the latter (Lucchini and Stoecklin 2020, p. 2).

According to De Moura (2002), the social construction of the category is in itself a form of intervention or "social action", which leads to a "naturalization" of deprivation. De Moura (2002) observes that the discourses about life in the streets is often based on macro-economic pushes and too much of a linearity process that ignores nuances and localities. The designation of a social problem becomes particularly ambiguous when a population is defined as a social category and the quantitative evaluations it is subjected to are multiplied. Indeed, the number of children in each country who can be situated within the highly flexible category of "street children" will determine the media response to the phenomenon, influencing social reactions and the debates over advocated solutions. This has led to today's paradoxical situation: there is a profusion of misleading descriptions of the living conditions of "street children", and a tendency to inflate their numbers. This in turn has led to passionate debates about what must be done, but there is no (perhaps *because* there is no) agreement on the very definition of the "street child".

The statistical approach is hence anything but neutral. It can even be misleading. It claims to measure the magnitude of a "social problem", but in fact it contributes to its construction: it signals a population that does not correspond to the norms, without questioning the latter. Within this function of naturalizing dominant social norms statistics are particularly effective, thanks to the nature of numbers: immediately graspable, but by the same token contributing to making the underlying stakes invisible. Their apparent neutrality establishes a vision of consensus about a social order that is in fact contested. We can see this process as a contemporary example of Plato's allegory of the cave: we are dazzled by our fascination with statistics to such a degree that we do not see the power that they incorporate. To shed light on this bio-power (Foucault 1991) we must hence read "behind" the numbers, and see its effects and its functions.

13.3 The Social Construction of Inequalities

The labelling of "street children" is a process by which the social construction of inequalities is reinforced (De Moura 2002). This is also observed by Lucchini and Stoecklin (2020), who have both worked about 30 years in this field. They still look at any definition (be it "street children", "street-connected children", or "children in street situations") as "a concept in search of an object". In their view, what is important is the object of study, the sociological process that can be observed by interviewing actors involved in it. It is in the nature of the sociological perspective to imply contextual analysis: social actors and their behaviours are seen as instantiations of the sociological dynamics. The sociological perspective sees the social categories in which people are falling (and one individual falls into several categories, according to age, gender, ethnicity, profession, etc.) as the result of a historical process of centralization of power.

One of the major issues for administrations based on a legal domination (Weber 1978), and hence conventional law, is to administer the population according to certain criteria. This is done through the qualification of each individual by of a certain number of attributes, to know how many individuals are actually concerned by a specific attribute. These attributes are constructed as meaningful, that is as having a "meaning" that is commonly recognized as uniform. Followers on the Foucaldian perspective have shown that categories used for statistical compilations are the categories that those in power have the means to impose as "taken-for-granted". The category is not put into question as long as the dominant view about what it "means to be" *x* or *y* is accepted. Hence, individuals may be classified as men and women, until some voices raise the issue of people not feeling comfortable with being classified in one sex or another. The gender issue is a typical example illustrating that taken-for-granted categories may be questioned at some point. The same goes for "street children". It was assumed for years that "street children" are, broadly speaking, children who live and/or work on the streets, until research giving more voice to such children showed that their reality is more complex than just "being on the street". Consequently, this change in the perspective, accompanied by more complex definitions, also affected to some extent the political-administrative practices and discourses. The hesitations and misunderstandings that are reflected in our present article are typical of this complex process of reciprocal adaptation between the perspectives of social sciences and the perspectives from administration.

So, we have a tension between the sociological perspective that is concerned with social processes, and the political-administrative perspective that is dealing with social problem. What is at stake is the response to the following question: how do social processes turn into public problems? In other words, what does it take for a process to be viewed as a problem? Not all social processes turn into public problems. For instance, ageing in traditional societies is not a problem, as traditional society are marked by high fertility rates and hence enough people can care for the elders. It however becomes a problem in modern societies, after demographic transitions (low fertility rates, low mortality rates): ageing becomes a problem of imbalance between working and non-working population, and the associated issue of transfer of social benefits. The same goes for childhood. To be a child is not a problem until a child is seen in a situation that is felt not appropriate for "people of this category", for instance the "street situation". This is linked not only to social representations regarding childhood but also to social practices that have constructed "streets" as they are, with specific characteristics and functions, opportunities and dangers, assets and shortcomings. It is the encounter of "street" and "children", with the connotations they have, that construct the social problem: the child (seen as good, innocent, vulnerable) and the street (seen as dirty, dangerous, noisy, etc.) form an oxymoron. The feeling of "strangeness" associated to an oxymoron hence makes the sociological process of "children living and/or working in the streets" a public problem called "street children". But now, since the research has shown how reductionist and stigmatizing it is to use this category, a movement of deconstruction of the category is gaining momentum.

In the last decades, many new approaches use the term "situations" to deal with "children in handicap situations, children in migration situations", and of course "children in street situations". They all reassess the social and material barriers that impede access to one's rights.

This movement is contributing to the "ontological turn" in childhood sociology (Alanen 2015, 2017; Spyrou 2019). While childhood is still seen as a social construction, along the "new paradigm" in childhood sociology (James and Prout 1990; James et al. 1998), the turn to ontology invites to reassess the link of social constructions with the material or physical aspects. It considers that symbolic constructions cannot be independent of the materiality of social life. Consequently, to come closer to the subjective meanings given by specific people to their situations, the very material conditions must be assessed as important components of experience".

This "ontological turn" makes sense in terms of human rights (foremost the right to be heard about one's own life or "situation") but it is challenging for practitioners. This is why the General Comment 21 of the Committee on the Rights of the Child on "Children in Street Situations" may be seen as puzzling, disorientating, to many people in charge of the administrative perspective of problem solving (sometimes even including researchers). They are somehow lost because the perspective of GC21 is closer to social sciences, regarding not just individuals, but their families, their contexts. Hence, the administrative habitus (Bourdieu 1998) of classifying individuals is jeopardized by multiple memberships and trans-sectorial thinking. The same can be observed with "childhood", once a sector in administration, becoming now a "transversal preoccupation" that should concern all sectors (economy, urban planning, education, security, etc.). Hence, the "ontological turn", departing from the administrative habitus of classifying individuals into boxes, is challenging in many ways. It challenges mainly traditional administration and power imbalances. And it also challenges essentialist views on human rights. These rights are "living" (Hanson and Nieuwenhuys 2013) as they gain signification only in context. Consequently, it is very important to understand the context of appearance of certain claims, turning into rights, and the negotiations that are then happening during implementation of guidelines based on certain interpretations of these rights.

Lucchini and Stoecklin (2020) proposed to contribute to "a renewed reflection on social processes that we must understand if we aim to link children's rights with their contexts of realisation" (p. 1). To do this, they advocate for:

> understanding the relationship to values (including children's rights) that govern observable behaviours. Illuminating the gap between claimed values and actual behaviours enables us to show the fundamental incompatibility between children's rights and the expression 'street child': this expression essentialises a human being whose situation in fact results from social processes that marginalise them. Access to their rights remains insufficient if we keep calling them 'street children', as this expression masks the very processes that must be changed to truly enable better access (Lucchini and Stoecklin 2020, p. 1).

With this perspective we can better illuminate the effects of the social construction of public problems through history. Most countries made or updated laws that criminalized vagrancy and idleness by the turn of the XIX to the XX century (Beier and Ocobock 2008). For instance, in Belgium a law was promulgated in 1891, sending many individuals to prison in case they were found in the streets, being idle or begging. In 1866, in the wake of the end of slavery in Virginia, the American State passed a law—"Act for punishment of vagrants" that targeted all those who were not working, especially the newly freed blacks of the State. Such was followed by many southern states. Anti- "vagabondage" laws appeared in Germany in 1871 and were later strictened during the Nazi Period (1933–1945). In the Early-Meiji Period, the Japanese efforts to "catch up" with the west included anti-vagrancy laws (Rusenko 2020).

Althammer found that the discussion about the "vagabond question" was much fostered through massive transnational expert forums, such as the International Penitentiary Congresses, the general meetings of the International Union of Penal Law and the International Congresses on Patronage (2014). The realization of those events worked as a form of "moral entrepreneurship" (Becker 1963) that stigmatized the streets as a place of a social evil. This walked hand in hand with the rapid advance of industrialization and capitalism that depended on the existence of mass of able-bodied free workers that demonstrated volition to work, and to which vagrancy and idleness was considered undesirable for the interests of a positivist and entrepreneurial state. It is not surprising that even children at the time were considered as a work force, and that their presence in the streets was seen as a threat. The First Code of Minors of Brazil (1927), for instance, subjected children found in the streets to institutionalization.

Much of the mainstream capitalist discourse of labour as a duty remains in developed and developing economies at large, now reinforced by neoliberalism. This is a counterpoint to the fact that many of the anti-vagrancy laws were repealed over time. Even though the criminalization is not in force, the social representations and the resulting stigma faced by the population in streets situation show the lingering effect of the discourse. Many studies of this population highlight the dichotomy of being perceived as lazy or being blamed by their own fate (Kidd 2007; Hopper 2015), and how homeless adults and children are placed against the capitalist notion of the self-made individual (Belcher and DeForge 2012, Bellingham 1983).

The evaluation of the street as a dangerous space stem from modernity, which, in its separation of private and public spheres, has reduced the street to the transitional function. Its differentiated use hence reveals social distinctions. This functional transformation of the street in countries of the Global South, via the middle classes, prompted a process stigmatizing the poor, as they cannot afford to make the street a mere transitional space. The transformation of the street with the rise of productivism in the nineteenth century marginalized many children who were taken away from their families by State intervention and placed in rural homes of Christian families, and in the case of U.K., in the colonies, as Belingham writes:

> Horatio Alger was a great booster. Most of his protagonists benefited from Children's Aid Society services, and one was sent West as an 'emigrant'. He prospered as a land speculator, escaping captivity by Indians. Brace personally oversaw the emigration of seventeen thousand city urchins. S. Wohl writes, 'By the end of the nineteenth century, a web of such benevolent enterprises, in which Brace's ideas and policies were flattered by almost slavish imitation, criss-crossed the nation'. The policy was adopted by the British beginning in the 1870s. The Barnardo Homes alone shipped over eighty thousand children from the slums of London to the colonies (Bellingham 1983, p. 304).

The social construction of inequalities is reinforced by normative frameworks built on implicit judgements on the depraved and immoral of people that are in fact marginalized the process of policing that is reducing psychosocial diversity to homogenous categories. We now turn more specifically to this normativity that is taken-for-granted or hidden behind layers of administrative proceedings.

13.4 The Hidden Normativity: The Entrepreneurial Mode of Action

For the interviews conducted with CSS in Brazil in 2022, we used the heuristic tool called "Actor's System" (Stoecklin 2013) and the theory of "transactional horizons" (Stoecklin 2021) that is built around the "actor's system" which assumes that one's system of action is the outcome of the recursive links between five components of personal experience, namely activities, relations, values, images of self and motivations (ARVIM). The actor's system is represented in Diagram 13.1.

The five components of the actor's system are found anywhere, as all human beings can refer to them when reflecting on what they do (activities), who they know (relations), what they believe (values), how they consider themselves (images of self) and what they want (motivations). These categories can embrace an infinity of experience. Social actors are not necessarily using these concepts (ARVIM) but they will nevertheless be the directions they look at when they want to communicate something about experience. These "directions to look at" are propositions with a content that is not given beforehand (Le Breton 2004, p. 41). These "sensitizing concepts" (Blumer 1969) are like containers of experience presiding over practical and discursive consciousness. Hence, to underline their function, they are called "transactional horizons", defined as "symbolic landscapes channeling social interactions" (Stoecklin 2021, p. 1).

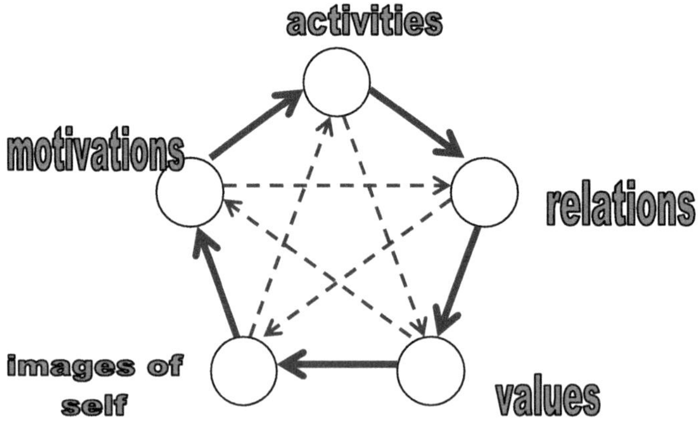

Diagram 13.1 «Actor's system» (Stoecklin 2013)

Social actors use transactional horizons (ARVIM) as daily routines (hence mostly unconsciously) to link together concrete items of perceived reality. Hence, ARVIM represents a symbolic matrix which is recursively involved in interactions, being simultaneously a medium for and an outcome of social practices. Accordingly, this duality of structure recalls the structuration theory (Giddens 1984) where "structure" is conceived as an abstract entity that is recursively constraining and enabling of practices. The transactional horizons (ARVIM) of the actor's system build up the structure that is mediating the relationship of social actors among them and towards the social system that they are co-constructing precisely by means of this symbolic structure. ARVIM are conducive to stabilized interactions in specific social systems. They are hence conducive to "modes of action" that are tinted according to the transactional horizon that is predominant: according to time and place, one can therefore act in a predominantly entrepreneurial, relational, moral, identitarian, or motivational mode of action. Of course, the fact that one mode of action is predominant doesn't mean that the others are absent, quite on the contrary: they are there but subordinated to the predominant mode of action. These modes of action are further described below:

> «The entrepreneurial mode of action focuses on activities that produce objects exterior to oneself (*poiesis*) and strategies believed to be the most efficient to achieve one's goals (corresponding to Weber's "rationally-purposeful action »). The relational mode of action puts emphasis on relational configurations (it is close to Weber's « traditional social action » when it favors habits and routines that reproduce the social status and positions of actors). The moral mode of action is based on the belief in the inherent worth of specific values (Weber's « value-rational action »). The identitarian mode of action bases on the intersubjective definition of self (it partly corresponds to Weber's « affective social action » as drives also inform subjective identity). The motivational mode of action is the most complex one. It has no correspondence in Weber's typology of social action, it is closer to inquiry (Dewey 1938)" (Stoecklin 2018, pp. 564–565).

The transactional horizons of the actor's system are quite illuminating when it comes to highlight the normativity, that is mostly hidden or taken-for-granted, in daily routines. For street life, the predominant entrepreneurial mode of action has massive impact. In the capitalist world, the individuals are reduced to what they can do and contribute—their entrepreneurial self, rather than what they are. This normative standard is pervading all settings, and this is how hidden frameworks are naturalized in daily interactions within the centres and between the centres and the administrative and political authorities. In the context of the functional transformation of street use along the globalization of market-oriented economy, any individual occupying the street in non-productive ways is stigmatized:

> The general trend of public places in big cities is towards a dichotomy between a majority of people using them as transit spaces between their major integrative spheres, primarily the family and the workplace, and a minority constituted of marginalized people who tend to occupy these places for some time. The social visibility of marginal people in the heart of cities therefore becomes even more evident. Social reactions to such visibility raise debates about 'policing' (Lucchini and Stoecklin 2020, p. 4).

Hence, institutional governance can be seen as the translation into the administrative language of the modes of action that are seen as the most relevant for this purpose.

Max Weber underlined this with his concept of bureaucratic domination, centred on the « instrumental-rational social action» (Weber 1978). Consequently, what eventually counts as good governance is made through the entrepreneurial mode of action. This entrepreneurial mode of action is therefore also pervading social interactions between CSS and the staff of care centres. It frames the forms that negotiation can take. As will be shown below, the entrepreneurial mode of action, with its focus on the "performative" individual, pervades the presentation of self of our respondents.

The results of our nation-wide qualitative research conducted with over 50 children placed in care institutions in Brazil focus on their street trajectories and points to a complex social process of virtues and stigmas associated with different aspects of the streets. One of the most striking elements is the pervasive stigma associated with the label "streets situation". Most of the children who responded to the questionnaire declined their relationship with the streets, but through other questions, such as their work and their leisure, the issue surfaced. This is the case of Layla,[1] 16, who responded: "I am not like the other teenagers [from this shelter]", and "I don't have the street profile". Later we found that she spent most of her weekends in the streets, using drugs, doing petty thefts, and staying with her boyfriend. Yet the fact that she first tries to escape the stigma associated with "street life" confirms the importance of the entrepreneurial mode of action: street life is perceived as antinomic to a performative person.

The stigma, deriving from this antinomy, is also present in the relationship with other persons of interest, such as the staff and public authorities. In another shelter, where the majority is composed of children in streets situations, a staff mentioned that stops and searches are frequent with the children in the shelters. In one episode they were threatened with a gun by a policeman, and they often heard: "do you defend those tramps[2]"? "Shame on you". The denomination of "tramp" is just one of the negative connotations associated with the street trajectories and idleness. The dichotomy work vs street was found throughout the different dimensions emerging from the data.

In one of those interviews, Neilton showed how much he was proud of being able to go to places and find his own way to work. He spent a long time in the streets working in land lots and other menial jobs, and as a result he skipped school. To him, being autonomous and working were of utmost importance, tracing a contrast with his "idle" and boredom-led life in a shelter in which he depended on the staff to find jobs. He did not associate his streets situation with a negative connotation of vagrancy, but rather of freedom.

Luan, 17, is a boy who declared just spending three days in the streets. He was found by policemen in the bus station of one of the capitals in Brazil after he ran away from a rehabilitation centre. When asked about work, he was tacit to say he was still selling items in the street. Interestingly, Luan had a very bad image of himself, possibly connected with his history of substance abuse.

Interviewer: what do you think about yourself?

---

[1] All surnames used in this chapter have been changed to respect the rule of anonymity.

[2] "Vagabundo" in the original.

Luan: I think it's shit.

Interviewer: why?

Luan: Because sometimes when I am far away, in some place, I feel useless.

Again, the performative self is the background on which this feeling of uselessness takes shape. This image could also be associated with a reinforcement of the stigma of the streets by the Tutelage Council and the staff. This is exemplified by the experiences of another boy in the same shelter. Nonato was removed from his family, according to him, because he had "progress" in street fights, he had progress of people seeing him in the streets selling torrones and that he was often seen cleaning some land lots. People used to comment that they would find him in the street and would call a shelter to get him. "I never believed that shelter existed, because I never saw a shelter". Nonato spent over six months in the streets, because he wanted to help his mom and his family, but now he was removed from them. In this case, the association of the working in the streets is less marked by autonomy and independence but leading to a dynamic of danger:

Nonato: I wanted to be in my home, but I can't.

Interviewer: Why can't you?

Nonato: Because there is no risk for me to stay home. But the shelter doesn't allow me, they say I will do the same things (...), selling torrone to help. So, I prefer to stay in the shelter, to not be in the streets and getting involved with those things.

Livia spent over a year in the streets. She was ordered by her mother to work outside, selling stuff such as socks or sweets in the streets, otherwise she would be beaten. She was discovered because of the bruises made by her stepfather. She describes the feeling of living in the streets:

> It was bad because I had to wait, I had to depend on everyone to eat and to sleep as well. There is a lot of evil in the streets. I had to keep waiting for food donations when I couldn't pay for it. I was selling my sweets. But I couldn't go to a pension because I was a minor, I had to sleep in the streets and wait for food donation, otherwise I would have hunger, and in the Sundays, there were almost no donations, I had to skip meals often until Monday. (…)In the streets I couldn't play with the kids [her siblings]. In the streets you can't even sleep properly, you have to keep one eye open and the other shut, someone can come and take advantage of you. However, I always had a lot of friends in the streets, nobody ever did bad to me.

The relationship between work and the streets was also evident in the way she saw herself:

Interviewer: What do you think it's most important in life?

Livia: To have a work. Without work, we are nobody.

It is interesting that Livia got pregnant during the time she spent in the streets. She mentions she "had to" leave the other shelters because her brothers and her were hurt by the "social parents". Food restrictions were common punishments. She preferred to stay in the streets. When she was about to deliver, she said she knew she would have to go back. Luckily for her, she stated, she was placed in the best shelter of the city, where she could take her child with her.

Bruno, 14, used to stay in the streets with his mom who was a drug user. Whenever she could, she would send him to day care and preschool. However, he remembers suffering prejudice there because of his situation—being the only boy that could only eat when someone shared their food with him, the teacher paid little attention to him and no one else wanted to play with him. "I only learnt how to count until three, and I thought all the cars were black because I only knew the name of the black colour". Lacking friends, he said, he had to create them: "I had over 200,000 voices in my head, they were guiding me to do activities".

His discourse raises the attention to an entrepreneurial mode of action, one that the child was considered less because he had not seen or his family as someone that could strive by themselves. His stay in the street and the needs of his mom led him to work picking up tin cans in the streets. More than a survival strategy, it seems that he also did it to work on his relationship with his mother. He often gave her the money so she could buy drugs. We can interpret that, due to his experiences and ambitions, to him giving more worth to his relations and values than to his activities. However, it seems he also have ingrained part of the entrepreneurial discourse. When asked how he sees himself, he states:

> I am a lazy person, I am very lazy person. (…) I look at myself in the mirror and think – 'wow'. (…) I was used to run, do exercises. Now I am like a family's old man. I am one of those husbands that have kids to take care, I only arrive from work and put the hand in my belly.

The image he has of himself is interestingly reflected on the way he portrays the dynamic of the shelter. Here the discourse of the institution becomes evident, and the result if an amalgam of the different lines of thought conversing about the role of the shelter, work and the streets.

> Interviewer: In your view, how does this house work?
>
> Bruno: It's an anthill. The workers, the mains, and the queen ant. We are the future of the anthill. We have those that got strayed from family, and so on. I only don't compare with bees because bees treat each other like family. In the anthill they take other ants, there are workers that take care of eggs, which are us, the workers, which are the educators that take care of the younger ones, the coordination are holders and the lady there is the queen ant.

One of the shelters took the concern of the streets and label to a different level: upon reading our questionnaire, they prevented the interviewer from asking a question about the reason they were institutionalized. They thought it was too invasive and could lead to the children to revive bad memories. In fact, it is the only shelter in its region that is focused on tending to children in streets situations. The refusal of the question was coupled with the fact that two out of three boys interviewed declined being on the streets and did not give more secondary information.

The excerpts above give light to the troubled status as subjects of rights gain according to different perspectives arising from their own images and acts and the discourses of practices of the shelters. Those children often felt like nobody, someone who has a lesser value to the society and whose rights don't count as much as the other. Often the result is that the shelter assumes a paternalistic stance that seeks to intervene on their life, at the cost of not listening to their voices. One example is the effects of the religious discourses. Leomar, a 15-year boy who spent a lot of time in the streets, speaks about the activities that the shelter asks the children to follow:

> Leomar: Sometimes the church come on Saturday, Sunday, I start to like, then we have a word, and we get to talk.
>
> Notes: The staff told me that generally the pastors organize outings with the teenagers and in the weekend there are Christian worships.
>
> Interviewer: How do you feel about that?
>
> Leomar: I always want to be alone, but it never works.
>
> Interviewer: Aham.
>
> Leomar: Sometimes I want to be alone, but it doesn't work, many screams, cries, of the people in the nursery as well, so much complaint.

The paternalistic stance also informed the discourse and practice of the intermediate organs, those persons and organizations that are responsible for first response in case of the found rights violation. It is the case of the Tutelage Councils, the police and also school officers. Often in the interviews it was found that the children or even the family had no participation in the process of placement/removal, which highlights the interventive facet of the public policies and reveals the hidden normativity of a culture of institutionalization (Rizzini and Rizzini 2005).

Ariane, the staff of another shelter, explains the process of institutionalization:

> The majority [of children don't participate in the process of placement] because it is only in the arrival that we can explain why there are here, because many arrive without … I had a kid that I received the person that didn't even say it's a shelter, they said he was just going for a stroll here in the city. He just left the kid here and we had to explain, so they don't …. Many families also complain about this, the mother was not aware of the event, someone came, took my son and brought him [to the shelter]… (…) when they arrive they normally don't know, they just fall here…

Irma, another staff in the same shelter:

> Irma: Some of those here had the talks about coming to a shelter, but others not, inclusively some of the children here, they were sought in their homes, by the police, the three of them, and the police told they were going somewhere else, and when they arrived it was here, a shelter. They didn't even know, were caught by surprise. Those teenagers, to adapt to the rules here and to continue in a new home, with to comply, it is hard.

However, the issue of inequality of rights is also presented on the way they strategize and capitalize on their situation. One of these events is that often the teenagers with streets trajectories know the rules of the system where they can find a place to stay or how to return to the shelter in case they evade. This is illustrated by the experiences lived by Mayana, a staff of a public shelter. When asked about how the children had notion of their rights, she was categorical:

> Mayana: They do, even so much, they talk a lot between themselves, because normally those who come to the shelter have a lot of experience outside, for instance, there are teenagers

who escape and say: "ah, tomorrow I am here, I will know what I want, and tomorrow I am here, I will just go to the police station.

Third person: I will just go out.

Mayanna: Yes, they know very well the rough paths, they know what to do, they stay here until they want, then after they don't want it anymore, they go to the streets and cause a mess, and it's something falling out of our control, as I said: we can't lock them here. We are not a prison, we are a shelter, the protection is for those who want to be protected. (…) For instance, sometimes we have a newcomer that is involved with drug cartels and has more experience than all of us here, then he gets two minutes chat with a teenager that we worked the whole month with, and he manages to get him to escape, got it?"

The inequality of rights of CSS is also reflected on the complexity of intersections instances of exclusions. The experiences reported by the girls were that they were often asked to step in in lieu of the mother to take care of younger siblings or were forced to do domestic work. In this scenario, the role of the street looked like a way out. It is the case of Nara, 17: her mother took her out of school so she could help take care of her siblings. She stated she didn't spend time in the streets, but the staff in her shelter told she spent a long time in the streets when she was not in the school. A similar story of Tabata, 16, who had to step up when her mother started to neglect their small sisters.

Interviewer: How was your life before you came here?

Tabata: Look, my life was good, I was studying, staying at home, I was going to my mom's home, taking care of my sister and that stuff, because my mother didn't take care of the small girls, only me when I was living with her, she didn't take care, was always parting out, did a lot of wrong stuff and complained.

When we see the statistics about children in streets situations and the gender ratio, data generally indicate that boys are the majority in the streets. This became evident with the research conducted by CIESPI and OPN in Brazil, in 17 different cities: in the streets, 75% of declared themselves boys and 25% declared themselves girls, while in the care institutions the numbers were respectively 70% and 30% (2020). However, the numbers can sure be misleading if we take the object as the result and forget about the causes. Similar to the approach of the Actor-Network-Theory (Latour 1996), if we ask where the girls stand in the process that led them to the streets, we will find that they are equally victims, but the destination is different. While boys are sent to the streets or rely on them for survival, girls recur and are used for the domestic work in the same family settings (Rizzini and Lusk 1995; Frangella 2000). The presence of girls in the streets is also an issue with the stigma of prostitution: girls are often assimilated with prostitutes, although a large majority of them does not take part in prostitution (Lucchini and Stoecklin 2020). This conflation is a consequence of stigmatization that affects women when they leave home for public spaces, and it is also reinforced by programmes for girls that focus on prostitutes.

The very nature of defining what street situations is in face of the evolution of the social phenomenon often made the staff confused on whether the children had trajectories in the streets:

Interviewer: Eventually, in this profile of children you tend to here, you said negligence, situation of violation of rights, etc. What about children in streets situations?

Marcia: Our profile also includes that, however nowadays we don't have anyone that is specifically from this "street situation". There are situations of street, like, in case they have some contact with the family, but they are exposed in the streets, like leaving home without warning and spending two, three days and returning, but always having that family reference.

Interviewer: So, it's a street situation that is not specifically living, but being exposed in the street.

Marcia: That.

Interviewer: being unprotected or unnacompanied by an adult.

Marcia: That

Interviewer: in a situation more of wandering…

Marcia: without approval from parents.

To complete the round of the different associations of children and care institutions, we had witnessed instances where the own staff let stigma surface. An educator of a public shelter once mentioned:

So [CSS] hinder a lot [of the process of the socialisation of other children] cos the children without experience in the streets start to have a link with those who have and then they start to go to the streets. And then it creates a vicious circle, because sometimes we even lose control [of the situation]. (…) Because they keep going on that path, I can say *escaping,*[3] you know? They spend a long time in the streets and when they go back they want to have the return like, how can I say, (…), an open arms welcome, like we were their family and that we had the obligation to accept them with all the vices of the streets, with all mistakes thy commit outside.

It is important to stress that such thought is not representative of the whole. In most cases the staff treated and considered CSS well. But traditional social representations still emerge here and there in the discourses of stakeholders that deal with this public. We find it relevant to conclude with this passage to show how even relationships are marked by the entrepreneurial mode of action. The relationships in the shelter are underlying in the concept of socialization as a task. "Like we were their family" is a thought that implies that the care institutions should weigh between tender, love, and care on one side and the procedimentalization of their passage towards an exit on the other side.

13.5 Conclusion

Observations made in Brazil show that the translation of social processes into an administrative logic that is eventually what counts as good governance is made through "transactional horizons" centred on an entrepreneurial mode of action. The

[3] Original verb « evadir», to leave the shelter without authorization. The "verb" evadir has a negative connotation as well. Like "runaway".

children and staff interviewed use labels that are recursively reinforcing the stigmatization of "street children". Yet, labels can only detach themselves from an implicit background: street life is detaching from the productive self. The entrepreneurial mode of action that forms the background on which intervention takes place reflects a future-oriented thinking (things to achieve in life) and detracts the observer from the factors that have caused the social exclusion experienced by our respondents. The analysis assesses how the inequalities are shaped and also shape the definitions, hence confirming the underpinnings of the structuration theory (Giddens 1984), and suggesting the recursive power of the entrepreneurial mode of action.

Despite the General Comment on Children in Street Situations, a paternalist and future-oriented intervention is still prevailing. This must lead us to consider the importance of statistics in the building of the Nation-State and its administrative regime of truth (Foucault 1991). Statistics are of primary importance for States which are based on legal domination (Weber 1978) and today all States use statistics, even those that are not guided by the rule of conventional law. The challenge remains for multilateral instruments of global governance, like the UNCRC, to redress the historical process of bio-power (Foucault 1991) that led to new forms of inequalities that are mostly linked to how people are labelled. The inequalities experienced by children in street situations are already stemming from the regime of truth of apparently neutral demographic data. The fate of millions of children is abstracted in numbers that are stratified along criteria that respond to the State's interest more than the children's best interest (Art. 3 UNCRC). This can be reverted only if more qualitative research is made whereby some reflections of the experienced situations can acquire some significance for those experiencing them. Numbers alone hardly do.

Acknowledgements For the redaction of this chapter, the authors benefited from the research grant no. 10001A_179098 of the Swiss National Science Foundation.

References

Alanen L (2015) Are we all constructionists now? Childhood 22(2):149–153

Alanen L (2017) Childhood studies and the challenge of ontology. Childhood 24(2):147–150

Aptekar L, Stoecklin D (2014) Street children and homeless youth: a cross-cultural perspective. Springer, Dordrecht

Belcher JR, DeForge BR (2012) Social stigma and homelessness: the limits of social change. J Hum Behav Soc Environ 22(8):929–946

Beier AL, Ocobock P (2008) Cast out. Ohio University Press

Bellingham B (1983) The "unspeakable blessing": street children, reform rhetoric, and misery in early industrial capitalism. Polit Soc 12(3):303–330

Blumer H (1969) Symbolic interactionism: perspective and method. Prentice Hall, Englewoods Cliffs

Bourdieu P (1998) Practical reason. On the theory of action. Stanford University Press, Stanford

Cowell F (2011) Measuring inequality. Oxford University Press

De Moura SL (2002) The social construction of street children: configuration and implications. Br J Soc Work 32(3):353–367

Forbes KJ (2000) A reassessment of the relationship between inequality and growth. Am Econ Rev 90(4):869–887
Foucault M (1991) Governmentality. In: G Burchell, C Gordon, P Miller (eds) The Foucault effect: studies in governmentality. Harvester Wheatsheaf, Hemel Hempstead, pp 87–104
Frangella SM (2000) Fragmentos de corpo e gênero entre meninos e meninas de rua. cadernos pagu 14:201–234
Giddens A (1984) The constitution of society: outline of the theory of structuration. University of California Press, Berkeley
Hanson K, Nieuwenhuys O (2013) Reconceptualizing children's rights in international development. Living rights, social justice, translations. Cambridge University Press, Cambridge
Hollander J, Howard J (2000) Social psychological theories on social inequalities. Soc Psychol Q 63:338. https://doi.org/10.2307/2695844
Hopper K (2015) Reckoning with homelessness. Cornell University Press
Kidd SA (2007) Youth homelessness and social stigma. J Youth Adolesc 36(3):291–299
Latour B (1996) On actor-network theory: a few clarifications. Soziale Welt 47(4):369–81. JSTOR. http://www.jstor.org/stable/40878163. Accessed 20 Nov 2022
Lucchini R, Stoecklin D (2020) Children in street situations. A concept in search of an object. Children's well-being: indicators and research, vol 21. Springer, Dordrecht
Rizzini I, Lusk MW (1995) Children in the streets: Latin America's lost generation. Child Youth Serv Rev 17(3):391–400
Spyrou S (2019) An ontological turn for childhood studies? Child Soc 33(4):316–323. https://doi.org/10.1111/chso.12292
Stoecklin D (2013) Theories of action in the field of child participation. In search of explicit frameworks. Child J Glob Child Res 20(4):443–457
Stoecklin D (2017) The general comment on children in street situations: insights into the institutionalisation of children's rights. Int J Child Rights 25(4):817–869. https://doi.org/10.1163/15718182-02503014
Stoecklin D (2018) Institutionalisation of children's rights: transformability and situated agency. Int J Child Rights 26(3):548–587. https://doi.org/10.1163/15718182-02603004
Stoecklin D (2021) The transactional horizons of Greta Thunberg. Societies 11(2):1–24, 36. https://doi.org/10.3390/soc11020036
Rusenko RM (2020) Homelessness regulation and neoliberalism's imperial past: the Janus face of anti-homeless urbanism and Tokyo's modern socio-spatial development. Antipode 52(6):1815–1836
UNCRC (2017) General Comment No 21 (2017) on children in street situations. CRC/C/GC/21.https://www.ohchr.org/en/documents/general-comments-and-recommendations/general-comment-no-21-2017-children-street. Accessed 20 Nov 2022
Waldman LK (1977) Types and measures of inequality. Soc Sci Q 58(2):229–241
Weber M (1978) Economy and society. In: G Roth, C Wittich (eds). University of California Press, Berkeley

Chapter 14
Challenges and Opportunities for Taiwanese Urban Indigenous People: Segregation, Assimilation, and Identity

Chien-Chia Liu

Abstract Indigenous people have lived in Taiwan for more than 5000 years, but the safe living space and opportunities of indigenous people have been seriously constrained since Han settlers migrated from the southeast coast of China to Taiwan in the late sixteenth century. Ruling regimes of the past, including the Ming and Qing Dynasties and Japanese colonial governments, adopted a segregation strategy for the indigenous people, which resulted in the indigenous and the Han people living relatively isolated spatially until a century ago. Driven by the sluggish tribal economy and the pull of employment opportunities in more urban areas, in the late 1960s indigenous people left their native tribes and resettled in metropolitan areas, emerging as a new group: "urban indigenous people." The term urban indigenous people generally refer to indigenous people whose household registrations relocated to metropolitan areas. Beginning in the early 1980s, a large number of indigenous people left the tribes; most of them lived in neighborhoods dominated by the Han people. However, there are still few illegal urban tribes in the northern and central metropolitan areas of Taiwan. Those ethnic enclaves protect new migrants from ethnic discrimination; urban indigenous people are also able to benefit from social and cultural networks within ethnic enclaves. This chapter has three sections. The first introduces the migration and residential history of the urban indigenous people through a review of historical literature. Based on official statistics, including population census data, government survey data, and data from the national indigenous database, the second section of the chapter examines the ethnic socioeconomic gap among urban indigenous, tribe indigenous, and Han Chinese people. The final section of the chapter elaborates on government policies and efforts to reduce ethnic inequality within cities, given the social and cultural challenges and opportunities for Taiwanese urban indigenous people.

Keywords Taiwan · Urban Indigenous people · Ethnic inequality · Ethnicity

---

C.-C. Liu (✉)
Department of Medical Sociology and Social Work, Kaohsiung Medical University, Kaohsiung, Taiwan
e-mail: chienchia@gmail.com

G. H. Tonon (ed.), *Urban Inequalities*, The Urban Book Series,
https://doi.org/10.1007/978-3-031-59746-6_14

14.1 The Migration History and Current Status of Taiwan's Indigenous People

Indigenous people lived in Taiwan, the birthplace of Austronesians, as the natural masters of the land for at least 5000 years (Chen 2004; Gray and Jordan 2000; Gray et al. 2009; Li 1993). When the Han Chinese arrived on the island at the end of the seventeenth century, they competed with the indigenous people for living space and survival resources. Conflicts between ethnic groups occurred and resulted in reduced living space for the Indigenous people, impeding their survival. This turmoil continued until the eighteenth century, when the Qing government adopted a new policy to manage the Indigenous people separately. The government created an aboriginal boundary to separate the Han Chinese and the Indigenous people. At the end of the eighteenth century, the Japanese colonial government continued this policy of separation. Indigenous people were confined to their established living areas and had little contact with the Han Chinese. In the middle of the twentieth century, control of the Nationalist government returned to Taiwan and the distribution of indigenous people changed drastically, which is also a characteristic of contemporary indigenous people.

The separation between the Indigenous people and the Han Chinese that began in the seventeenth century and continued through the period of Japanese occupation found most Indigenous people living in the reserve land in the mountains by late in the occupation period. Those who moved to the plains lived primarily in Hualien and Taitung in the eastern part of the island; fewer than a hundred people moved to the cities in western Taiwan (Wang 1967). When Taiwan was colonized by Japan at the end of the nineteenth century, the colonial regime enacted policies, including mass relocation that halted the free movement of indigenous people; overall, the areas where the indigenous people lived and were active were fixed (Chen and Sun 1958). Demographers Chen and Sun (1958) analyzed the distribution of the Indigenous people using population data from the period of the Japanese occupation, demonstrating that the distribution of the indigenous people remained limited to a specific area.

When the Nationalist government took over Taiwan in the 1950s, it continued the Japanese colonial government's administrative structure for managing Indigenous people in the mountains. However, the Nationalist government did not prohibit Indigenous people from relocating; the number of Indigenous people who left the tribes to work in cities increased year by year. The large-scale migration of indigenous people in modern times can be traced back to the 1970s. The weakness of the economy in the mountains and modernization process made it increasingly difficult for the indigenous people to maintain their lifestyles in native tribes, as those tribes depended on the natural mountain forest resources. Some indigenous people lost their land to land nationalization and private trade. Indigenous people who did own land could not borrow from banks[1]; those who wished to use reserve-land forest

[1] Reserve-land cannot be used as collateral for bank loans.

resources were restricted by the Forestry Act. Indigenous people were also affected by the structural transformation of the economy and the development, and convenience, of communication technology. Modern life in cities and the charm of the capitalist society led to the mass migration of Indigenous people from their homeland.

Migration was driven by the unequal distribution of resources in Taiwan. Most Indigenous people moved for schooling and employment reasons. They had a higher rate of migration compared to mainstream society (Liu 2021; Liu and Lin 2010). Many of them moved more than once.

Since the twenty-first century, the trend of Indigenous people migrating to metropolitan areas has become more evident. The migration status of Indigenous people can be demonstrated by housing and population censuses. According to census estimates, approximately 39.9% of Indigenous people left their tribes to live in metropolitan areas in 2000; by 2010, the rate had increased to more than 60% (Liu and Lin 2010). Relocated indigenous people clustered around the metropolitan areas of western Taiwan, with the majority (50%) residing in areas around the northern metropolitan areas, less urbanized satellite towns near the metropolitan areas[2] (Liu 2011, 2021).

14.2 Inequitable Social Opportunities for Indigenous People in Cities

Currently, more than half of the indigenous people in Taiwan have left their tribes. These relocated indigenous people are called "urban indigenous people," a term the Council of Indigenous Peoples defined as those who have moved away from the 55 statutory traditional townships of indigenous people. The classification of urban Indigenous people is based on individuals' household registration; those who are not registered in their original hometowns are regarded as urban Indigenous people while those who are registered in their original hometowns are regarded as tribal Indigenous people. Indigenous people who live in rural areas outside of the tribes are also urban Indigenous people by definition.

According to Council of Indigenous Peoples statistics, in 2002 approximately 32.8% of the Indigenous people in Taiwan were urban Indigenous people. That percentage increased to more than 40% in 2007 and continued to increase year by year to reach nearly half (48.69%) in 2021 (Fig. 14.1). At the same time, the proportion of people remaining in their original tribes decreased yearly (Council of Indigenous Peoples [CIP], 2022a). Since 2002, the population of urban indigenous people has increased by 0.8% annually; the number of urban indigenous people is expected to soon exceed half of the total indigenous population.

[2] Liu's (2013) investigation of the areas in northern cities with higher densities of Indigenous people found a strong negative correlation between the density of the Indigenous population and the socioeconomic index in the areas. Hot spots of Indigenous people were often the areas with poor socioeconomic attributes.

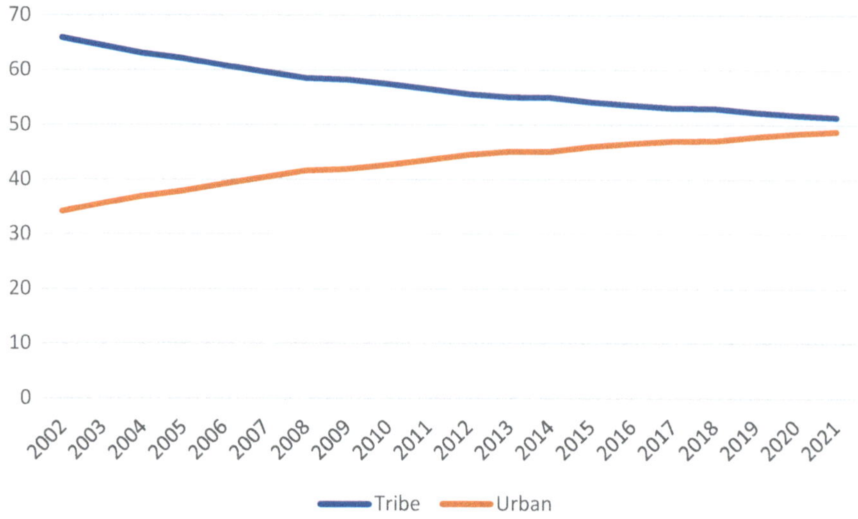

Fig. 14.1 The proportion of Indigenous people in Taiwan, 2002–2021. *Note* The graph shows changing proportion of Indigenous people in Taiwan from 2002 to 2021. *Source* Council of Indigenous Peoples Database, 2022

Existing studies on urban Indigenous people worldwide have focused on the disadvantages faced by urban Indigenous people in job-seeking, schooling, and social interaction, discussions focused primarily on social issues, welfare, and charity. These studies have shown that the situations faced by urban Indigenous people are very similar across the globe. For example, most Indigenous people in the United States and Canada who have left reserve areas struggle to survive in urban areas, working hard but barely maintaining basic living standards. Some have become urban slum dwellers, thus facing numerous new problems. Australia and Canada have formulated various policies to protect indigenous people, but have not yet succeeded in eliminating or even improving the plight of urban indigenous people who have left the reserve areas and migrated to metropolitan areas.

As traditional cultures have declined and tribal lifestyles changed completely in Taiwan, Indigenous people have moved to areas dominated by Han Chinese, either actively or passively, and have lived according to the customs of the Han communities. Compared with Han Chinese, these Indigenous people are underprivileged in terms of socioeconomic status; have higher unemployment rates, shorter average life expectancies, and lower academic achievement.

Although they remain underprivileged in terms of socioeconomic status, when urban indigenous people in Taiwan move to cities, more resources and job opportunities are available to them. Compared with indigenous people who remain within tribes, urban indigenous people achieve higher levels of education and higher socioeconomic status. Tribal Indigenous people receive 9.8 years of education on average, with a socioeconomic score of 36.8; urban Indigenous people receive 10.3 years of education on average and have a socioeconomic score of 38.9 (Liu 2011). Due to structural constraints, including a lack of regional resources and underdeveloped industries, tribal indigenous people have a lower average education level; their low employment rates are a result of the limited employment opportunities in their home areas. Agriculture is still the primary form of production in tribes; Indigenous people also perform odd jobs in nearby towns during the slow seasons of farming. Some Indigenous people within the tribes are passionate about forming their own businesses. Many engage in creative cultural businesses and tribal ecological tourism with the help of the government, but relatively few are successful financially.

While tribal Indigenous people live with tribal community members, most urban Indigenous people who have migrated to metropolitan areas live with and among Han Chinese. Financial constraint, primarily housing expenses lead most transplanted indigenous people to scatter across urban areas and reside in towns that surround metropolitan areas, towns with lower costs of living. Studies have confirmed that Indigenous people who live across urban areas tend to choose to live in districts surrounding metropolitan areas, and that those districts have lower housing prices and lower consumer price indices (Liu 2013).

However, some Indigenous people continue to live together in a tribal-settlement form after relocation to metropolitan areas, retaining familiar living and working habits. Using northern Taiwan as an example, Indigenous people who moved to the Taipei metropolitan area in the 1970s were engaged in "3D" (dirty, dangerous, and demanding) jobs. Their jobsites could be categorized into the 3 "mosts"—the highest (construction site), the furthest (ocean-going fishing vessels), and the deepest (mines) (Yan 2005). Most of the urban Indigenous people who lived in the Taipei metropolitan area at that time were engaged in the construction industry. Those workers lived in temporary dormitories next to the construction sites primarily and did not have fixed residences. When they were between jobs, they had to find a new place to stay within the city.

Indigenous people who moved to urban areas before the 1980s faced high rental costs, but had lower demand for accommodation. Working on construction sites, they had easier access to construction materials and created their own lodging. Accustomed to living with other tribal members, urban Indigenous people developed several urban tribal settlements in northern Taiwan. They built simple dwellings either in uninhabited marginal areas of cities or along the banks of the streams of northern Taiwan.[3] These self-constructed settlements became known to tribal members through word of mouth and gradually developed into larger urban settlements in metropolitan areas (Yan 2006, 2008).

These urban settlements provided spiritual and physical sanctuary for the Indigenous people working in the metropolitan areas before migrated tribal members could gain firm footholds in cities. However, since most of the urban settlements were

[3] Well-known settlements along riverbanks were built by the Amis people primarily; they were used to living near water and tended to settle in an environment close to the hometown of their tribes. Right-to-residence conflicts developed in several Amis cities in northern Taiwan, including one involving the Xiaobitan tribe and the Xizhou tribe by the Xindian River and one among the Sanying tribe, the Saowac tribe, and the Kanjin tribe by the Dahan River (Chen 2010; Yang and Lin 2021).

located on state-owned land, the land was generally not suitable for living (the districts were either intended for military use or were new riverside land). Settlements along riverbanks flooded every time a typhoon or torrential rain occurred. As residents lived in the settlements illegally, the safety of illegitimately connected water and electricity was a question. Most urban settlements faced disputes about their legality. Before 2000, government policy was to eliminate any illegal settlements, forcefully demolishing all buildings. However, the Indigenous people rebuilt their residences again and again. Later, after residents protested and negotiated with the local government, issues were solved primarily by the creation of legitimate settlements in the same area or resettlement elsewhere.[4]

The indigenous people living in urban areas were not only underprivileged in terms of housing, but experienced other huge disparities in education, labor and health. Statistics from the Ministry of the Interior show that the proportion of Taiwanese people older than 15 years educated at the high-school level or above increased from 28.3% in 2004 to 40.6% in 2021. The number of Indigenous people educated at the high-school level or above increased, but only from 13.9% in 2004 to 29.0% in 2021. As Fig. 14.2 demonstrates, among people older than 15 years with tertiary educational attainment, the proportion of Indigenous people has remained lower than the general proportion nationwide. Although ethnic group differences in tertiary education are narrowing, the gap has remained approximately 10% to 14% for more than ten years (CIP 2022b; Ministry of the Interior 2022).

The institutional educational capital gap is also reflected in employment performance in the labor market, thereby widening the overall class-status gap between Indigenous people and Taiwanese society as a whole. In a sample survey conducted across Taiwan in 2010, approximately 16% of the indigenous labor force was estimated to be unemployed and a high proportion of those who were employed were at risk of losing their jobs within a year (Chang et al. 2010). As shown in Figs. 14.3 and 14.4, Council of Indigenous Peoples employment surveys over the years have shown that indigenous people had a higher labor force participation rate than the general population, but their unemployment rate was also higher than the general labor force (CIP 2022b). Most Indigenous workers cope with unstable working hours and wages. Employed or unemployed, Indigenous workers experience ethnic-group differences in the workplace.

The Council of Indigenous Peoples survey data indicate that the majority of Indigenous workers were employed in the construction and service industries, holding positions that require lower professional and technical competencies. According to survey data, the defense and public administration industries were the industries where a higher proportion of Indigenous workers held supervisory or professional positions. Generally, only a very small number of indigenous workers were shown by the data to hold supervisory or professional positions; instead,

[4] To solve urban settlement issues, the government often built public housing near the settlements. For example, residents of the Sanying tribe were resettled in the adjacent Long'enpu public housing, and the residents of the Bachimen tribe were resettled in the Haibing public housing. However, the strategy of collective resettlement could not overcome the Indigenous people's traditional desire to live near water (Chiang and Chi 2011; Yu 2019).

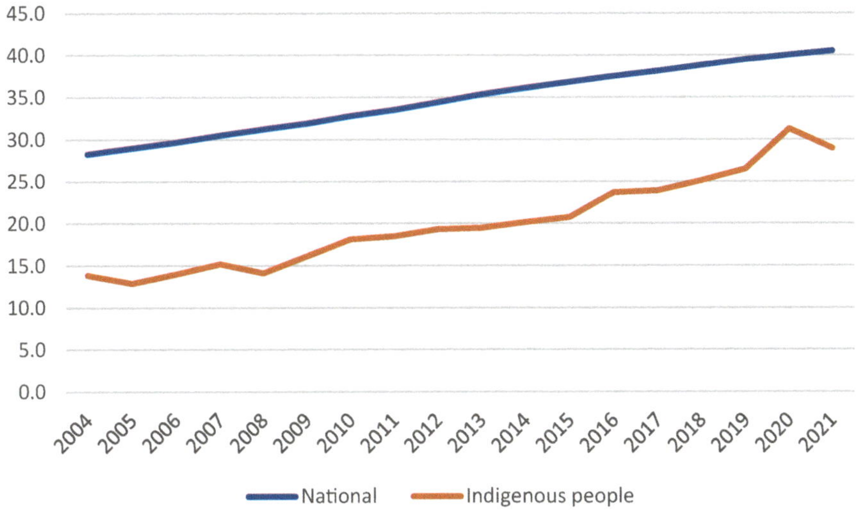

Fig. 14.2 Civilian population with tertiary educational attainment in Taiwan (% of population aged 15 + years), 2002–2021. *Note* The graph shows the proportion of Taiwanese people and indigenous people with tertiary educational attainment among older than 15 years in Taiwan from 2004 to 2021. *Source* Ministry of the Interior Database, 2022

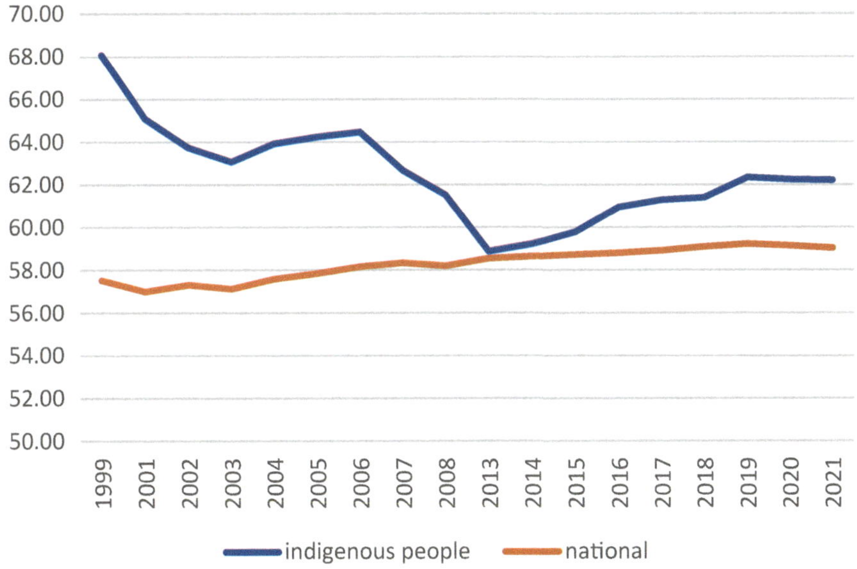

Fig. 14.3 Labor-force participation rates in Taiwan, 1999–2021. *Note* The graph shows labor-force participation rates of indigenous people in Taiwan from 1999 to 2021. *Source* Council of Indigenous Peoples Database, 2022

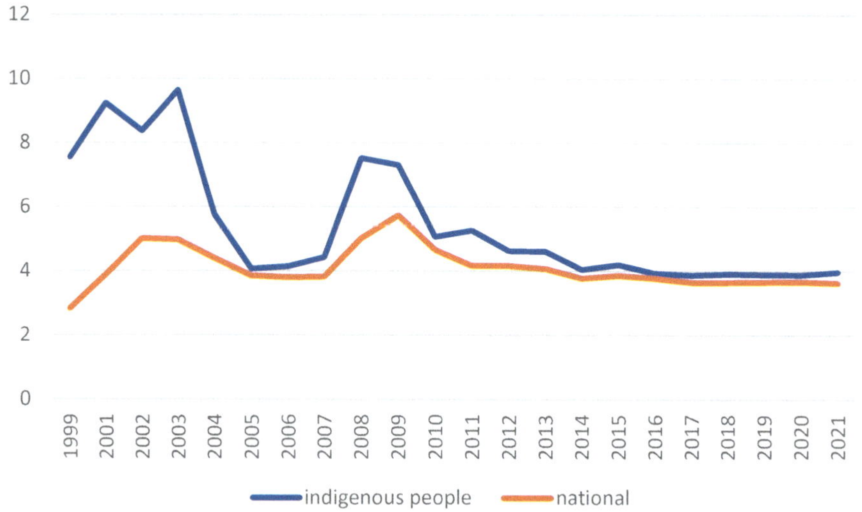

Fig. 14.4 Unemployment rates in Taiwan, 1999–2021. *Note* The graph shows unemployment rates of indigenous people in Taiwan from 1999 to 2021. *Source* Council of Indigenous Peoples Database, 2022

they were shown to work within three occupation groupings primarily: technicians and assemblers, mechanical equipment operators, service workers and salespersons. Based on survey data, workers tended to stay within these three occupation types when they switched to other industries (Liu, 2009).

These employment characteristics are related to the criteria of Indigenous people in the labor market. When Indigenous people leave their tribes and move to metropolitan areas, they have limited employment opportunities. First, their human capital is lower because the level of education of indigenous people is lower than that of other people. Second, indigenous people are attracted by wages that are higher and paid at once. Their job stability is low, and they focus less on the development potential of their careers. As a result, many change jobs frequently (Chan 1991; Chang 2001, p. 223; Chiu 2007; Chu 2001, pp. 105–107; Lin 1981, p. 17; Xie 1991, pp. 89–90, 176–177, 206–207). Indigenous people tend to take unstable and low-skilled jobs, often unrelated to any education and training that they have received.

The underprivileged social status of Indigenous people is related to their human capital, but also results from limitations in their social networks and limits to their capability to obtain information. Indigenous people face an information gap in their new hometowns, as they are accustomed to searching for employment information through acquaintances. They are usually introduced to employers by people from the same tribe or the same hometown; resultingly, the majority of Indigenous people work in specific industries, including the construction industry and personal-service industries. The highly overlapping networks of Indigenous people limit the information that can be obtained; Indigenous people are led to similar jobs and cannot maximize employment opportunities and benefits. Jobs that Indigenous people are

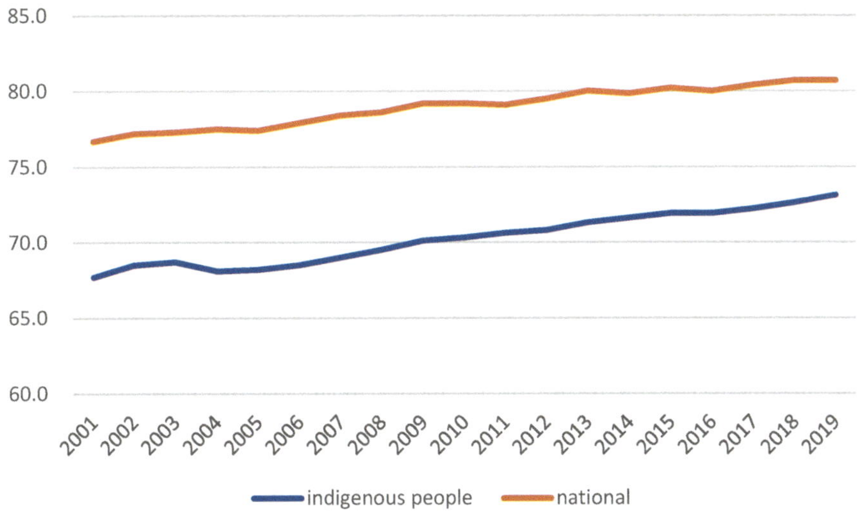

Fig. 14.5 Average life expectancy in Taiwan, 2001–2019. *Note* The graph shows average life expectancy of indigenous people in Taiwan from 2001 to 2019. *Source* Council of Indigenous Peoples Database, 2022

introduced to are usually jobs at a lower professional level, making them inconducive to improving socioeconomic status.

The socioeconomic disadvantages of Indigenous people are also reflected in health inequalities. The average life expectancy of Indigenous people is lower than that of the general population. Based on annual health reports, the life expectancy of Indigenous people was 67.7 years in 2001, 9.0 years shorter than the 76.7 years average of all Taiwanese people. In 2019, the life expectancy of Indigenous people increased to 73.1 years, still 7.6 years shorter than the 80.7 years expected for the Taiwanese population generally (CIP 2022c). Figure 14.5 demonstrates both the 5.4-year increase in life expectancy from birth of indigenous people since 2001 and that indigenous life expectancy has remained lower than the national level over the years. In that period, ethnic-group differences in life expectancy decreased, yet the gap between the two remained between 7.6 and 9.4 years, also demonstrated in Fig. 14.5.

14.3 The Identity of Urban Indigenous People and the Embodiment of Contemporary Indigeneity

If the first wave of Indigenous people who migrated to metropolitan areas in the 1970s is considered the first generation of urban Indigenous people, after half a century the second and third generations have emerged. The second generation of urban

Indigenous people enjoys higher human capital and socioeconomic status than the first generation (Liu 2011). Beginning with the second generation, urban indigenous people have had more and better opportunities, as their parents were successful in raising the family social status by moving to metropolitan areas in years past. Thanks to the parents' socioeconomic status developing and improving in metropolitan areas over elapsed years, the children of those parents' are in a higher socioeconomic status than indigenous people who only recently left the tribes.

Although Indigenous affairs and policies that guarantee education, employment, and medical treatment are tailored to the status of Indigenous people generally, urban Indigenous people are a special group. In recent years, tribe members who reside in urban areas have considered themselves urban Indigenous people, a new identity type (Hu 2005; Tsai 2005). Shih (2008) concluded that urban Indigenous people who have left their native tribes will have more complicated and nuanced feelings about their ethnic identity. The close contact with the Han ethnic group in larger society often leads indigenous people who have migrated to metropolitan areas to develop a new pan-indigenous consciousness and identity that transcend the boundaries between ethnic groups.

Many second- and third-generation Native Americans in the United States have no experience living on Indian reservations. The indigenous identities that they recognize and perceive are different from the collective identity of the first generation and of their ancestors, which was formed when Native American people lived together in tribes (Lobo 2001; Straus and Valentino 2001). The newer urban identity is more prominent among schoolchildren who belong to the younger generations (Tsai 2005). The Native Americans in North America who have left reservations differ significantly from tribe members living on the reservations. They strive to develop ethnic identities and discourse based on their lives and childhoods in the metropolitan areas (Jacobs 2014; Weaver 2012).

Urban Indigenous people and tribal Indigenous people are not fixed social categories in Taiwan. Within household registration changes, the identities of urban Indigenous people and tribal Indigenous people may change. Once a household registration is changed to indicate that someone is living outside of the native tribe, the designation becomes urban Indigenous people; when or if the household registration is changed back to a tribe, the designation is again tribal indigenous people.

People who were born and raised in a native tribe but left their hometown as an adult and settled in another place, people who were born in a native tribe and moved with their parents who work outside of the tribe, and young tribe members who worked, got married, gave birth to children outside of the tribe, and had never lived in the tribe are all regarded as urban indigenous people; however, they have different stories and life experiences.

Conversely, tribe members who were considered urban indigenous people, including second-generation indigenous people who worked outside the tribe in the early days and returned to the tribe for retirement although they had never lived in the tribe (Hu 2005; Tsai 2005), people who had moved from place to place outside of the tribe and decided to return to the tribe to start anew, and the indigenous elderly who had worked outside the tribe for many years and returned for retirement (Chan

2014; Liu and Lin 2010), are no longer considered to be urban indigenous people once they change their household registration back to their native tribes.

The migration of Indigenous peoples is often interim, repeated, and back-and-forth (Chen 1990; Liu 2009, 2017; Sun, 1991). Urban Indigenous people and tribal Indigenous people cannot be considered two separate social categories. Urban Indigenous is a fluid and changeable status. The category demonstrates the series of responses that tribe members offer in different environments and the life-stage changes that Indigenous people make after weighing the capital they own. Chan (2014) explained the wandering and returning of contemporary indigenous people from the perspective of "modern nomadism/hunting." In his study, the hometown is likened to the "main house," while the migration destination is likened to a "work hut (cottage) like the ones built away from the communities in the past for accommodation during farmland reclamation" (Chan 2014).

In addition to the volatility that frequent migration brings to the experience of urban Indigenous people, intermarriage blurs the social boundaries between ethnic groups. Since the 1970s, the migration trend has reshaped the activity space of Indigenous people and contact between Indigenous people and Han Chinese has become more frequent. Fu (1994) found that in areas where Indigenous people had frequent contact with the Chinese Han, the sense of estrangement between the two groups was weaker. As urban Indigenous people age, intermarriage between Indigenous people and Han Chinese becomes more common. Census data show that in 2000 approximately 39% of married Indigenous people were married to Han Chinese; in 2010, that percentage rose to 60% (Liu and Chang 2012, 2017). The intermarriage trend has been even more pronounced among urban Indigenous people. Survey data show that among the population aged 18 to 64, approximately 43.6% of urban Indigenous people married Han Chinese; the intermarriage rate among the tribal Indigenous people was 14.4% (Liu and Chang 2017). With the reorganization of ethnic groups' living space, there is more frequent contact between the ethnic groups, and the more harmonious the relationship between the Indigenous people and Han Chinese becomes.

14.4 Urban Indigenous People Reflect Ethnic Inequality in Taiwan

Adhering to race-based positive actions and the principle of equality, in Taiwan Indigenous people are protected in terms of education, employment, medical treatment, and housing. People with an indigenous identity all enjoy the protections and benefits provided in today's Taiwan. However, the specific rights and benefits that they enjoy may vary based on different areas of residence. For example, the use of resources, as well as the co-management and use rights of land, in traditional territories and reserve land are available only to Indigenous people who remain in tribes.

Indigenous people who leave tribes to live in cities not only lose the right to participate in the collective resolution of tribe issues (e.g., tribal meetings), but also face a fundamental problem of living in metropolitan areas: the lack of connected communities like those in tribes. However, it is precisely because of the cultural shock and "detribalization" experienced by urban Indigenous people that they develop both a collective identity of pan-indigenous people and an urban indigenous people identity. These new identities appear only within urban life (Lobo 2001).

In the 1980s, the government of Taiwan successively proposed supportive acts for Indigenous people who leave their native tribes (including the Taiwan Province Mountain Aborigines Social Development Program in 1989 and the Life Counseling Program for Urban Indigenous People in 1992). In the early years, the programs focused on supporting and alleviating the challenges of urban Indigenous people by offering housing and employment assistance and unemployment counseling. In recent years, programs have shifted to focus on the subjectivity of the urban indigenous people, strengthening efforts to continue and embed culture (schoolchildren learning ethnic languages, establishment of indigenous universities and indigenous cultural centers in the metropolitan areas), encouraging the horizontal social network connection of the indigenous tribes, and helping urban indigenous people set up communities (clansmen associations, societies, and associations) in metropolitan areas. Efforts focus on the lost opportunity to participate in the connected communities of tribes, the relative isolation of urban indigenous families in the metropolitan areas, and the differences in the development of social safety nets and cultural care for urban indigenous people and the tribes. An important current issue is providing care for the families of urban Indigenous people, specifically care for children and the elderly.

In multi-ethnic societies such as the United States, Canada, and Australia, the migration of Indigenous people in society has always been the focus of national ethnic and human rights discussions. Census data from the United States, Canada, and Australia show that indigenous people tend to leave traditional territories and migrate to urban areas. The percentage of urban indigenous people in Canada increased from 49% in 1996 to 56% in 2011 (Indigenous and Northern Affairs Canada 2017).[5] In the United States, approximately 67% of Native Americans lived in metropolitan areas in 2000, rising to 71% in 2010 (U.S. Census Bureau 2010). In Australia, approximately 27% of indigenous people lived in urban areas in 1966, increasing to 73% in 1996 (Ross 1999); by 2011, 79.6% had left their traditional territories (Australian Bureau of Statistics 2013). Judging from these reports, Indigenous people leaving their tribes and reservations is a global trend. However, Indigenous people who have moved to metropolitan areas struggle to survive in the marginal districts of metropolitan areas. Important challenges include how to improve and protect the well-being of urban indigenous people; how to prevent them from falling into cultural, financial, and social underprivileged situations and statuses; and how to build understanding of the

[5] The cities with the largest indigenous populations in 2011 were Winnipeg (78,420), Edmonton (61,765), Vancouver (52,375), Toronto (36,995), Calgary (33,370), Ottawa-Gatineau (30,570), Montreal (26,280), Saskatoon (23,895), and Regina (19,785).

living conditions of and inequalities faced by urban indigenous people in metropolitan areas in terms of employment, economic stability, and health.

Since the Indigenous people's rights movement in the 1990s and the promulgation of the Indigenous Peoples Basic Law in 2001, the stigma attached to Indigenous people has faded in the larger society of Taiwan. Ethnic discrimination has also been eliminated as much as possible through institutional education. Migration is an important topic in contemporary indigenous studies, a topic that can be approached with perspectives that include demography, regional geography, economic and labor forces, and social contact. Wang (1997) proposed the "edge studies of ethnic history" approach that suggested that ethnic relations be examined from the points of view of identification and ethnic groups, on these groups' edges. The approach suggested that:

> ...in ethnic relations, once the fringe of an ethnic group is defined by a certain subjective standard, it is not necessary for the people within the ethnic group to constantly emphasize their cultural connotations, but the ethnic characteristics are emphasized at the edges of the ethnic group. Therefore, the edges become the best place to observe and understand the phenomenon of ethnic groups. (Wang 1997, p. 77)

The indigenous articulation concept proposed by Clifford (2001) examined the existence of contemporary indigenous people from the viewpoint of the connection and disconnection between politics, society, economy, and culture. It asserted that it is possible to articulate, disarticulate, and re-articulate, which is a hanging, non-permanent form (Clifford 2013), in particular, how to get rooted in and routed through the cultural colonization and decolonization encountered in the migration process and how to integrate the elements of urban life into a person's own ethnic culture to generate unique cultural assets. Furlan (2017) found sketches of Native Americans in the corners of cities among the graffiti that can be seen everywhere in the streets of American metropolises; the sketches emphasized the existence of Native Americans and declared the legitimacy of their activities in urban space. Similarly, cultural articulation can often be seen in artistic creation and dance performances as well as in small items such as architectural totems and large items such as landmarks (Chen 2013, 2016; Wei 2009). The migration and back-and-forth movement of Indigenous people have added more elements of modernity and contemporary life to indigeneity; contemporary indigeneity has gone beyond traditional discourses. Descriptions of the inequality experienced by contemporary Indigenous people in metropolitan areas require continuous assessment of the expansion of ethnic inequality in cities. To assess ethnic life in contemporary urban contexts fairly, these descriptions also necessitate acknowledgement of the possibility that a novel kind of indigeneity may develop, which would then generate a corresponding discourse.

References

Australian Bureau of Statistics (2013) Estimates of aboriginal and Torres Strait islander Australians. http://www.abs.gov.au/ausstats/abs@.nsf/mf/3238.0.55.001

Chan H-S (1991) A study on the living conditions and employment needs of urban youth in Taipei report of a special study commissioned by the youth counseling committee, executive Yuan. Youth Development Administration
Chan Y-C (2014) The issue of return migration in the elderly indigenous. Soc Policy Soc Work 18(2):37–72 (in Chinese)
Chang C-F (2001) History of urban aboriginal social issues. In: Tsai M-C (ed) History of Taiwan aborigines-urban aborigines. Taiwan Historica, pp 207–240 (in Chinese)
Chang Y-H, Lin J-P, Liu C-C (2010) Relocation and migration of indigenous peoples in Taiwan and the changes and current situations of their socioeconomic status. In: Huang S-M, Chang Y-H (eds) Government policy and social development of Taiwanese indigenous people. Institute of Ethnology, Academia Sinica, pp 51–120 (in Chinese)
Chen C-H, Sun TH (1958) Population distribution and Chang in Taiwan. Geogr Rev Jpn 31:644–654 (in Japanese)
Chen C-B (2013) Hybridity in Taiwanese aboriginal cassette culture. Guandu Music J 13:27–50 (in Chinese)
Chen C-B (2016) From "singing school songs" to "singing our songs" articulations of modernity and aboriginality in the musical on the road. Taiwan Radical Q Soc Stud 103:1–49 (in Chinese)
Chen Y-L (2010) Pangcah tribal settlements along the riverbanks the social-security significance of rebuilding a self-reliant livelihood in lieu of social welfare provisions. Taiwan Radical Q Soc Stud 77:135–175 (in Chinese)
Chen Y-C (1990) Migration and occupational mobility: a study of migration and inter- and intra-generational mobility among Taiwan's Mountain aborigines. Unpublished doctoral dissertation. Tunghai University, Taiwan (in Chinese)
Chen Y-M (2004) Is Taiwan the homeland of the Austronesian? Artifacts, culture, and people: a discussion of south island languages as an example. Hist Monthly 199:34–40 (in Chinese)
Chiang W, Chi C-C (2011) Aboriginal tribal identity and reconstruction under urban marginalization: a case study of the new Sanying tribe in New Taipei City. Taiwan Indigenous Stud Rev 10:125–156 (in Chinese)
Chiu J-N (2007) Employment difficulties of the Taiwan indigenous peoples and its policy solutions: from social exclusion perspective. Unpublished doctoral dissertation. National Chi Nan University, Taiwan (in Chinese)
Chu J-J (2001) Urban aboriginal labor history. In: Tsai M-C (ed) History of Taiwan aborigines-urban aborigines. Taiwan Historica, pp 91–120 (in Chinese)
Clifford J (2001) Indigenous Articulations. Contemp Pac 13(2):468–490
Clifford J (2013) Returns: becoming indigenous in the twenty-first century. Harvard University Press, Cambridge
Council of Indigenous Peoples (2022a) Population of indigenous ethnic groups in urban areas of Min County, Taiwan—metropolitan ratio. Council of Indigenous Peoples. Retrieved January August 1, 2022 from https://www.cip.gov.tw/zh-tw/news/data-list/940F9579765AC6A0/F01A654848F03AFF6F51CB2375B3D43B-info.html (in Chinese)
Council of Indigenous Peoples (2022b) Employment status survey of indigenous peoples, 2021. Council of Indigenous Peoples (in Chinese)
Council of Indigenous Peoples (2022c) Health statistics of Taiwan indigenous peoples, 2019. Council of Indigenous Peoples (in Chinese)
Fu Y-C (1994) The socioeconomic plight of Taiwan's native Austronesian-speaking peoples: squaring the accounts of Han Chinese and native peoples. Inst Ethnol Acad Sinica 77:35–87 (in Chinese)
Furlan LM (2017) Indigenous cities: urban Indian fiction and the histories of relocation. University of Nebraska Press, Nebraska
Gray RD, Drummond AJ, Greenhill SJ (2009) Language phylogenies reveal expansion pulses and pauses in pacific settlement. Science 323(5913):479–483
Gray RD, Jordan FM (2000) Language trees support the express-train sequence of Austronesian expansion. Nature 405:1052–1055

Hu H-X (2005) Urban experience and identity of the second generation of urban aborigines. Unpublished master's thesis. National Tsing Hua University, Taiwan (in Chinese)
Indigenous and Northern Affairs Canada (2017) Urban indigenous peoples. https://www.aadnc-aandc.gc.ca/eng/1100100014265/1369225120949
Jacobs M (2014) Urban American Indian identity: negotiating Indianness in Northeast Ohio. Qual Sociol 38(1):79–98
Li P-K (1993) New data on three extinct Formosan languages. Bull Inst Hist Philology 63(2):301–323
Lin JP (1981) The urban aboriginal in Northern Taiwan. China Forum 12(7):24–28 (in Chinese)
Liu C-C (2009) Migration of Taiwan aborigines: climb up or stumble in life course? Unpublished doctoral dissertation. National Chengchi University, Taiwan (in Chinese)
Liu C-C (2011) Intermarriage among Taiwan's urban indigenous: the social boundary of different generations. J Population Stud 42:115–153 (in Chinese)
Liu C-C (2013) The spatial distribution and migration pattern of urban indigenous people in Northern Metropolitan Areas of Taiwan. Dept Geogr 69:1–22 (in Chinese)
Liu C-C (2021) Migration and distribution of indigenous people in contemporary Taiwan: long-term migration pattern of urban indigenous people in household registration administrative data. J State Soc 22:1–38 (in Chinese)
Liu C-C, Chang Y-H (2012) The application and merit of census data in the study of ethnic intermarriage between the indigenous and Han people in Taiwan. Survey Research 27:7–51 (in Chinese)
Liu C-C, Chang Y-H (2017) Generational change and gender difference in the ethnic intermarriage between indigenous and Han people. Taiwan J Indigenous Stud 10(1):71–106 (in Chinese)
Liu C-C, Lin J-P (2010) Migration of Taiwan indigenous peoples in response to the effects of location-specific factors and individual life course. City Plann 37(3):305–342 (in Chinese)
Liu K-Y (2017) We wish a better life: the lived experiences of second-generation urban male indigenous workers. Unpublished master's thesis. National Yang-Ming University, Taiwan (in Chinese)
Lobo S (2001) Is urban a person or a place? Characteristics of urban Indian country. In: Lobo S, Peters K (eds) American Indians and the urban experience. Altamira Press, pp 85–94
Ministry of the Interior (2022) Population of 15 years and over by educational attainment, Excel, 2.02
Ross K (1999) Occasional paper: population issues, indigenous Australians. Australian Bureau of Statistics
Shih C-F (2008) Constitutional-making in Taiwan: substantive and procedural issues. J Law Politics 21:121–140 (in Chinese)
Straus T, Valentino D (2001) Retribalization in urban Indian communities. In: Lobo S, Peter K (eds) American Indians and the urban experience. Altamira Press, pp 85–94
Sun R-X (1991) Social mobility of urban aboriginal. National Chengchi University. Unpublished master's thesis. National Chengchi University, Taiwan (in Chinese)
Tsai CL (2005) The ethnic identity of urban aboriginal descendants: the case of twelve urban aboriginal descendants. Unpublished master's thesis. National Dong Hwa University, Taiwan (in Chinese)
U.S. Census Bureau (2010) Census 2010 American Indian and Alaska native summary file: Table: PCT2
Wang M-K (1997) Hua-Xia Margins—historical memory and ethnic identity. Asian Culture (in Chinese)
Wang R-Y (1967) Population change of Formosan aborigines. Institute of Ethnology, Academia Sinica (in Chinese)
Weaver HN (2012) Urban and indigenous: the challenges of being a Native American in the city. J Community Pract 20:470–488

Wei Y-C (2009) Taiwan indigenous literature's Scheherazade-the dialective formation and literature practice of the authors in the indigenous movement generation. Chin Mod Lit 16:67–96 (in Chinese)
Xie G-Q (1991) A study of the adaptation problems of Taiwan's mountain people after moving to the city. National Development Council (in Chinese)
Yan S-F (2005) Pits, oceans and scaffolds. Tonsan Publications (in Chinese)
Yan S-F (2006) New homes for the Amis in the city: a study of the history of aboriginal urban community making in Taipei County in the last fifty years. Tonsan Publication (in Chinese)
Yan S-F (2008) Drifting tribe: a social history of the aboriginal urban homeland along Xindian Creek for nearly 50 years. Tonsan Publications (in Chinese)
Yang F-R, Lin P-S (2021) Floods information transmission and evacuation in urban riverside Pamatangan tribe. J Commun Work Commun Stud 11(2):1–34 (in Chinese)
Yu T-M (2019) Retrospection and discussion on the research on the urban indigenous policy. Nanhua J Soc Sci 6:3–25 (in Chinese)

Chapter 15
The Impact of Globalization on Citizen's Political Autonomy in Cities: The Case of London

Jack Simpson

Abstract Due to the dramatic rise in populations in the twentieth century, and early twenty first, cities are increasingly where people pursue their goals. In the late twentieth century and early twenty first, we have also seen a dramatic rise in technological development. These changes suggest huge implications for the ways humans live together. In particular, with regards to the colegislation of citizens in cities, this chapter situates these changes in the context of London. Globalization has made global trade much easier in a way which has increased the cost of space and thus living for many in the city. Further, the private sphere has seen a dramatic rise in the use of data on people. To what extent are democratic processes keeping pace? This chapter frames these phenomena with the capability approach, arguing that current approaches to housing and technology use in city making are failing to protect the political autonomy of London's citizens. In this context, we see cities as increasingly subject to forces which create inequalities in the decision-making space for citizens. As such, we can see growing inequalities within cities like London, with regards to the political autonomy of its citizens.

Keywords Globalization · Cities · Capability approach · Political autonomy · London · Housing

15.1 Introduction

What are cities for? The question has been asked since at least the time of the ancient Greeks. Closely related to the question, *who* are cities for? For Aristotle (1999, p. 63), the city had a clear teleological purpose; it "…exists for the sake of a good life, and not for the sake of life only". The city may have been founded due to people's natural social drive to be around others, further their own needs and better their physical environment, but, ultimately, the city, for Aristotle, has a guiding aim: the

J. Simpson (✉)
University of Leeds, Leeds, UK
e-mail: J.simpson3@leeds.ac.uk

G. H. Tonon (ed.), *Urban Inequalities*, The Urban Book Series,
https://doi.org/10.1007/978-3-031-59746-6_15

good life. The city is an important political space in human culture. More recently, some, like Henri Lefebvre, have observed that the city seemed to have developed an alternative purpose, the development of the value of capital. Lefebvre uses the Marxist distinction of use and exchange value, to make clear how this plays out. The use value in a city being the way space was valued for its use, exchange value the way space is valued for its value in commercial exchange. Cities are increasingly serving the needs of capital, rather than their citizenry.

This chapter aims to show that London has been long at the mercy of such forces. That is, aspects of the city are, at best, not focused on developing space for the development of its citizenry. At worst, the goals of the city could be seen to be in opposition to such development. This chapter will focus on two aspects of citizen's development: access to space for housing and decision making. Both of these will be shown to be integral to an account of human development which has a focus on political autonomy; our ability to effectively colegislate with others.

If cities are increasingly at the mercy of outside forces, we can understand such forces as essentially representing the political autonomy of others, those from outside of cities, which creates a correlative reduction of the political autonomy of those within cities. The city space then is a key battleground of inequality, where citizens of cities are locked in a battle, with those with the capacity and capital to affect the city from outside.

Throughout the twentieth century, increasing numbers of people moved to cities and urban spaces to pursue their goals and attempt to live good lives. Over the last century, urban populations have been dramatically rising (Harvey 2013; WHO 2020; UN 2018; IMF 2007); this is true both in absolute terms and relative to rural populations. Though cities have existed for over 7000 years, the modern city, as seen from around 1750, has witnessed unprecedented growth. Many towns and villages have developed from regional marketplaces to vibrant centres in their own right (Lampard 1955). This expansion of urban living has been global; currently, 23% of the world's population live in a city with a million inhabitants or more. By 2030, that is predicted to reach 27% (UN 2016, p. 3), and this trend is expected to continue.

Cities are then increasingly the sites from which we pursue our goals, from which we attempt to make sense of life and live in accordance with those thoughts. The rise in the importance of city spaces can be seen in the shift to the sustainable development goals (SDGs), which added cities as a key focus in the renewal of the millennium development goals (MDGs). The explicit aim of Goal 11 is to make "cities and human settlements inclusive, safe, resilient, and sustainable" (United Nations 2022a, b). In Sect. 15.2, I will explore how globalizing forces have affected space for housing in London.

As city populations developed at pace in the twentieth century, so did technology. Since at least the 1950s, the expansion of computing, urban development, and the digital realm have become ever more intertwined (Cardullo et al. 2019, p. 2). Computers were increasingly used to store and process city administration through the second half of the twentieth century. Academics also used systems to model and look at areas of urban living, such as transportation scenarios. In the late 1960s, the idea arose of the city as a system of systems that might be optimized (Forrester cited in Cardullo et al. 2019). As personal computers became more widespread in the 1980s and 1990s, cities utilized computers more and more to administer and deliver services. As the Internet grew in the 1990s and 2000s, these computers became increasingly networked. This led to a significant investment in e-governance (Castells cited in Cardullo et al. 2019) and the widespread proliferation of networked infrastructures such as traffic management and surveillance systems (Lyon cited in Cardullo et al. 2019).

In Sect. 15.3, I will explore the potential expanded capacity such technological change brings for deliberation in city spaces. Arguing that whilst London's citizens are being monitored, their data gathered and their political processes have done little to expand space for political participation. Barcelona, Calgary, and Madrid's online forums are used as examples of how the city could improve.

15.2 Political Autonomy and the City

In terms of the city providing space for citizens' development, this chapter shall focus on one aspect of that development: the development of citizens' political autonomy. The importance of political autonomy is framed using the capability approach, a normative framework used across philosophy, economics, and various applied fields in social science. As is common practice, the framework is used to analyse the justness of social arrangements as well as to determine the wellbeing of individuals (Robeyns 2017, p. 24). The capability approach is predicated on the notion that, when evaluating an individual or society's wellbeing, what counts is what people are able to do and be (Robeyns 2017, p. 7).

In a social context, our ability to do and be things relies to a large extent on our ability to colegislate with others. To think about the ability to exercise such freedom in a social context, I adopt the idea of "effective autonomy", which, according to Brown (2009, pp. 156–157), is: "The ability to reasonably achieve the intended or desired result of one's conscience and self-legislation in a social context". This ability is influenced to a large extent by one's "aspect capacity". Aspect capacity is "the ability to formulate, hold, and act upon an individual's social conscience based on external elements (aspects or facets for external freedom) affecting the development of one's colegislative capabilities" (2009, p. 157). We find a clear relationship here between innate human potential (the ability to reason and self-formulate ends) and the interaction between the agent and the environment in which that capacity can be expressed. In other words, an aspect capacity is, in effect, the capacity for autonomous self-reflection within a particular environment where aspects of autonomous action are hindered, permitted or passively, and/or actively advanced.

We can understand the city's role for developing the citizen's aspect capacity, and thus ability to colegislate in two stages. Initially, the city provides the environment for the individual's development. Through its material and social conditions, its opportunities, limitations, and space for development. Secondly through the practising of such skills; by taking part in the city in a colegislative manner. The citizen and city are then involved, ideally, in a state of codevelopment. As a result, in terms of political autonomy, the city has two roles that are key to living a flourishing life:

a. Firstly, the city provides a space, the cultural and material conditions that foster or inhibit flourishing, while,
b. Developing the city offers the chance to take part in one's community, contributing to the development of the city in which the individual also continues to develop.

The investigation into the effects of globalization on London citizens' political autonomy will focus on these roles through two important spaces: opportunities for housing, space and decision making.

15.3 London, Global City

This rise in city populations was accompanied by and interwoven with a period of dramatic globalization. Especially in the late twentieth and early twenty-first century, in many ways, the development of London, like other global cities, can be seen as symbiotic with the development of the global moment. Alongside other global cities such as New York and Tokyo, the development of London in the late twentieth century and early twenty first is a part of the movement of globalization (Sassen 2001). Such cities gather global capital and power practices: global brands and companies' staff live and work there; consumers of globally produced products and services consume there. London is very much in this mould. Its global ambitions and reality are clear; this can be explicitly heard in the rhetoric of politicians like former UK Prime Minister and then London Mayor Boris Johnson, who said, "London must retain and build upon its world city status as one of the three business centres of global reach" (Therbon 2017, p. 309).

When thinking of major cities at the centre of globalization, few stand out like London. Until the end of the Second World War, it was the capital of the British Empire, the largest in the world. It remains to this day a world centre of financial transactions (Therborn 2017, p. 304). It was said that, at the peak of the British Empire, the sun did not set on it. But the impact of London on the world now is perhaps even more relentless. Alongside New York, Tokyo, Shanghai, Frankfurt and Hong Kong, London is where a significant share of world banking occurs. Canary Wharf, where many of those transactions occur, displays various facets of globalization: built with foreign capital, materials and techniques by foreign designers. Foreign workers from the Global South and Eastern Europe clean the offices (Therborn 2017, p. 306). London has, in this period, become a truly global city, both part of the system which developed this globalizing period and subjected to globalizing forces.

For the most part, through the twentieth century, cities were treated as a sub-national phenomenon, held within national territories and hierarchies. But, in the late twentieth century, cities like London became supranational, for instance, in the way that they handled a disproportionate amount of the world's business. These cities have become known as global cities, a development of the term "World City", a term coined by Patrick Geddes and developed by Peter Hall (Keil and Ren 2018, pp. XXIV–XXV). We can understand such cities as "spatial materializations" of the system they are in *and help reproduce*—processes associated with capitalist modes of production; in particular, accumulation and class struggle (Keil and Ren 2018, pp. XXIV–XXV).

15.4 London, Globalizing Forces, and Property Prices

In this section, I will look at how global capital has shaped the ability of the citizens of London to afford space. In Sect. 15.2.2.1 I will show the relationship between affordable space and political autonomy. I will examine two areas as reasons for added demand:

a. London as a centre of global trade has attracted firms that service this network. This has had the effect that companies with bases there have bought up space. Such companies and industries bring their staff, and salaries, which then bring their own culture.
b. London being a space for foreign investment has meant that space has been bought up for investment purposes.

London attracts firms that occupy space in the city. Such firms represent a new spatial organization of the global economy: the development of firms expressly set up to service globalizing industries such as finance and service. The growth of many firms that service and build the globalizing economy contributes to the rise in property prices. This is due, in part, to the way that certain industries such as financial and service firms concentrate. Much of the shift in property prices is due to internationally owned property by industries such as finance. This has, in part, led to London being described as the most concentrated and specialized international property market in the world. We can see part of this in the high rate of financial-service firms owning the buildings they occupy (Baum and Lizieri cited in Sassen 2001, p. 193).

These radical shifts of the global economy in the late twentieth century were transformative for major cities like London making them key sites of production, servicing, marketing and innovation (Sassen 2001, p. 87). While governments played a role in approving and legitimizing this movement, the deregulation of financial services and foreign investment created a space for transnational economic activity. A second related way such movement of industry affects local economies is by gathering large numbers of people on salaries above the norm for the area. These high-income workers bring a demand for residential space, luxury shops, art galleries, etc. (Sassen 2001, pp. 190–195). The economy pivoting towards those with more consumptive power.

A further reason for the rise in prices for London property is that the city has been used as a safe house for capital. Assets have been bought with foreign money. This is not a phenomenon restricted to London, of course. In recent years, governments worldwide have conducted enquiries into the impact of foreign investment in property. From Chinese investment, Barcelona and Berlin rents, Canadian house prices, foreign investment is having a real impact on housing costs for residents worldwide. Some countries wish to attract more attention and create Visa schemes specifically to target potential investment. Spain, Greece, Turkey, and Cyprus have all done this to attract investment from Russia, Asia, and North America (Koh and Rogers 2017, p. 2). Cities like London have been significantly affected by such investment. One explanation for the influx of foreign direct investment is that global cities like London present relative safety for money from less politically stable economies (Badarinza and Ramadorai 2018, p. 1). On this model, a rise in political risk in Russia, parts of Africa, and the Middle East manifests as a rise in property prices in London. It is not merely the promise of a return on investment that acts as a pull. The *push* of capital from risk is also at play (Badarinza and Ramadorai 2018, pp. 3–4).

The new global environment of the 1980s created an increasingly active environment for foreign investment. As the network of global cities grew, so did foreign investment. This gave rise to phenomena such as London land prices becoming increasingly unlinked to the conditions of the national economy. Such high prices were also not spread around the city equally but confined to certain areas of central London. This led to the development of run-down areas, such as the old docks (Sassen 2001, pp. 190–191). Foreign capital was reshaping the city to provide a return on investment. Wealth created away from London is deposited in London, with space and buildings in the city used as assets, stores of value. London has been linked so clearly to global political strife that one property executive described how "every mortar bomb in Libya adds £1/sq foot to investments here" (Minton 2017, p. 23), leading to it being said that the exchange value has had its link to the use value entirely broken (2017, pp. 22–24). The use of space in this way has some, perhaps to be expected, results. Land banking and the preservation of buildings as long-term assets are very different from a space being used locally by the people who need to see and experience it daily. In 2016, Sajid Javid, then UK Secretary of State for Communities and Local Government, spoke of how he could not:

> …look the other way when I see land banking holding up development. Some of you have conceded to me, in private, that it happens. Some of you still deny it's an issue. But there's clearly something going on. The number of plots approved for residential development each year rose by 59 per cent between 2011 and 2015. But the number of buildings starts rose by just 29 per cent. (Minton 2017, p. 66)

Shelter, a UK housing and homelessness charity, estimates there are almost 500,000 sites in the UK with planning permission that have not been acted upon. Banking the land helps control the price—the exchange value. A further validation of Lefebvre's concerns. Even the perceived abandonment of land can be a political strategy used by the interests of private capital to pressurize councils and planning departments (Hackworth 2015).

Land banking relies on having the resources in the first place to sit and wait for the right time to cash in an investment. But this is not how people locally behave with space. It shows a divergence from the interests of the locals, the use value of space. Barcelona's PAH translates as the Platform for People Affected by Mortgages. Ada Colau, a leading PAH activist, argues that the right to the city ought to be taken seriously: "The point is to be open; to keep innovating, listening to citizens" (Minton 2017, p. 169). Land banking and the use of space and buildings as assets make such space move to the tune of a *global* song, not a *local* one. It is fundamentally to shift the relationship between space and its relation to the local and make it part of a new system of movement, where such space looks to the constellation turns of stock markets and wide investment cycles, not the needs of the local populous.

Cumulatively, the movement of capital characteristic of the late twentieth and early twenty-first century can be seen to do two things. Firstly, it has pushed up prices as firms have become based in global cities. Such firms have brought with them staff who have pushed up prices. Secondly, London has been used as a safe space for capital in the property market. When considered in the context of local political autonomy, it can be seen that the period has witnessed a diminishing of such autonomy. Decision making and power over space have moved away from the citizens of many areas of London.

15.5 London House Price Rises

Data from three key areas (house prices, house price to earnings, and salary growth in the lowest quartile of the London population) from the mid-1990s to 2020 indicates a marked rise in both absolute housing costs and housing costs relative to earnings. House prices, especially house prices to earnings (as shown in Table 15.1), tell us something important about the cost of living and impact on people's ability to spend on other things that, to place within a capability approach framing, contribute to expanding their capabilities and functionings. An upward curve in the ratio would suggest increasing numbers for whom housing costs will be difficult.

The ratio between house prices and earnings in Table 15.1 shows striking growth across the city between 2002 and 2018. In 2002, the average house price was 6.9 times average annual earnings. This was compared to the UK equivalent of 5.11. By 2018, this had grown considerably in both London and the UK more broadly. But, in London, this had developed to 13.09 compared to the UK-wide rate of 8.04. Furthermore, within this data, we can see more pronounced unequal growth in particular areas. Hackney in this period went from 7.37 times to 16.34. Hammersmith and Fulham, from 8.55 to 20.04 (London Datastore 2021a). Between 2002 and 2018, house prices outstripped earnings across London. This, of course, makes house buying less accessible to more people.

Whilst UK housing has had increases in costs above earnings, this has been more pronounced in London and the Southeast. These areas have an above-average spend on housing in the UK, and, since the 1970s, housing in these areas has grown faster than in the rest of the UK (Holly et al. 2010). Further, the rapid rise in house prices in London has been accompanied by rapid rises in rents. October 2019 to September 2020 saw the highest-ever recorded median monthly rent in England,

Table 15.1 Ratio of house prices earnings, residence based

Area	2002	2004	2006	2008	2010	2012	2014	2016	2018
City of London	6.29							14.83	16.55
Barnet	7.42	8.63	9.12	9.16	9.7	10.05	12.14	14.05	14.45
Brent	8.1	9.24	9.85	9.89	10.31	11.13	12.39	15.21	16.32
Camden	9.71	10.4	10.38	11.97	12.63	13.3	17.43	18.53	18.23
Ealing	7.09	8.68	8.88	9.66	9.61	9.94	12.28	14.73	14.17
Greenwich	5.99	6.86	7.7	7.88	7.61	7.96	10.09	11.87	11.95
Hackney	7.37	8.7	9.48	9.06	9.63	10.41	12.82	16.67	16.34
Hammersmith and Fulham	8.55	9.83	10.8	11.76	12.36	12.48	16.95	19.21	20.04
Hounslow	7	8.65	9.03	9.02	8.54	8.67	10.3	12.62	12.75
Islington	8.71	8.67	8.91	10.84	10.95	10.64	13.57	16.51	13.09
Kensington and Chelsea	10.68	10.04	11.19	15.26	15.93	21.32	25.28	30.46	33.43
Lambeth	8.02	8.38	8.98	9.49	8.97	9.85	12.43	15.01	13.82
Richmond upon Thames	7.26	8.35	9.18	9.85	10.35	10.35	12.79	14.38	14.88
Tower Hamlets	7.81	7.97	8.56	9.12	8.88	8.7	10.88	12.82	12.73
Westminster	11.01	10.57	12.32	12.93	12.14	14.09	20.38	23.48	23.71
London	6.9	8.02	8.37	8.52	8.75	9.15	10.77	12.91	13.09
South East	6.12	7.33	7.68	7.89	7.74	7.64	8.2	9.43	9.92
England	5.11	6.6	6.95	6.96	6.85	6.77	7.09	7.72	8.04

Source London Datastore (2021a)

£725. London, at £1435, was almost double (ONS 2020). Perhaps unsurprisingly, the greater difficulty of buying houses is accompanied by higher housing costs for those renting.

As discussed in Sect. 15.2.1, one reason for raised house prices and rents is the sheer added demand that a global city like London has. This can be seen when we look into the data of areas like Spitalfields. Between 1997 and 2016, the population of Spitalfields grew by around 40%. Spitalfields is in the Tower Hamlets borough, a borough which saw 7.81 times average earnings in 2002 jump to 12.82 times average earnings by 2016 (London Datastore 2021a).

Table 15.2 shows salary growth amongst the lowest quartile in London between 1999 and 2019. Hackney, for example, shows a rise from £17,391 annually to £24,990. In 2015, the Greater London Authority (GLA) published figures updating household income estimates. In 2001/2, Hackney had a figure of £22,000, which by 2012/13 was £35,140. At the same time, Hackney, like most areas, had complex arrangements of people living there. A 2014–2015 report found that 31% of households with children were workless, more than double the national rate of 14% (Hackney Learning Trust 2015). So, in areas like Hackney, we can see a pronounced element of the phenomenon we are discussing: rising housing costs. Alongside ratios of house prices to earnings, there are those with lower earnings who live in an area where house prices and rent are increasing. Table 15.2 shows how, between 1999 and 2019, earnings in the lowest quartile went from £17,391 to £24,990.

Table 15.3 shows how in almost the same period, between 2000 and 2020, average house prices went from £121,135 to £555,696. This demonstrates then that, whilst average earnings may show us something important, something else going on: housing in the area is becoming increasingly out of reach for those on the lowest incomes, both relative to earnings and in absolute terms. This is especially important when we consider that many new people to areas like Hackney will be on higher incomes within the price to earnings data. Those already in the area on lower incomes are effectively a drag on the average earnings. Understanding that tells us something significant about the increasing difficulty of Hackney residents to afford housing: some are moving to the area, whereas those already there are faring much worse on this equation. What can be seen is that the poorest are increasingly priced out of areas like Hackney.

That London is an expensive city to live in and that this expense is partly due to high land values is widely accepted. The London Assembly (an elected body, which scrutinizes the activities of the Mayor of London) recognizes that the city as a whole is an expensive city to live in, in part because of high childcare and transport, as two key costs. Part of this is attributed to the high land values. Of course, if the cost of living goes up, other costs like childcare must too (London Assembly 2021).

We can see that this massive rise in land value has created wealth for some. But that rise in wealth has been accompanied hand in hand with poor living standards for many. The figures of poverty in the city do not show a city that is a wealthy one. Over 1 in 6 Londoners, including 3 in 10 children, live in persistent poverty; 13% of pensioners in the city live in material deprivation, double the UK rate. Between 2005/6 and 2019/20, the number of rough sleepers in the city almost quadrupled, according to records (London Assembly 2021).

This section has shown clearly escalating property prices. These property prices have outpaced earnings growth. But, further, in the data is also the story that, whilst property price to earnings ratio has risen, within those earnings will be people whose earnings have barely grown. Some of the earnings growth will relate to those who have moved in. All of this suggests that areas like Hackney have become increasingly expensive for those who live there. In the following section, we will see how access impacts upon political autonomy.

15.6 House Prices and Political Autonomy

By focusing on what people are able to do and be, the capability approach poses a challenge to more standard ways of evaluating a societal arrangement. Regarding housing policy, the capability approach can be used in this arena in two ways: firstly, to evaluate a given situation: people's access to housing. The second is by attempting to understand the nature of sources of capability deprivation and inequity. The capability

Table 15.2 Salary growth—London 1999–2019–lowest quartile

Area	1999	2002	2004	2006	2008	2010	2012	2014	2016	2018	2019
City of London	22,515	27,016	29,427	32,000	33,251	34,535	36,659	35,816	35,999	39,021	38205
Hackney	17,391	17,400	20,485	22,653	22,844	25,668	23,772	23,676	24,553	25,745	25000
Hammer-smith and Fullham	18,664	21,214	21,672	22,898	25,260	26,657	26,118	25,221	25,097	32,230	28048
Westminster	17,562	20,872	22,986	23,899	25,584	26,604	27,080	27,340	27,752	30,000	30,787
London	16,090	18,689	20,496	21,817	23,441	24,417	24,547	24,760	25,381	26,930	27,381
Inner London	–	–	22,804	24,049	25,833	26,998	27,300	27,108	27,873	29,884	29,927
Outer London	–	–	17,711	19,036	20,290	20,868	20,846	21,121	22,000	22,345	23,465
England	12,852	14,852	15,948	16,734	18,078	18,577	19,011	19,390	20,252	21,263	22,052

Source London Datastore (2021b)

Table 15.3 Average house prices, Hackney, London, and England

	Hackney	LONDON	England
	E09000012	E12000007	E92000001
Jan-95	61,297	74,436	53,203
Jan-00	121,135	130,411	75,219
Jan-05	230,551	232,422	158,572
Jan-10	295,824	279,724	174,458
Jan-15	469,391	402,847	202,856
Jan-20	555,696	475,918	247,970
Jan-21	602,507	503,308	267,711

Source London Datastore (2021b, c)

approach allows for a shift away from the focus of housing that has traditionally been on material and monetary resources (Kimhur 2020, pp. 262–263).

Housing and the spatial elements of people's lives also appear key to most capability theories. After all, people exist somewhere. As such, housing and access to space can be seen as key to other elements of living. The UN links housing to work, education, health, and security (UN-Habitat 2021). The provision of adequate housing speaks to living dignified lives (OHCHR 2021).

In Nussbaum's central capabilities, we find both "adequate shelter" and "being able to hold property" (2003, p. 42). As discussed, the capability approach evaluates the beings and doings that people have. These beings and doings are split into capabilities and functionings. As Kimhur (2020, p. 266) describes, we can understand housing as both: housing can be seen linguistically to straddle both noun and verb, material object and the activity of housing. Yet, housing has become subject to increasing financialization. It has become "a commodity, a means of accumulating wealth and often as security for financial instruments" (UNHRC cited in Kimhur 2020, p. 257). In London, house prices have grown rapidly in both absolute terms and in terms of the ratio to earnings. Booming house prices can be seen as a sign of a market in demand, a positive thing. But with rising prices comes an impact on those who do not benefit from those rises in wealth.

These two ways of viewing housing speak to alternative visions of what cities, economies, space are for. Expanding capabilities or the city as node of global capital. This speaks to that which Lefebvre showed in terms of the Marxist distinction between use and exchange value was explored in the Right to the City literature. As described in the introduction, exchange value is the value of something as it is being exchanged, as with commodities. Use value is the value that something has in its use, like a playground or a good view. The financialization of property shows how much housing is being used for its exchange value, creating pressure on its use value.

As the exchange value of space in London rises, it bestows benefits, but as has been described, not equally. There are some who are in a position to take advantage of this. But others are more likely to be at the mercy of it. As such we can see the use of space in this way as exacerbating inequalities. The use value for those who do not own space, under threat.

In Sect. 15.2.1, the rise of the global city was argued to be accompanied by a rise in firms and individuals who owned property in big global cities like London. Likewise, London and other cities' property was used for their stability in property markets; London property, including housing used for its exchange value. We can see from Table 15.3 that London housing is doing well on this understanding. Absolute prices are rising. Housing on the exchange value metric is, in London, performing well.

On the other hand, for the dweller of the house, on the other hand, there is use value. As Kimhur demonstrates, using a capabilities framework, we can understand many things that are valuable in providing people with the space to live good lives, that exchange value will struggle to pick up, including the ability to live and do so in a healthy environment. Housing is also linked to being able to work and generate income, as well as income levels after paying for housing (Kimhur 2020, p. 269). Between the UN-Habitat report extolling the reasons for enshrining a right to adequate housing and the UNHCR report on how housing is increasingly part of a financializing process, we can see the tension between the two. The exchange value of housing is rising, being pushed by London attracting foreign investment and global companies. This makes using housing for all the capability-enhancing reasons cited by Kimhur (2020) more challenging to achieve.

There are various ways that we might understand housing prices relating to political autonomy:

People's ability to have control over their own space.

The economics of paying more for housing: we either work more to keep the same preference set, in which case we have less time or work the same amount and have less to spend on other things. We have less freedom in a market-governed space.

The capability approach is a valuable framework for a number of reasons. In part, it highlights how it is not merely what exists, what architectural assets, or new flats, buildings, or space in general. Instead, we are invited to ask how different individuals have access to it. In this instance, citizens in areas of London like Hackney might find it increasingly difficult to achieve a reasonable living standard, where property prices and rent are being driven up by the use of housing as an asset. The capability approach can help highlight differing individuals' abilities to access such space. Alongside this, we are also invited to look at the relationship between space, justice, and injustice, and ways in which spatial segregation occurs in the urban environment (Frediani 2015, pp. 65–66). One way we can understand the relationship between housing and political autonomy is simply in a just arrangement of space. In Sect. 15.2 I used Garrett Brown's model of effective autonomy where, "The ability to reasonably achieve the intended or desired result of one's conscience and self-legislation in a social context" (Brown 2009, pp. 156–157).

Effective autonomy was shown to be reliant on aspect capacity, "the ability to formulate, hold, and act upon an individual's social conscience based on external elements (aspects or facets for external freedom) affecting the development of one's colegislative capabilities" (Brown 2009, p. 157).

The ability to legislate is critical to the individual's development. When people need to move away from their friends and family, and when these things disproportionately affect groups of people with low income, we can see this as an unfair distribution of space in which people have to live. In the context of cities, it is hard to think of an understanding of political autonomy that does not in some way rely on an ability to have some space to live. If one has to move out of the neighbourhood, or the neighbourhood becomes increasingly expensive, then it, of course, impacts the resources we have, in a way which inhibits our ability to legislate.

Further, if we attempt to stay in an area with rising costs, our political autonomy is curbed because we must cut back on other things, or else we have to work more to achieve the same basket of options. Either way, our political autonomy, understood as our ability to legislate and have some kind of control over the space in which we live, is diminished when the movement of global capital creates rises in rent and house prices in the areas in which we live. This diminishing comes hand in hand with an expansion of political autonomy, for those in a position to benefit from the rise of exchange value of London space.

15.7 Politics, Technology, Data, and Decision Making

The impact of global capital and development on London property prices seems clear. A second, related, issue will now be examined: how decision making in London kept pace, as technological development enhanced the ability to express and monitor preferences. As described in Sect. 15.1, political autonomy is reliant on the ability to colegsilate in a social context (Brown 2009, pp. 156–157). Decision making in the city might range from how taxes should be spent, which politicians should represent us and how, to trees being felled, playgrounds built, sports opportunities, access development; all kinds of decisions which impact upon those who live there. Technological development offers unprecedented capacity to gather, store, and manage information around people's preferences. As such, the potential for deliberation and colegislation in cities has expanded accordingly. This expansion could ameliorate some of the issues described in cities. Referring to Lefebvre, we could expand the space for decision making, which would expand space for use value. It would do this, by allowing a greater array of voices communicating what they would like from their political space.

That technology has developed in such a way as to allow for greater information handling and processing is clear. Companies like Acxiom, Experian, and Datalogix have accrued vast amounts of information. One database alone has 1.4 billion data points (Moore 2018, pp. 59–61). This gathering of information by private companies is now well documented. Companies like Facebook and Google have become some of the biggest companies ever, powered by their understanding of user preferences. This has led to micro-profiling in consumer behaviour and political parties in elections (Bartlett 2018). This shows that technologies exist that can gather preferences at unprecedented levels. Further, this is being used by private companies and political parties to pursue their goals; expand revenue and become elected. This expanded ability to gather and utilize information is, undoubtedly, producing real change. In 2016, the UK's decision to leave the European Union sent shockwaves around the world. It has been described as a precursor to the Trump election win in the US. Leave campaign funder Aaron Banks claimed it was AI that won it for them. Few deny that the US and EU campaigns were seriously impacted by technology (Moore 2018, p. 67).

Expanded technological capacity could be used in cities like London for greater political participation. Later in this section, I will show how this is the case in cities such as Barcelona, Calgary, and Madrid. Such participation could even help counter the some of the issues discussed in Sect. 15.2, allowing more significant decision-making input to local preferences. Effective autonomy and, in turn, aspect capacity require that an agent take part in a process, which allows for the expression of that which they can "formulate, hold, and act upon" (Brown 2009, p. 157). The technology available could, in part, allow for citizens to engage in meaningful deliberation about the space in which we might consider their need for legislation to exist.

But UK democratic institutions have not kept pace with these developments. These democratic institutions were in part shaped by another period of technological development: the Industrial Revolution. They were shaped in this period, in part because the Industrial Revolution created an environment in which labour could organize. The new supply of energy, which powered the revolution, accompanied a radical shift in societal organization. The wave of technological development in recent times draws comparisons with the Industrial Revolution due partly to how seismic the changes could be to many areas of life. The new revolution is in part fuelled by data and information, which are both consumed and exchanged. Information, of course, flows much more quickly now. With these dramatic shifts created by technological development, one might expect that our institutions follow suit. The Internet could be seen as providing an opportunity to help develop democratic participation and fill this void (Contucci et al. 2019, pp. V–VI). This sentiment can be felt in the following statement, shared by London borough Lewisham council, which describes how, though we can bank, study, and communicate globally with others: "One area that seems to have remained impervious to these benefits is our model of democratic governance, which has remained largely unchanged since it was invented in the twentieth century" (Nesta 2021).

Even how we vote seems at odds with how we live much of the rest of our lives (Nesta 2020). Though that which surrounds our democratic system has been radically reshaped, the processes themselves have not. This has been argued to have led 47% of British citizens to describe feeling like they have no say over local politics (Nesta 2020). Goal 16 calls explicitly for "responsive, inclusive, and participatory and representative decision making at all levels" (Stockholm Environment Institute 2019), but in London and the UK, the digital revolution has so far not materialized in participatory decision making.

A more democratic participation through digital engagement is possible. This is due to the low entry barrier and the vast scalability of programs (UNICEF 2020, p. 3). Support for a greater enabling of digital democracy comes from a wide range of places, receiving enthusiastic support from sources as opposing as government officials and libertarians, from Liberal individualistic digital democracy, deliberative digital democracy, counter-publics digital democracy and autonomist Marxist digital democracy. The possibilities of digital democracy to offer a more deliberative space have been discussed for some time and have been linked to Habermas's work on deliberative democracy. In part, the promise of the Internet, and contemporary computing, is the possibility of a two-way, low-cost space for information sharing (Dahlberg 2011, pp. 2–9).

Managing information flows is integral to any democratic environment. Since the late twentieth century, information and communication technologies have affected society much more in general. The term Digital transformation (DX) has been used to describe such changes. Looking at information management has drawn comparisons with how Athenian democracy is depicted based in part on an aspiration for transparent and inconclusive knowledge management processes. Digital transformation affects communities, allowing the development of both local and global organizing. One reason to look to Athenian democracy for examples of pursuing a better digital democracy is how Athens excelled at "organizing useful knowledge". Delivering democratic self-governance has been argued to be possible, in part by using "…openness, transparency and inclusivity across multiple inter-dependent knowledge management processes". (Diaconescu et al. 2017, pp. 38–44)

A UK parliament enquiry was set up to look at the Internet, digital voting, and issues around the digital divide. In 2015, the Digital Democracy Commission (DDC) report said that the digital revolution had disrupted and posed a serious challenge to representative democracy.

The report had five aims:

1. By 2020, the House of Commons should ensure that everyone can understand what it does.
2. By 2020, Parliament should be fully interactive and digital.
3. The 2015 newly elected House of Commons should create a new forum immediately for public participation in the debating function of the House of Commons.
4. By 2020, secure online voting should be an option for all voters.
5. By 2016, all published information and broadcast footage produced by Parliament should be freely available online informants suitable or re-use. Hansard should be available as open data by the end of 2015 (Digital Democracy Commission 2015).

These are the aims set by the House of Commons and are for the most part yet to be met. When it comes to formal legislation, in the form of democratic processes, London and the UK have barely moved forward. Yet the citizens of London are, for the most part, both: (a) expressing preferences regularly when using phones and laptops, and (b) subject to forces that are harnessing such technology. One way that technology could be harnessed to ameliorate some of these concerns is by embracing technology which enhances the possibility of citizen participation. Next, I will introduce the idea of technopolitics as a way of framing how.

Technopolitics: tech for management or deliberation?

Much thinking about the interface between modern ICT and urban spaces is encapsulated in the idea of the Smart City. The notion of engaging modern technologies in cities is key to ideas of the Smart City. Modern technology and technology companies have, through the Smart City, been utilized globally in urban government agencies and planning scenarios. They are often presented as ways to increase efficiencies and economic development and increase participation (Levenda et al. 2020, p. 2). At the same time, uses of big data, AI, and the Internet of Things have moved beyond technologies we merely access into regular features of life (Hitachi UTokyo 2020, p. xi). Discourse around the Smart City has dominated discussion around digital technologies and governance in research, policy and practice. At the heart of Smart Cities is essentially the search to improve city life through the application of digital technology. This is through applying such tech to "the management and delivery of city services and infrastructures and solving urban issues" (Cardullo et al. 2019, p. 2).

Technopolitics is a discourse that represents the movement of digital democracy platforms, which contrast with those modes described by the Smart City. It is informed by a sociological understanding of technology and seeks more democratic developments of the uses of technology in political space (Kurban et al. cited in Martín and Smith 2021, p. 312). In cities, citizens are often participants, testers who may provide feedback but are rarely in the role of proposer or cocreator, not leading in any meaningful sense. Technopolitics invites questions of control, representation, and participation. Democracy is a key starting point. Technology here is to be critiqued and expected to carry normative elements. Its use is strategic, and aimed at improving the arena of the democratic, with emancipation and decentralization as goals (Martín and Smith 2021, p. 313).

The roots of technopolitics are not in opposition to the Smart City but provide an alternative vision of a contemporary role for technology and the Internet, a hacking-free environment of open-source software (FOSS) and the possibilities of culture and knowledge which could be reproduced at almost no cost. A parallel is the rise of corporations' approaches to digitizing the city. But technopolitics is not merely an urban practice, more one predicated on how we should direct this expansion of digital capabilities. Technopolitics is a recognition that tools and tech are never neutral; that assumptions about society are embedded in them (Martín and Smith 2021, p. 313).

Technopolitics then aims not at obscuring the political but raising it to its explicitly democratic state; to use technology to advance democratic aims. This may be through enhanced activism, platforms for citizen debate, coordination, and decision-making tools. The technology aims to create space for voice, recognition, and to make possible action, including developing the tech itself. It is inspired by Manuel Castells, Yochai Benkler, and others (Martín and Smith 2021, p. 314). Such space aims to provide a network capable of uniting different forms of knowledge and expertise—space for deliberation, but also space for the expression of diverse lives and professional knowledges. The hope is that, rather than see democratic deliberation diminish, it is possible at a scale never before seen—democracy as process, without outcome (Martín and Smith 2021, p. 314). We must embrace the fact that participation proper is messy. It is political. For this reason, we cannot take a cookie-cutter approach to it (Hammock cited in Biggeri and Frediani 2019, pp. 40–45), but, instead, have some clarity on the commitments involved.

The commitment is to try and maintain some of the messiness of democracy and deliberation that advocates of technopolitics espouse. The logic of computational language persists and must be acknowledged. These technologies are placed and understood as a site amidst a broader political struggle (without end). The aim is to place the contested notions and ways of democracy into existing frameworks of power (Martín and Smith 2021, p. 314).

15.8 Calgary, Madrid, and Barcelona: Three Examples of Technological Development and the Expansion of Space for Deliberation

In the following section, I will use examples from Barcelona, Calgary, and Madrid to show how London might look to the possibilities provided by technology, to create more participatory frameworks. In terms of developing platforms, look first to Calgary's Engage project. The project strapline is "Meaningful Dialogue. Informed Decisions" (Engage 2021). The website explains that:

"This site, the Engage portal, was established to provide Calgarians with an online space to learn about and participate in City projects and initiatives that are open for public input.

Within the Engage portal, you can:

- Provide feedback on projects that are currently open for input.
- Learn about upcoming in-person public engagement events.
- Find out what we heard from participants on projects previously open for input.
- Check the progress of ongoing projects.
- Discover what final decisions were made on projects that had public involvement".

The project has led to upwards of 50,000 pieces of engagement, compared to 200 attendees consulted at a typical event. By going online and building a bespoke platform to create space for participation, Calgary's Engage project opens up issues and space for dialogue and views from new perspectives. In this way, it has been argued that being online can create a more diverse set of views for Engage (Levenda et al. 2020, pp. 6–7).

In a second project in Calgary, Levenda et al. studied participatory budgeting in which Calgarians were asked how they might help manage city finances via an online app (2020, p. 7). This type of project demonstrates the exciting inroads that could be made in the creation of a more participatory civic space. Many citizens are used to utilizing such technology for inputting preferences concerning personal spending. To invite them to use such a method to join in city spending is an exciting prospect.

In September 2015, Madrid launched its citizenship participation panel, Decide Madrid. Barcelona followed shortly after, with Barcelona Decidim. Both have similar features, such as discussion threads, scoring and ranking of discussion and points, and the ability to share files. They also both have the ability for users to follow each other and create space for discussion of proposals and budgets for their respective cities. Anybody can contribute to discussion, but a higher level of access is afforded to those who can verify residency. This level allows residents to vote on topics. Common to both are space for citizen debates, citizen proposals, citizen budgets, and citizen plans. Any citizen can initiate a debate, which then allows others to comment. Any citizen can propose policies, propose how council budgets should be spent, and participate in development and planning discussions for their city. In September 2019, 655,559 users created over 27,000 proposals, making over 200,000 comments. Decidim Barcelona had launched 40 participatory processes, with almost 15,000 proposals and 1100 citizen initiatives. Roundtables, local meetings, walks, and online interactions all combine to add the ability for citizens to input on decision making in their city. In Barcelona, citizens added 17 proposals to the council's 45, including improved pavements, neighbourhood cleaning, bus routes, cycling facilities, and sports facilities. Participatory budgeting has been active in Madrid, with 60 million Euros of funding for citizen proposals. Safe houses for women, extra trees, drinking water, noise monitoring, urban allotments, and many more local projects were funded. The local nature of these projects requires a level of input from citizens; if it were merely the behavioural observation of the city, surely some of these preferences would be lost (Martín and Smith 2021, pp. 317–318).

Such work is onerous. Whilst technology may be used to expand the space for deliberation, humans are still needed to work through proposals and negotiate between the expectations and desires of citizens and the bureaucracy of the council. This has led to some problems in Madrid, where projects have moved more slowly than anticipated. This has led to frustration amongst campaigners working hard to increase engagement (Martín and Smith 2021, p. 321).

The commitment of platforms like these is a commitment to connecting the decision making of councils to the lived experience of citizens. This is an ongoing process. Experiments have been made to add in physical workshops. The design of sites inevitably relies on human assumptions about cities, which can be proved wrong (Martín and Smith 2021, p. 323).

Technopolitics here highlights the difference between itself and Smart Citizenship: firstly, with an open commitment to direct democracy and a sense of collective intelligence. A commitment to technopolitics and the expansion of space for deliberation accompanies representative democratic institutions. Leaders of the spaces discussed were committed to new kinds of urban democracy. Platforms then are not merely contracted to technology companies by those without a commitment to widening participation (Martín and Smith 2021, pp. 324–325).

The platforms are also committed to a sense of collective intelligence. Professional planners and administrators respond to citizens. This authoritative intelligence is, however, open to criticism and discussion via the platform. This is a reversal of the situation in which many proposals take place, where proposals exist and then citizens are invited to comment on them (Cardullo and Kitchen cited in Martín and Smith 2021).

Technopolitics offers space to understand how London might look to develop space for greater digital participation. Engage, Decide Madrid, and Decidim Barcelona are thus examples of how we might encourage greater activity in the space for urban deliberation, planning and budgeting (Martín and Smith 2021, p. 326). Technopolitics sees technology as a common requiring participatory development. City councils often have to deal with low engagement, power imbalances and a lack of diversity in voices: an often strong free-market private sector and weakened civil society (Chatterton 2019, p. 90). Technopolitics can help ameliorate this by expanding the voices heard in city decision making, space for civil society and engagement. Further, technopolitics offers a fruitful avenue of critique for fields looking to incorporate participatory work into technical solutions. In the pursuit of expanding space for the political autonomy of London's citizens, a technopolitical approach to new technology and deliberation could prove beneficial.

15.9 Conclusion

In London, we can see a city with much of its space prioritized by exchange rather than use value. To preserve London as both a global city and a livable city, a balance must be struck between market forces and the needs of citizens living locally. This chapter has shown that global markets have radically shaped London's housing market in a way that has undermined how livable the city is for its residents. Such a process exacerbates inequalities within cities, as those with less financial sway and access to decision making see their political autonomy diminished, whilst those who can profit from higher land prices, and access to decision making see their political autonomy increase.

This has happened due to several phenomena: space is being bought up and used as assets for global companies; the culture and salaries these companies bring, and the distortions this adds to local markets; the use of London space as a safe asset.

These three phenomena have combined to produce a London housing market increasingly out of reach for many London citizens. This has had the effect of: (a) making it harder to stay living in an area and (b) where people can have space, making it more expensive for them to cover basic living costs. This, in turn, diminishes their choices beyond paying for housing. Cumulatively, this can be seen as a diminishing of citizens' political autonomy and, in turn, an increase in the ability to affect London space by others, which are not based there, using, to take Lefebvre's distinction, the city for its exchange rather than use value.

This was described alongside the rise in the ability of companies to gather people's private information, preferences, and data. This has increased such companies' ability to run international businesses and affect elections. In London and the UK, this has not been accompanied by increased transparency, space for deliberation or political participation. All of this means that the relationship between London citizens and global capital has been an uneven one. Many London citizens have seen a diminishing in political autonomy, whilst the city has developed global capital. Whilst others within, and outside of the city have expanded their political autonomy through the same movement. To mitigate this, it is argued that decision-making processes ought to be reconfigured, taking advantage of contemporary technology and offering space for radically different decision-making processes. Such processes would be developed around the needs of the citizens living in London. This would involve the transparency of information. It then ought to be recognized that such global capital movements threaten citizens' ability to have control over their own space. As such, steps should be taken to empower citizens and mitigate against the threats described caused by roving global capital.

Acknowledgements Reproduced from *The role of the city in expanding and preserving political autonomy: A capability theory* by Jack Simpson with permission from Jack Simpson.

References

Aristotle (1999) Politics. Batoche Books, Kitchener, Ontario

Badarinza C, Ramadorai T (2018) Home away from home? Foreign demand and London house prices. J Financ Econ 130(3):532–555

Bartlett J (2018) The people vs tech: how the internet is killing democracy (and how we save it). Random House, London

Biggeri C, Frediani (2019) The capability approach, empowerment and participation. Palgrave Macmillan, London

Brown G (2008) Globalization is what we make of it: contemporary globalization theory and the future construction of global interconnection. Polit Stud Rev 6(1):42–53. https://doi.org/10.1111/j.1478-9299.2007.00144.x

Brown G (2009) Grounding cosmopolitanism. Edinburgh University Press, Edinburgh

Cardullo PD, Feliciantonio C, Kitchin R (2019) The right to the smart city. Bingley, Emerald Publishing

Chatterton P (2019) Unlocking sustainable cities. Pluto Press, London

Colau A (2016) After Habitat III: a stronger urban future must be based on the right to the city. The Guardian. https://www.theguardian.com/cities/2016/oct/20/habitat-3-right-city-concrete-policies-ada-colau

Contucci P, Omicini A, Pianini D, Sirbu A (2019) The future of digital democracy: an interdisciplinary approach. Springer, Cham

Dahlberg L (2011) Re-constructing digital democracy: an outline of four positions'. New Media Soc 13(6):855–872

Diaconescu A, Ober J, Pitt J (2017) Knowledge management processes and design principles for self-governing socio-technical systems. In: IEEE 2nd international workshops on foundations and applications of self* systems (FAS*W). Tucson, AZ, USA, pp 97–102. https://doi.org/10.1109/FAS-W.2017.127

Digital Democracy Commission (2015) Foreword. http://www.digitaldemocracy.parliament.uk

Engage (2021) Calgary engage. https://engage.calgary.ca

Frediani AA (2015) Space and capabilities: approaching informal settlement upgrading through a capability perspective. https://discovery.ucl.ac.uk/id/eprint/1520857/1/Space%20and%20Capabilities_AF.pdf

Hackney Council (2019) Child poverty. https://hackney.gov.uk/cyps-needs-assessment
Hackney Learning Trust (2015) Hackney childcare provision assessment 2014/15. https://www.hackneyservicesforschools.co.uk/system/files/extranet/Hackney%20Childcare%20Sufficiency%20Assessment%202014-15%20Summary.pdf
Hackworth J (2015) The normalization of market fundamentalism in Detroit: the case of land abandonment. In: Smith and Kirkpatrick O (eds) Reinventing Detroit. Transaction Press, New Jersey
Harvey D (2013) Rebel cities. Verso, London
Hitachi UTokyo (2020) Society 5.0. A people-centric super-smart society. Springer Open, Tokyo
Holly S, Pesaran MH, Yamagata T (2010) Spatial and temporal of house prices in the UK. https://ftp.iza.org/dp4694.pdf
IMF (2007) Population shift to cities. https://www.imf.org/en/News/Articles/2015/09/28/04/53/sores104a
Keil R, Ren X (2018) The globalizing cities reader. Routledge, Abingdon
Kimhur B (2020) How to apply the capability approach to housing policy? Concepts, theories and challenges. Hous Theory Soc 37(3):257–277. https://doi.org/10.1080/14036096.2019.1706630
Koh SY, Rogers D (2017) The globalization of real estate: the politics and practice of foreign real estate investment. Int J Hous Policy 17(1):1–14. https://doi.org/10.1080/19491247.2016.1270618
Lampard EE (1955) The history of cities in the economically advanced areas. Econ Dev Cult Change 3:81–136. https://doi.org/10.1086/449680
Levenda AM, Keough N, Rock M, Miller B (2020) Rethinking public participation in the smart city. Can Geograph/Le Géographe canadien 64. https://doi.org/10.1111/cag.12601
London Assembly (2021). https://www.london.gov.uk/what-we-do/housing-and-land/tackling-londons-housing-crisis
London Councils (2021) The mayor of London. https://www.londoncouncils.gov.uk/who-runs-london/mayor-london
London Datastore (2021a) Ratio of house prices to earnings, Borough. https://data.london.gov.uk/dataset/ratio-house-prices-earnings-borough
London Datastore (2021b) Earnings by Workplace, Borough. https://data.london.gov.uk/dataset/earnings-workplace-borough
London Datastore (2021c) UK house price index. https://data.london.gov.uk/dataset/uk-house-price-index
Martín PP, Smith A (2021) Going beyond the smart city? Implementing technopolitical platforms for urban democracy in Madrid and Barcelona. J Urban Technol 28:1–2, 311–330. https://doi.org/10.1080/10630732.2020.1786337
Minton A (2017) Big capital, who is London for? Penguin, London
Moore M (2018) Democracy hacked, how technology is destabilising global politics. Oneworld books, London
Nesta (2020) A framework for understanding democratic innovation in the UK. https://media.nesta.org.ukdocumentsA_Framework_for_Understanding_Democratic_Innovation_in_the_Nesta_UK_June_2020.pdf
Nesta (2021) The rise of digital democracy. https://councilmeetings.lewisham.gov.uk/documents/s63778/Appendix%20I%20-%20NESTA.pdf?
Nussbaum M (2003) Capabilities as fundamental entitlements: Sen and social justice. Fem Econ 9(2–3):33–59
OHCHR (2021) Special rapporteur on the right to adequate housing. https://www.ohchr.org/en/issues/housing/pages/housingindex.aspx
Office for National Statistics (2020) Private rental market summary statistics in England: October 2019 to September 2020. https://www.ons.gov.uk/peoplepopulationandcommunity/housing/bulletins/privaterentalmarketsummarystatisticsinengland/october2019toseptember2020
Robeyns I (2017) Freedom, wellbeing and justice. Open Book Publishers, Cambridge
Sassen S (2001) The global city. Princeton University Press, Princeton, NJ

Shelter (2018) Learning from international examples of affordable housing. https://assets.ctfassets.net/6sxvmndnpn0s/1byrK8fEQut88x9kk4w6cP/0d3452ab160befa27c832f1e1ae83f8e/International_examples_of_affordable_housing_-_Shelter_URBED_Trust.pdf
Stockholm Environment Institute (2019) Making space: how public participation shapes environmental decision-making. https://www.sei.org/wp-content/uploads/2019/01/making-space-how-public-participation-shapes-environmental-decision-making.pdf
Therborn (2017) Cities of power. Verso Books, London, New York
UNCHR (2021) Access to secure and affordable housing. https://www.unhcr.org/handbooks/ih/housing/access-secure-and-affordable-housing
UN-Habitat (2021) Housing rights. https://unhabitat.org/program/housing-rights
UNICEF (2020) Digital civic engagement by young people. https://www.unicef.org/sites/default/files/2020-07/Digital-civic-engagement-by-young-people-2020_4.pdf
United Nations (2016) The world's cities in 2016. http://www.un.org/en/development/desa/population/publications/pdf/urbanization/the_worlds_cities_in_2016_data_booklet.pdf
United Nations (2018) 68% of the world's population projected to live in urban areas by 2050 says UN. https://www.un.org/development/desa/en/news/population/2018-revision-of-world-urbanization-prospects.html
United Nations (2022a) Make cities and human settlements inclusive, safe, resilient and sustainable. https://sdgs.un.org/goals/goal11
United Nations (2022b) Around 2.5 billion more people will be living in cities by 2050, projects new UN report. https://www.un.org/en/desa/around-25-billion-more-people-will-be-living-cities-2050-projects-new-un-report
World Health Organisation (2020) Urban planning crucial for better public health in cities. https://www.who.int/news-room/feature-stories/detail/urban-planning-crucial-for-better-public-health-in-cities

Chapter 16
How the Theory and Practice of Wellbeing Advances Equity and Civic Leadership in Santa Monica

Catalina Langen and Julie Rusk

Abstract In this chapter we will explore the ways in which the community of Santa Monica has engaged with community-level wellbeing measurement and action to address issues of inequality and social isolation. We will begin by contextualizing Santa Monica's local action for wellbeing, including how the measurement tool of the Santa Monica Wellbeing Index was developed. We reflect on how disaggregated community-level wellbeing data highlights inequalities and opens up opportunity for targeted interventions to reduce inequalities. We will explore the evolution of how we practice community wellbeing, through the pandemic period of collective disruption to a new perspective within a community-based nonprofit organization. We also show how the pandemic served as a crucible moment for community leadership toward collective wellbeing. This includes impacts from the Wellbeing Microgrant Program, and how the Program seeks to measure, understand, and improve community wellbeing. And we will share the work underway to analyze what is needed to build the future of a wellbeing movement—in Santa Monica and across the United States, with learnings from the rest of the world. We conclude with reflections and recommendations based on the lessons learned.

Keywords Civic wellbeing · Equity · Leadership · Innovation · Systems change

C. Langen (✉)
Santa Monica Bay Area Human Relations Council, P.O. Box 1307, Santa Monica, CA 90406, US
e-mail: catalina@civicwellbeing.org

J. Rusk
Civic Wellbeing Partners, 2218 21st Street, Santa Monica, CA 90405, US
e-mail: julie@civicwellbeing.org

G. H. Tonon (ed.), *Urban Inequalities*, The Urban Book Series,
https://doi.org/10.1007/978-3-031-59746-6_16

16.1 How the Theory and Practice of Wellbeing Advances Equity and Civic Leadership in Santa Monica

A resident sits at a desk in the common area of a Santa Monica apartment building. She is registering voters. A community organizer knocks at the door of another residential building. He is canvassing to encourage fellow members of his local Ethiopian community to get informed and vote at the local and national levels. Meanwhile, another resident is out at Veterans Row in Brentwood handing out 155 solar-powered phone chargers to unhoused people, a need identified at a community dialogue. A member of the senior commission knits prayer shawls and sends them to those who have lost loved ones to COVID-19. And a recent college graduate prepares two cameras—her webcam angled at her face, her phone propped up above her desk to show her hands—and demonstrates how to sew cloth face masks to a group that has gathered over Zoom.

With the pandemic front and center for these October 2020 projects, a cohort of individuals met the moment with leadership in Santa Monica, supported by a program called the Wellbeing Microgrant Program. In a moment of dire uncertainty, this group demonstrated the power of social connection and community action to drive local wellbeing. What began as a project to measure and actively improve community wellbeing from within the Santa Monica city government is now a nongovernmental entity, Civic Wellbeing Partners (CWP). CWP stewards the science and practice of wellbeing through shared learning, research, and consulting services. CWP brings practices for taking action to advance local wellbeing to the local nonprofit Santa Monica Bay Human Relations Council (HRC), including the Wellbeing Microgrant Program.

This chapter offers a case study of Santa Monica, California. Santa Monica's award-winning Wellbeing Framework and Index has provided unique and meaningful insights into the community's strengths and disparities, which when disaggregated reveal specific opportunities for introspection and action. Through the Wellbeing Microgrant Program, we offer small-scale action and promising practices with the potential for future growth. The Wellbeing Microgrant Program challenges community members of all ages to pitch an idea for leadership in an area of wellbeing—health, learning, economic opportunity, place and planet, and community. The Microgrants support small-scale, short-term action with equity, impact, and the community at their core. Supported by the CWP team, Microgrant leaders report increased feelings that they can make a difference in their community, they learn and engage more deeply in community issues, and they access opportunities for economic success. In this chapter we will explore the ways in which Santa Monica has engaged with community-level wellbeing measurement and action to address issues of inequality and social isolation.

16.2 Implementing Wellbeing: Development and Measurement in Place

Santa Monica, California is a small beach city surrounded on three sides by the City of Los Angeles, with a western border of the Pacific Ocean. The city has long been an urban laboratory developing, testing, and advancing policy and practices that go beyond what many US municipalities have tackled. Over the past decade, Santa Monica has become a pioneer in the wellbeing movement and is seen nationally and internationally as a beacon for local action. Situated in the large metropolitan area of Los Angeles County, Santa Monica has long toggled between its small-town identity and big city aspirations and resources. With a resident population of approximately 93,000 and a daytime population of 250,000, Santa Monica is a tourist and workforce destination. Within local government circles, the City of Santa Monica is described as a "full-service city" anchored by thriving health care, technology, retail, and hospitality sectors along with distinctive neighborhoods, vibrant arts, and culture. The community and its government have a long-standing commitment to education (at all levels), rent control, affordable housing, social services, environmental sustainability, and more. The history of the community reveals a trajectory of success and consolidation of wealth and capital in the downtown area, with tourism thriving at peak times in the areas like the world-renowned Santa Monica Pier and Third Street Promenade. The city maintains excellent facilities for its residents, and the natural beauty of the beaches to the west and mountains to the north provide plentiful sites for accessible recreation.

For longtime residents, however, rising rents and historical housing policies illuminate a picture of a city grappling with issues of inequality. Housing prices vary widely by neighborhood, and rental prices have skyrocketed as Santa Monica has built out its tourism powerhouse of the Downtown area (Peters 2016, pp. 94, 103). From the bifurcation of Santa Monica by the 10 Freeway in the 1960s to the more recent "subway to the sea" public transit from downtown LA to the Santa Monica Pier, residents have long debated and felt the brunt of changes that have excluded or marginalized many. These changes have disproportionately impacted the Pico neighborhood, Santa Monica's historically multiethnic neighborhood (Pfeiffer 2007, p. 41). Homelessness, social isolation, mental health, and concern over the increasing cost of living are all top-of-mind issues for people in Santa Monica. Several public incidents of youth violence and suicide in the early 2010s brought community leaders together to try to understand how the abundant resources in this community were not adding up to outcomes of a thriving community for all.

These community leaders sought to redirect efforts toward improving community-level wellbeing, which came to be understood as the state of the community as informed by the individual outlook of its members, the collective sense of place and planet, access to lifelong learning opportunities, access to economic opportunities, health, and community. The Santa Monica approach focused on wellbeing as the self-organizing term over happiness, as described in the following selection from the

2013 whitepaper, commissioned by the City of Santa Monica in the early phase of its wellbeing work:

> "Happiness" communicates its appeal across a spectrum of audiences. It's easy for the general public to understand it. However, the definition of happiness is a moving target, depending upon one's background and perspective. The word "happiness" can be easily confused with the emotion of pleasure. In contrast, the word "wellbeing" draws upon the traditions of Aristotle and the ancient Greek philosophers as well as contemporary positive psychology to marry two traditions together: hedonia –a state of pleasure – and eudaimonia – a life experienced as meaningful and engaging. Wellbeing, as it is used in measurement and social change, deliberately emphasizes this Aristotelian concept of living well and reaching our full human potential. At the individual level, wellbeing may be defined as living a meaningful life, characterized by feeling empowered to make a change, be happy, healthy, and connected to one's environment and community (Kern and Warner 2013, p. 5).

A wellbeing lens helped the community to look at local resources in a different way, holding institutions accountable for outcomes. Beginning with youth wellbeing and a collective impact initiative, Santa Monica Cradle to Career (Santa Monica Cradle to Career 2022), the concept expanded to the development of a community-wide Wellbeing Index. The City of Santa Monica won a 2013 Bloomberg Mayors Challenge prize to develop the Index and eventually create the Office of Civic Wellbeing. Through data from the Wellbeing Index, issues facing the community would be brought to light and contextualized. The use of data from the Index has the power and potential to inform decision-making, reduce and prevent inequality, and transform the central goals of local government and affiliated institutions and organizations.

16.3 Multi-dimensional Community Wellbeing to Reduce Inequalities

Wellbeing as it is understood in Santa Monica is multi-dimensional—spanning individual outlook, health, learning, economic opportunity, place and planet, and community. "Civic wellbeing" is defined as the use of governance, social, and political structures to foster equitable access to services, programs, and amenities in ways that positively influence resident and community wellbeing through ongoing use of data, participatory decision-making, and collaboration across departments and sectors (Fig. 16.1).

Community or collective wellbeing is relational, encompassing both the individual and community levels. Communities come together not by chance, but because they want the place they call home to improve their lives. That place needs to provide for their most basic needs, allow them to earn a living, and be a haven for their children. It also needs to allow them to relax, be exposed to new ideas, and feel contentment. Community wellbeing is a positive state where the wellbeing of any one individual depends highly on the wellbeing of that individual's relationships and on the wellbeing of the community in which he or she lives.

Fig. 16.1 Framework for community wellbeing in Santa Monica. *Source* Made by the authors

Research on community-level wellbeing suggests that levels of social wellbeing correlate with outcomes in terms of mental health (Van Lente et al. 2012). As well, community infrastructure intervention approaches can boost social relations and wellbeing in communities; importantly, the way these interventions are conducted determines how impactful they will be (Bagnall et al. 2018). A sense of belonging and connection to one's community is tied to individual wellbeing, and Santa Monica's wellbeing work seeks to foster a sense of community by providing opportunities for individuals to come together and work toward a common goal. From this perspective, cultivating wellbeing is not the job of the collective or the individual. It is a shared responsibility; a process that is dynamic and interactive, from the personal level to the group level and back again.

Strategies for improving community wellbeing seek to reduce inequalities, create greater access to opportunities across the dimensions of wellbeing (economic, health, learning, and more), and improve individual outlook. Research shows that the relationship between income inequality and subjective wellbeing is complex (Ngamaba et al. 2018, pp. 577–596). Research has also shown that if we seek to improve mental health outcomes for communities, using a subjective wellbeing approach acknowledges and accounts for structural inequalities that can present very real barriers to empowering individuals and communities to thrive (White et al. 2016).

To better understand the state of wellbeing in Santa Monica, the City's Office of Civic Wellbeing team oversaw the production of three rounds of the Wellbeing Index, with data collected in 2015, 2017, and 2019. In partnership with researchers at RAND Corporation (headquartered in Santa Monica) and the UK-based New Economics

Foundation, the team surveyed the community, collected data from across city departments, and shared the results widely (RAND Corporation 2023; New Economics Foundation 2023). The Index measured key areas of social connection, civic action, and community leadership. The Wellbeing Framework and Index provide unique and meaningful insights into the community's strengths and its inequalities, which when disaggregated reveal specific opportunities for introspection and action.

16.4 Collaboration Across City Departments

The Office of Civic Wellbeing worked to take action on findings from the Wellbeing Index through collaboration across city government departments. Civic Wellbeing staff used wellbeing data to inform and act on interdepartmental initiatives including equity, urban planning, performance management, citywide messaging, and a Wellbeing Summit.

Through the Index findings, disaggregation of results was prioritized across income, gender, age, zip code, and race, with the goal of using data to target action to address inequalities.

In particular, ethnic breakdowns consistently demonstrated inequalities in Santa Monica. For example, in 2017 findings, Hispanic/Latino residents reported feeling more stressed than their peers in other ethnic groups and reported lower levels of trust in their neighborhood. In contrast with higher rates of economic worries among Hispanic/Latino and African American and African diaspora respondents, 75% of white respondents reported *not* being worried about housing costs. White residents also reported the highest use of community amenities. White residents reported lower job satisfaction and poorer work/life balance than other ethnic groups national. The wellbeing team used targeted interventions to take action on reducing inequalities in these areas and more.

The overall picture of the 2017 findings informed an energized group of staff who joined the Government Alliance for Race and Equity (GARE), a national network of government working to achieve equity, eliminate disparities, and advance opportunities for all (GARE 2022). The Office of Civic Wellbeing played a coordinating role in the equity efforts within city hall and in the community, reflecting the understanding that race and wellbeing are inextricably linked in Santa Monica, in the state of California, and across the United States more broadly. The Santa Monica GARE staff organized an action plan, staff affinity working groups, and a playbook for equitable community engagement.

The Santa Monica Downtown Community Plan, approved in 2017, featured an analysis of the results of the Wellbeing Index for zip code 90,401 (City of Santa Monica 2017a, b). The Plan highlights: "Downtown residents benefit from many positive dimensions measured in the Wellbeing Index, such as access to healthy food, proximity to culturally enriching experiences and institutions, use and enjoyment of Downtown's outdoor spaces, and frequent use of the many mobility options provided. However, the Index also points to areas of concern that highlight the growing pains of a residential neighborhood. The metrics for stress, stability, and work-life balance all register slightly higher Downtown than in any other neighborhood in the city. One indicator of particular concern to the planning team reveals that Downtown residents do not feel connected to one another in the way that fosters community, trust among neighbors, and social interaction" (City of Santa Monica 2017a, b).

The work of the Wellbeing Index was further embedded into practice through the Pico Wellbeing Project, a collaboration of the city's planning department and the Office of Civic Wellbeing, in response to community needs. Pico Wellbeing Project served the 90,404 Pico neighborhood of Santa Monica, where increasing unaffordability for longtime residents and gentrification have had lasting effects on community networks (Pfeiffer 2007, p. 41). Pico residents continue to face displacement, reporting the highest rates of considering a move out of Santa Monica due to housing costs. Pico residents reported the lowest life satisfaction, higher economic worries than the city as a whole, and highest rates of difficulty addressing medical needs due to costs (Fig. 16.2).

The Pico Wellbeing Project provided a community wellbeing approach to an update to the neighborhood's zoning code, engaging hundreds of neighbors in the process. After a year of focused engagement from staff through the Pico Wellbeing Project, the Pico neighborhood zip code 90,404 reported the highest rates of agreement with the Wellbeing Index question, *I can influence decisions affecting Santa Monica* (Fig. 16.3).

Citywide priorities shape work plans and performance metrics for staff; in the City of Santa Monica, performance management brought wellbeing center stage to improve outcomes for the community. The performance management team developed a dashboard to organize outcomes around dimensions of civic wellbeing. Progress toward outcomes was quantified by work streams with metrics to improve wellbeing of all people in Santa Monica (City of Santa Monica 2019a, b, c).

In the February 2019 State of the City entitled *Shaping a Changing World*, Santa Monica's Mayor framed her speech around the future of the city (Youtube

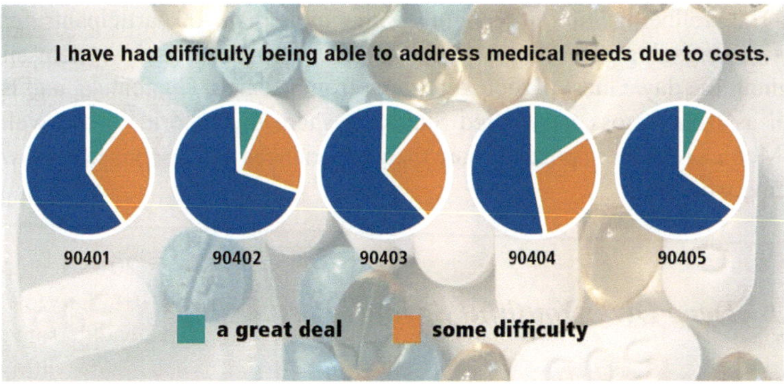

Fig. 16.2 2019 finding showing residents' capacity to address medical needs due to costs (Wellbeing Index Preview). *Source* Made by the authors

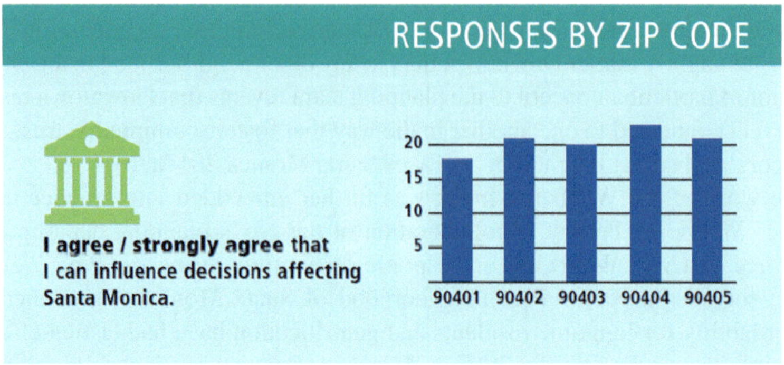

Fig. 16.3 2019 finding about residents' perceptions of their ability to influence decisions in their community, by zip code (Wellbeing Index Preview). *Source* Made by the authors

2019). At the time, Santa Monica had recently been awarded "2019 Wellbeing City" from NewCities (NewCities 2019). The Mayor outlined her speech around pillars of community wellbeing—affordability, keeping neighborhoods safe, homelessness, climate change, mobility and access, and creating an engaged and thriving community:

> "Santa Monica is recognized throughout the world as a sustainable city of wellbeing. This is not just a pie-in-the-sky idea about a vague notion of happiness. As Santa Monica demonstrated when it won the Bloomberg Philanthropies grant for creating our Wellbeing Index, wellbeing can be measured. And what can be measured can be improved… Our wellbeing index is showing some improvements, including gains in kindergarten readiness, and a reduction in the sense of isolation among our older residents. But we are committed to doing more… We will continue to look towards more innovative ways to improve our collective and individual wellbeing so we all can thrive" (Youtube 2019).

Already embedded and spreading within city government, by November 2019 the city hosted its first international Wellbeing Summit at the Santa Monica College Center for Media and Design, an innovative new community college campus (Cole 2019). The Wellbeing Summit brought together more than 900 participants, featuring dozens of engaging workshops, free local food, vendors, and keynote speakers throughout the day. Guest participants came from Mexico, Guatemala, and Bhutan, as well as from across the United States, to share their stories about wellbeing. Santa Monica residents spoke to share their experiences and lessons learned with the wellbeing work.

16.4.1 Deep Dive: Youthful Savings Partnership

As an interdisciplinary approach, the wellbeing work is about collaboration at its core. By initiating partnerships, and investing in the purpose and possibility of individuals, partnerships can thrive. One key partnership stemmed from our Wellbeing Microgrant

Program, through a grant awarded to the CEO of Youthful Savings. (Youthful Savings 2022).

Members of the Youthful Savings team tested their economic empowerment learning modules, part of their educational platform the Learning Marketplace, through a partnership with a class of middle schoolers at John Adams Middle School in Santa Monica. In a session with the students, they engaged the class in a foundational conversation on positive and ethical money management. They provided a number of learning modules to the class for future use.

Through Youthful Savings, young entrepreneurs learn skills of ethical business ownership, while creating networks of mutual support via the Youthful Savings Marketplace, an active Slack channel, and opportunities to meet virtually to continue advancing their skills and connecting with other youth leaders.

Youthful Savings Marketplace practices the following: "Youthful Savings Marketplace believes that youth deserve to partake in the global economy at a young age. All youth vendors are graduates of the My Own Business Challenge program and receive socioeconomic education that increases their mental wellbeing, financial literacy and community development skills. Our philosophy is that if all youth can receive the proper education and initial investment, they can create businesses that economically empower themselves and their communities. Our impressive youth vendors contribute to positive gross domestic product (GDP) and provide necessary sales tax revenue to their neighborhoods. Youthful Savings Marketplace aims to disrupt the growing wealth gap through the power of entrepreneurship and education. Together, let's create a kinder and more inclusive global economic system." (Youthful Savings 2022).

A youth Microgrant leader participated in a My Own Business Challenge through Youthful Savings. She demonstrated that investing in youth can be an impactful strategy for local pandemic economic recovery. Through a Microgrant project of her own, she has grown her online business. She has also grown her leadership skills, developing a curriculum to teach her peers about financial literacy and entrepreneurship.

Youthful Savings works at the intersection of youth, wellbeing, and entrepreneurism. They teach youth to harness the systems around them to thrive well into the future.

16.5 Wellbeing in Transition

At the start of 2020, the COVID pandemic sent economic shock waves through city hall as well as neighborhoods and institutions across the community. Santa Monica's tourism revenue projections bottomed out, austerity measures led to the elimination of 421 staff positions; libraries were shuttered, bus routes were reduced, some recreation activities were eliminated, city government underwent restructuring, and the Office of Civic Wellbeing was closed in June of that year (Haselmayer 2020; City of Santa Monica 2021).

At this critical juncture, we, staff from the Office of Civic Wellbeing, worked quickly to transition the City's groundbreaking work to advance community wellbeing out of city hall, forming Civic Wellbeing Partners (CWP) as a startup entity to steward the work. CWP advances wellbeing through shared learning, research, and consulting services, and brought practices for action to the Santa Monica Bay Human Relations Council (HRC).

Crucially, the HRC stepped up to ensure the continuation of the Wellbeing Microgrant Program, which fosters local action and leadership for wellbeing. A local nonprofit board founded in 2004, the HRC had previously financially supported the Wellbeing Summit, and is a strong proponent of equity and social justice through action, education, advocacy, and resolution of local conflicts. On July 2, 2020, the HRC signed an agreement with the City of Santa Monica to act as fiscal sponsor and engaged supporter of the Wellbeing Microgrant Program.

Geographically, the HRC serves the Santa Monica Bay area, particularly those residents facing systemic inequities and gaps in services and resources available to them. Pre-pandemic measurements found that 67% of Santa Monica residents said that they don't think their children will be able to afford to live here as adults, pointing to soaring housing costs and an inability to access pathways for economic mobility (Wellbeing Index Preview 2019a). 28% of Santa Monicans worried about paying their rent or mortgage, 19% worried about credit card bills, and 17% worried about losing their jobs. And most did not feel they can influence city decisions (Fig. 16.4).

The board members of the HRC represent a network of leaders with their fingers on the pulse of community life. They include executive directors of social service providing nonprofits and affordable housing providers, Santa Monica College faculty and staff, community leaders, and local government staff. Their vision is as follows:

- Believes in the value of a socially just community—a community that:
- Views diversity as an asset;
- Advances understanding and cooperation between diverse perspectives;
- Respects the rich heritages, experiences, and perspectives of the people who live and work here;

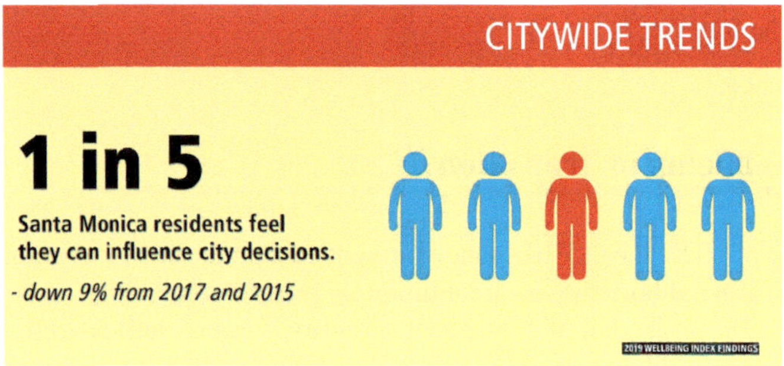

Fig. 16.4 2019 finding that feelings of influence over city decisions were down overall that year. *Source* made by the authors

- Seeks to include the voices and equal participation of all of its members, and
- Fosters individual responsibility for the benefit of all. (Santa Monica Bay Human Relations Council 2023).

The HRC fosters civic engagement, community conversations, policy solutions, and special events. Their primary focus is on the residents of the Santa Monica Bay area, with special priority on those least likely to be engaged by local government and institutions. The Wellbeing Microgrant Program advances the mission of the HRC by creating a pipeline for resident leadership and civic engagement and promoting community wellbeing.

As Civic Wellbeing Partners (CWP), in addition to bringing the Wellbeing Microgrant Program to the HRC, we see ourselves as connectors of wellbeing research to practice within communities. This has been a gap in the wellbeing space—a challenge to connect research and action in local communities. Many communities have emphasized the value of mutual aid, connection, culture, and care as essential for thriving together. The research and science of wellbeing are relevant to this, but communities have different entry points even though they are doing the same work.

16.6 The Practice of Community Wellbeing Through Collective Disruption

Through the rocky global context of 2020, Santa Monica's wellbeing work moved in conversation with the multiple layers of international pandemic and social upheaval. A national 2020 timeline of COVID in the United States created by the *American Journal of Managed Care* marks various phases of uncertainty and social distancing (AJMC 2021). Lockdowns began in California on March 19, and reopenings came in waves based on COVID case numbers in the following months. Then, on May 25, 2020, George Floyd was murdered by police officer Derek Chauvin in Minneapolis, Minnesota (MPR News 2022). The weeks afterward marked a shift—millions of people across the U.S. and the world took to the streets to demand justice for Floyd's death as well as the deaths of countless other African Americans at the hands of police. Further, the 2020 election resulting in President Biden's win marked another national call for systemic change in the U.S.

Through this time period, the ripple effects from national issues layered onto Santa Monica's local conditions to supercharge grassroots agitation for systemic and political change. Six days after George Floyd's death, protestors in Santa Monica faced a police response largely viewed as disjointed and mishandled—a Los Angeles Magazine article cited outside experts who called it a "largely preventable catastrophe" (Gumbel 2020). As looters busted the storefronts of big box shops and small businesses alike in the downtown area including the Third Street Promenade, police focused attention on protesters in a chaotic scene that resulted in significant property damage and physical injuries to protestors from tear gas canisters and rubber bullets. The events of that day ultimately heightened tensions between police and the local resident and business communities. It led to the formation of a police oversight commission that has yet to make a substantial change to policing in Santa Monica beyond presenting its initial set of recommendations. Stanford Law School recently profiled the issue, and cited current and former commissioners, who shared they have experienced "delay, interference, undue influence… [The commission has] been starved for resources, and there was a shift in city council, which changed the leadership's interest and motivations to do the things that the city council instructed the city to do" (Stanford Law School 2022). The spring and summer of 2020 marked a time of triple crisis through the unparalleled convergence of the COVID pandemic, economic crisis, and long overdue national reckoning.

A wellbeing lens asks us to analyze and respond holistically to take action through social confusion, pain, economic worry, and institutional change. For the wellbeing work in Santa Monica, 2020 was a time of adapting, continuing, and ultimately strengthening in new ways. The Wellbeing Microgrant Program continued throughout this time period and demonstrated the power of a wellbeing frame to help local people see the ways that issues interconnect, serve as an antidote to toxic local discourse, and generate civic leadership. Micrograms served as an important platform to channel energy around the collective calls for ethnic justice, in particular through projects that served African American communities and communities of color generally, projects that lifted up local history, and projects that brought people together across race through dialogue. The HRC board adapted to an online format in March 2020, and the wellbeing work supported their success in moving forward as well.

Through this period of time, information about how people are doing, in what ways they are hurting or thriving, was both hard to measure and vital for the moment. Equally important was combating narratives of being collectively stuck, mired in confusion, and lacking leadership. Through a wellbeing frame, we recognize that generating change requires intentional effort and support networks to make an impact. To do this, we take action for improved local dialogue, civic leadership, creative solutions, and empowerment through local programming like the Wellbeing Microgrants.

16.7 Wellbeing Microgrant Program as Leadership Pipeline

The Wellbeing Microgrant Program is a promising intervention that can help prevent and overcome inequalities by promoting social connection, civic action, and community leadership. Its emphasis on equity, impact, and community participation can help build a more just and equitable society by empowering individuals to take action and work toward a common goal. (Wellbeing Microgrant Program 2022). The projects and businesses funded through Microgrants are proposed by residents, and upon award of the grant, leaders receive direct support from program staff throughout the project period to bring their visions to life. Leaders also receive support through an engaged cohort of fellow project leaders. The program staff helps project leaders to level up their leadership skills. They connect leaders to community resources, businesses, and other cohort leaders; and they provide technical assistance to leaders, secure event locations and in-kind donations, and support the design of promotional materials and flyers.

Since June 2018, eight rounds of Wellbeing Microgrant cohorts have successfully generated energy and results for community wellbeing. Through 88 projects led by 91 individual leaders, upwards of 3000 participants have been engaged across all projects and rounds of Wellbeing Microgrants. Dozens of community partners, from local grocery stores and shops, social service agencies, city departments, religious and interfaith organizations, affordable housing providers, and more have contributed their time, resources, and platforms to support Microgrant leaders in achieving their goals.

Wellbeing Microgrant program staff are bilingual in English and Spanish, and one staff member previously applied for, received, and led a successful Wellbeing Microgrant project herself. Our program manager is well-versed in issues of immigration, housing and homelessness, education, and inclusive economic and community development. Our staff's lived experience informs our understanding of and ability to proactively address the immediate needs of the community. The core of our work is led by, for, and within the community.

In a recent round of Microgrants, applicants spanned diverse demographics. 5% identified as transgender or nonbinary, 66% identified as women, and 29% men. We reached diverse age groups with a strong group of high school applicants. One third of applicants were teens to 24 years old, one third aged 25–44, and another third aged 45–74. In terms of ethnic and ethnic diversity, populations represented include 14% who identified as mixed race; 5% American Indian or Alaska native, 14% Asian, 29% African American and African diaspora, 33% Hispanic/Latino, and 29% White. Microgrant leaders include monolingual Spanish speakers, and bilingual Ethiopian immigrants (Amharic and English). The diversity of Microgrant leaders is a strength of the program and a result of outreach efforts and skills of the team members.

The community most centrally served by the Wellbeing Microgrant Program has been the Pico neighborhood of Santa Monica. Pico neighborhood is the Santa Monica neighborhood with the highest use of programs and services, and the most likely to spend time in other neighborhoods of Santa Monica (Wellbeing Index 2017). Pico residents consistently apply for and lead Microgrant projects, a reflection of the energy that this program both channels and amplifies.

Findings from their Microgrants illuminate the impact of individual interventions for community wellbeing through civic leadership and belonging. An analysis of pre- and post-surveys from the fifth cohort of Microgrant leaders (early 2022) found:

- Most (73%) Microgrant leaders said that participating in the Microgrants program impacted their feelings that they can make a difference in their community and world.

- A 25% increase in those Microgrant leaders who ranked their overall life satisfaction 7 or higher on a scale of 0 (0 = not at all satisfied) to 10 (10 = completely satisfied).
- A 25% increase in those Microgrant leaders who said they feel like they belong to their neighborhood.
- A 20% increase in those Microgrant leaders who said that Santa Monica has offered them good opportunities to get ahead economically.
- No changes in frequency of socializing with friends, relatives, or work colleagues.
- No change in those who say they regularly stop and talk to people in their neighborhood.
- Opinions about whether their neighborhood was a healthy place to live became somewhat more polarized (responses grew stronger in both positive and negative directions).

These findings demonstrate the strengths of the program in terms of measurably improving individual empowerment and sense of belonging. The Wellbeing Microgrant Program organizes residents to work together, alleviate stress, and enhance creativity through the upheavals of the pandemic and beyond. Through their involvement in this program, grantees develop their voices as leaders and channel their skills and passions to impact the systems around them. Microgrant leaders report increased feelings that they can make a difference, they learn and engage more deeply in local issues, and they access opportunities for their own economic success.

Material benefits for Micrograms leaders also demonstrate effectiveness of the program in opening previously unforeseen opportunities. Microgrant leaders whose projects focused on increasing community access to culturally supportive beauty supplies, teaching skills for healthy Mexican cooking, and teaching craft paper flower creations all went on to create small businesses building from the platform they received with the Microgrants. We remain connected to these leaders for years after completion of their projects, hiring them for events and services or supporting their businesses through continued promotion through our social media networks. Through connections to our broader network of allied wellbeing organizations, still more Microgrant leaders have gone on to present at and attend convenings at venues ranging from the Santa Monica Rotary Club to the Biden Administration's Convening on Advancing Equity and Wellbeing through Data, to the Coalition for Social Isolation and Loneliness' Action Forum.

Additional program benefits include the inner work and development that has consistently emerged for project leaders. The experience of being recognized and empowered can change the internal belief system of individuals in this program. One leader said, "I learned it takes one person or one idea to bring change in a community." Another shared, "We learned about what trust means. During the event, we were able to really see the connections happening to the community, and even set up and breakdown helped establish a sense of community." These testimonials bring together the importance of seeing oneself as a leader with evidence of increased connections and mutual trust.

There has been a shift since the Microgrants began: the emergence of new, energized, and connected cohorts of like-minded leaders across diverse interests and backgrounds. What began as a focus on individual projects has developed through the COVID pandemic to center an interconnected cohort model. By noticing and attending to organic connections in the first few cohorts, we have adapted the program to promote mutual support among the leaders. The program commits to supporting access to local economic opportunities, decreasing isolation and loneliness, and advancing wellbeing. The following examples feature outcomes of a few Microgrant leaders' projects, in their own words.

16.8 Community Engagement: Ethiopian Community in Santa Monica

The Wellbeing Microgrant leader of this project aimed at bringing Ethiopian Community members residing in Santa Monica together to organize as one community. A member of Santa Monica's local Ethiopian community himself, the project leader planned to convene and strengthen the local influence and issues of his community.

After an extensive door-knocking campaign, 53 Santa Monica-based Ethiopian families were engaged and alerted to the formation of a new group by and for their community. The first in-person gathering on August 7, 2021 brought members of the Ethiopian Community under one umbrella. Accordingly, the committee proceeded to formally establish a group, Ethiopian Community in Santa Monica, on Saturday, November 6, 2021, at Santa Monica's Virginia Avenue Park.

Through the regular convening of this group, members have reported increased opportunities to support one another. Members share information using social media and e-mail exchanges. Through this formalized network, they have also been able to provide financial and spiritual support to individuals with personal and health problems.

Through the regular convening of this group, members have reported increased opportunities to support one another. Members share information using social media and e-mail exchanges. Through this formalized network, they have also been able to provide financial and spiritual support to individuals with personal and health problems.

The group plans to continue to register all members of the Ethiopian community residing in Santa Monica; to find a common ground with other similar leaders in Santa Monica; to meet with the city officials and civic and nonprofit organization leaders; and to plan community events to share and exchange traditional and cultural assets with other communities, etc.

The project leader reported that this project and the process he went through to accomplish its goals impacted him in several ways. It helped him to understand his community's needs, meet more people, and make new friends. He discovered community members who had a great desire to have an organized community but did not know how to start. He learned that it takes one person or one idea to bring change, and he discovered that there are individuals and organizations in the community who can step up to support good intentions and community-supporting agendas.

Participants in the establishment of the Ethiopian Community group saw real-time benefits, including substantial donations to a health fund for one community member, the uplifting of Ethiopian cultural heritage and tradition, and greater community power for advocacy and organization. Through testimonials, the project leader demonstrated the impact of this newly formed community group on many participants. One participant said, "This is huge; I did not expect all these members to love the idea of community organization. Now, I need the community more. I received contributions for my sick friend all at one time. You can not ask for better than this!" Another participant shared, "This gathering is way more than I expected. It is energizing! This is the best way to teach our children about our cultural heritage. That is why I am motivated to spare my time to drive door to door to reach out for more members." A third participant said, "This is the way to go. Look at what we have done in just a few months; I learned that if we unite and are well organized, there will be nothing we can not accomplish."

The project leader also reported a few challenges that he overcame, including the challenges of finding potential members, issues with coordinating with partners to perform outreach, finding common meeting times, and distributing invitations. However, he reported that the project accomplished its goal and that the overall outreach, participation, volunteering of members, and support from the Microgrants team was a success.

16.9 Entrepreneurial Pipelines: Pop-Up Beauty Supply Store and Hair Parlor

In a local economy, a wellbeing framework calls for centering outcomes for people (rather than endless growth) as the goal of economic activities. This topic has been the focus of community conversation within the Pico neighborhood for years, with one upcoming solution being the development of a holistic economic hub in the neighborhood, a multi-use project of local affordable housing provider Community Corp (threesixtyCITY Podcast 2022). Wellbeing Microgrant leaders were well-poised to apply for this opportunity. One Microgrant leader who is a multi-talented business owner demonstrates the ways a Wellbeing Microgrant can serve as a launching pad for entrepreneurs.

Her Microgrant project produced a pop-up beauty supply store and hair parlor in the Thelma Terry Room at Virginia Avenue Park, a key community gathering space. "As a Santa Monica resident for over 23 plus years, African Americans and people of mixed cultures have to travel very far out of the community to find a beauty supply store," the project leader said. "As a woman, our hair is very important and dear to us. It's a part of our culture and identity. This is an ongoing conversation in high demand from the people of the Pico Community of Virginia Park. Beauty supplies serving African Americans and people of mixed cultures are in high demand, and this business can serve local weddings, proms, school graduations, birthday parties, and special events. Recycling dollars back into the community via this Microgrant makes a positive impact."

She gathered 50 of her neighbors and community members for an energizing afternoon event which included a visit from the Mayor of Santa Monica and Chief of Police. Santa Monica College beauty school students styled the hair of attendees seated in barber chairs, and women discussed the best products for different hair types, all surrounded by colorful displays of hair products and elaborate hats. In Kimberley's post-event survey, 100% of respondents felt their hair care needs were acknowledged during the event, and 96% said that having products of their culture available in the Pico neighborhood was very important to them.

During her project, the Wellbeing Microgrant Program organized and financially supported a group of Microgrant leaders to attend an Entrepreneurial Training Program through Pacific Asian Consortium in Employment Los Angeles. Kimberley participated in that training, which provided the actionable tools and skills needed to launch a small business.

A woman of many talents and entrepreneurial interests, Kimberley also operates a Creole soul food business and has gained traction at the Saturday morning Pico Farmer's Market, selling beignets and other fresh treats. She is among the local vendors applying for a food business space within the in-development economic hub, 1819 Pico Marketplace.

16.9.1 Healthy Soil, Partnerships, and Policy: The Soilist

The Soilist's Wellbeing Microgrant journey has taken her through multiple local connections, leveraging her deep knowledge of urban soil and compost health to positively impact, educate, and generate sustainable solutions in her community. She connects people with the living soil around them and supports action for healthy communities. Through her Microgrant, the Soilist's objective was to disseminate knowledge about "living/biological compost" beyond the boundaries of her previous scope of managing the garden of a local, community-engaged church, the Church in Ocean Park (CIOP) (Fig. 16.5). She filmed and published videos of her process supporting a local homeowner in organizing his composting system (Youtube 2021).

The Soilist further practiced her studies with Soil Food Web School through biological assaying of local compost at various sites around town (Soil Food Web 2022). As she says, she "ascertained a serious lack, across all samples, of numbers and diversity of micro-organisms necessary for healthy soils and composts. The assays confirmed what I knew theoretically: that our urban soil environs are ailing."

During her project period, she participated in a compost giveaway by Santa Monica City Yards. "As I drove into the Santa Monica Yards, an unpleasant smell alerted me that something was wrong. Upon examining the sample under the microscope, I found

Fig. 16.5 The soilist in the garden at the church in Ocean Park in Santa Monica. *Source* made by the authors with permission

a lack not only of the diversity of microbes needed for the material to be considered as compost, but it lacked microbes, period: not one of any of the needed nematodes, protozoa, or fungi was present. As the public, we think we are applying compost to our gardens when in fact we are adding, at best, virtually lifeless, anaerobically produced material suitable as mulch or for bulking up the soil."

Knowing what she now knows about Santa Monica's soils and landscape waste, the Soilist grew determined to channel her knowledge into supporting a better solution for the local ecosystem. She had made valuable contacts in the Santa Monica Parks system through her Microgrant project. Her new goal is to educate the community to generate awareness so that Santa Monica's green waste might instead "be converted into 'live' compost that would promote full nutrient cycling in soil and plants, carbon sequestration, a reduced carbon footprint while working to convert Santa Monica's soils into a water-retaining sponge."

Her efforts since receiving a Microgrant have led to her own increased civic engagement and awareness of the health of the soil and compost systems around her. As well, her efforts have generated awareness of soil health among diverse populations of people—at various points in her work she has presented on composting for local residents of affordable housing buildings, seniors, homeowners, unhoused people, and more.

She reported that receiving the Microgrant has "kept her motivated and energized around a daunting task—working to change mindsets and systems in the hopes of converting organic waste into a viable product that will assist and encourage urban growers, gardeners, and property owners to contribute to the healing of our environment by assisting in the regeneration of our soils."

16.9.2 The Future of Wellbeing in Santa Monica and Beyond

Local action, as illustrated above, ties directly to greater systems change and growing grassroots calls for the government to take seriously the issues that contribute to (or else degrade) a community's sense of collective wellbeing. As demonstrated through the ongoing and recent projects through the Wellbeing Microgrant Program, action for wellbeing is alive and well in the Santa Monica area. The lessons from Microgrant leaders in this place offer possibilities for replication in other areas. Pairing wellbeing research with action in a community offers a path forward for individuals to deepen their awareness of the systems around them. It also opens space for the possibility of individual creativity to have a platform for greater positive community impact. In a space like Santa Monica with serious social inequality and a challenging political climate, giving people a platform (like the 90 past and present Microgrant leaders) offers a different approach for advancing community wellbeing.

Through connecting grassroots and research perspectives, crossing sectors, and forming new coalitions, the wellbeing work and the nascent movement forming around it offer a more interconnected path forward for measurement and action in communities and countries around the world.

16.9.3 Wellbeing, Hope, and Models for Replication

Wellbeing initiatives have an opportunity to look holistically at the types of metrics we use, the communities we engage, and the channels for effective impact and systems change.

By leveraging partnerships with research-based organizations, the Wellbeing Microgrant Program has continued to expand the kinds of information it measures. In a recent partnership with Dr. Carol Graham of the Brookings Institution, we explored the impact of Microgrants as interventions to support hope and wellbeing, with an emphasis on the ways Microgrants support the arts and artistic expression. The questions we asked were as follows:

> These questions are about your future and your hope for success.

- I trust that I will be able to do everything that I want to do in my future.
- I believe that the things I am doing now are preparing me for what I want in the future.
- I believe that I will be successful even when there are difficulties in my life now.
- There are people who can help me when I need guidance to achieve what is important to me.
- Are you still hopeful about your future since the onset of the COVID-19 pandemic? (Graham 2023, p. 147)

We found that during the short three-month measurement window, hope for the future increased among some of the Microgrant leaders, and life satisfaction increased more broadly. The leaders' confidence and management skills also improved.

The Wellbeing Microgrant Program has a record of localized impact including community engagement, entrepreneurship, and policy. The program offers a promising practice for bridging research and practice in the wellbeing space and is a replicable model.

Additionally, the disruption of the COVID pandemic put a pause on active data collection through the Wellbeing Index, but as evidenced by the multiple layers of fallout from the pandemic, information about our collective wellbeing is more important now than ever. A next iteration of the Wellbeing Index would provide vital information about how people in Santa Monica were impacted by the pandemic (across the wellbeing dimensions of health, economic opportunity, place and planet, community, and learning). It would also potentially show where residents of this community remain resilient and flourishing in spite of the disruptions of 2020 and beyond. The tools and systems developed in Santa Monica are primed to serve others in communities around the United States and the world.

16.9.4 Systems Change for Wellbeing Across the U.S. and the World

Organizations, local governments, and communities across the U.S. and the world are reimagining what is possible for tomorrow in the midst of significant challenges today. Despite, or perhaps because of, inequities across ethnic, gender, socioeconomic, and geographic lines, an opening has emerged for collective wellbeing.

Civic Wellbeing Partners is actively engaged in partnerships across the United States and the world. We have been actively engaged in bringing the work of the Wellbeing Economy Alliance to the United States, serving as founding partners on the steering committee of the WEAll California Hub. We bridge spaces of sustainability, democracy, and equity with a wellbeing lens to orient various initiatives around shared outcomes. We see the wider work as both informative for and deeply informed by the local Micrograntss work.

In December 2020, CWP connected with co-creators of the Madison, Wisconsin-based Institute for Collective Wellbeing. The Institute for Collective Wellbeing's goal is "to share stories, ideas, practices, research, and frameworks that help all of us continuously evolve our understanding of collective wellbeing and how it is practiced and seen in our communities and organizations" (Institute for Collective Wellbeing 2022). Our teams are connected as actors in the wellbeing space with deep roots within systems, branching out anew to form our respective organizations. Together, our teams have extensive experience working across community government, nonprofit, academic, and private sectors, and have developed and implemented collective wellbeing initiatives, programs, and scholarship. We have experience in policy and program development, innovation, research and evaluation, public and private sector organizations, community engagement, and cross-sector partnerships. Our teams work in collaboration to advance the field of collective wellbeing.

With this connection to ground us, CWP and Institute for Collective Wellbeing embarked on a partnership for a project called Mapping Collective Wellbeing: Advancing Opportunities for Systems Change, a project to map and understand the system of collective wellbeing initiatives in the U.S., while applying lessons from around the world (Mapping Collective Wellbeing 2023). With funding support from the Robert Wood Johnson Foundation (RWJF), the team worked to map collective wellbeing efforts that are emerging locally through initiatives across the U.S. This project responds to growing demands for progress, across the interconnected areas of social and climate justice.

At the beginning of our project, we understood collective wellbeing as an approach that connects social and ethnic equity, sustainable ecologies, and the ability of individuals to realize their full potential. It is characterized by authentic, generative relationships and reimagining the purpose of society. We have continued to iterate upon this definition with reflections from over 80 thought leaders in this space—and growing.

An RWJF program officer said: "In the United States, our policies, programs, and budgets are more often oriented around economic indicators than the wellbeing of people. For several years, the Robert Wood Johnson Foundation has been looking at other countries that put wellbeing as their North Star to learn how their approaches might inform policy and practice here in the United States. We're excited to see how these lessons can now be applied to help accelerate the collective wellbeing efforts that are emerging in the U.S."

Through interviews and research, the team identified leadership skills, resources, and mindsets needed to make a significant shift toward collective wellbeing practices. The project, completed in 2023, has engaged and supported dozens of collective wellbeing system stakeholders across community, local government, nonprofit, academic, and for-profit sectors. The team collected and shared lessons from place-based initiatives in other parts of the world, recommending actions to advance the collective wellbeing system in the U.S.

Key learnings and recommended next steps emerged through a co-sensing process which employed the Theory U process of change management developed by Otto Scharmer of MIT (Scharmer 2009). These include recommendations for learning and action. Within learning spaces, collective wellbeing practitioners need support through peer-to-peer connections that foster authentic relationships and facilitate greater understanding within and across sectors. A key function of wellbeing work is identifying and acting strategically to advance equity in the pursuit of wellbeing for all. There is also a major role for those with access to capital in fueling the collective wellbeing movement. Stakeholders asked, what is possible when we adequately fund equitable, innovative, and human-centered initiatives? What impedes progress? How might we reimagine an economy where wellbeing drives decision-making? Practitioners also seek activities, tools, and resources that promote equitable change and evaluate success and failure. Offerings are continually needed to support applying lessons from wellbeing initiatives across the globe. Innovative initiatives are necessary to bridge the gap between the collective wellbeing system today and where it needs to be to affect deep systems change.

16.10 Conclusion

To improve community wellbeing, Santa Monica's experience demonstrates that there is a need for translating measurement into action, fostering strategic partnerships, and advancing equitable programmatic design. From its origins as a data project started with Bloomberg Philanthropies Mayors Challenge prize funding, to its life as the operating system of city government, to its current iteration as a community-based organization, the wellbeing work in Santa Monica persists and grows in new ways. Learnings and partnerships picked up along the way inform the landscape of possibilities for growth and deeper impact.

Currently, the Wellbeing Microgrant Program sheds light on the assets and skills of community members and connects them to one another for mutual support. The program crosses disciplines and dimensions of wellbeing to create interesting connections, opportunities for learning, and leadership development. The program impacts both project leaders and their participants, increasing civic engagement, leadership skills, belonging, and overall life satisfaction. The examples shared by Microgrant leaders all highlight the varied ways Microgrants make an impact. We have learned that applying the framework of wellbeing in specific contexts and situations empowers individuals to see themselves and their role in their community through a more holistic frame.

The future of wellbeing work is still being written. The many assets, relationships, and action projects undertaken up to this point could converge within the context of replication opportunities in other communities. They have the potential to serve as a hub for research to practice pipelines, as evidenced through the example of embedding questions about hope for the future. Civic Wellbeing Partners and our regional allies are well-positioned to continue to lead in the wellbeing space through partnerships with global organizations like the Wellbeing Economy Alliance and the Mapping Collective Wellbeing project, an exploration of the larger system of collective wellbeing across the U.S. The Mapping Collective Wellbeing project outlines recommendations for learning and action for the nascent wellbeing system in the U.S.

The U.S. is at a point in its history where wellbeing as a framing can serve to unite across a fractured national dialogue by returning to the first principles of government. Entrenched narratives about the economy, race, health, and democracy are being challenged in new and evolving ways. Ultimately, what we ask with a wellbeing framework is: How are the people in a community really doing? How do we collectively define a good life and thriving community? And what actions can be taken to make measurable improvements?

What we demand with a wellbeing framework is accountability based on results, centering people, and the planet, with mindfulness of place. We measure and take actions for better outcomes in order for all of us to thrive. Hope, trust, and wellbeing come together when we work collectively.

References

AJMC Staff (2021, January 1) A timeline of COVID-19 developments in 2020. AJMC. https://www.ajmc.com/view/a-timeline-of-covid19-developments-in-2020

Bagnall A, South J, Di Martino S, Southby K, Pilkington G, Mitchell B, Pennington A, Corcoran R (2018) A systematic review of interventions to boost social relations through improvements in community infrastructure (places and spaces)

City of Santa Monica (2019a, March 9) Wellbeing index preview. https://santamonicawellbeing.org/Media/Default/docs/wellbeing-preview-03092020.pdf

City of Santa Monica Cradle to Career. (2022). https://santamonicacradletocareer.org/.

City of Santa Monica (2017, July) Downtown Community Plan. https://www.smgov.net/uploadedFiles/Departments/PCD/Plans/Downtown-Specific-Plan/FINAL%20DCP_web.pdf#page=26

City of Santa Monica (2021, May 10) Introducing the proposed FY 2021–23 Biennial Budget. https://www.santamonica.gov/blog/introducing-the-proposed-fy-2021-23-biennial-budget

City of Santa Monica (2019b, April 9) New performance dashboard shares progress on community outcomes with the public. https://www.santamonica.gov/press/2019/04/09/new-performance-dashboard-shares-progress-on-community-outcomes-with-the-public

City of Santa Monica (2017) Wellbeing index findings

https://santamonicawellbeing.org/Media/Default/findings/2017_WellbeingIndexFindingsSummary_SaMo.pdf

City of Santa Monica (2019) Wellbeing index 2019 snapshot. https://drive.google.com/file/d/1n3xcXmRsZW8GJZqCYxXydJL9y60JZg4a/view?usp=sharing

Cole R (2019, October 4) Wellbeing Summit: Santa Monica's vision to define what matters most. City of Santa Monica. https://www.santamonica.gov/blog/rick/wellbeing-summit-santa-monica-s-vision-to-define-what-matters-most

Government Alliance for Race and Equity (2022). https://www.ethnicequityalliance.org/about/who-we-are/

Graham C (2023) The power of hope how the science of well-being can save us from despair. Princeton University Press

Gumbel A (2020, November 17) Inside the Santa Monica Police Department's Botched Response to May's Looting Spree. Los Angeles Magazine. https://www.lamag.com/citythinkblog/santa-monica-police-may-31/

Haselmeyer S (2020, June 18) Santa Monica's austerity plan turns the city away from the future. Medium. https://medium.com/states-of-change/santa-monicas-austerity-plan-is-turning-the-city-away-from-the-future-bd955276a38

Institute for Collective Wellbeing (2022). https://www.instituteforcollectivewellbeing.org/

Kern M, Warner K (2013) A city of Wellbeing: the what, why & how of measuring community wellbeing. https://use.metropolis.org/system/images/1606/original/MeasuringCommunityWellbeing_Whitepaper2013.pdf

Killing of George Floyd (2022) Minnesota Public Radio. https://www.mprnews.org/crime-law-and-justice/killing-of-george-floyd

Mapping Collective Wellbeing (2023). https://mappingcollectivewellbeing.org/

NewCities (2019) Wellbeing City Award. https://newcities.org/wellbeing-cities-2019/

NewCities (2022, March) Wellbeing in action with Julie Rusk and Tara Barauskas. ThreesixtyCITY Podcast. https://open.spotify.com/episode/6qyHaj8seBRCFuC5jonYQj?si=605ff9888ff1465b&nd=1
New Economics Foundation (2023). https://neweconomics.org/
Ngamaba KH, Panagioti M, Armitage CJ (2018) Income inequality and subjective well-being: a systematic review and meta-analysis. Qual Life Res 27:577–596
Peters D (2016) Density wars in silicon beach: the struggle to mix new spaces for toil, stay and play in Santa Monica, California. In: Protest and resistance in the tourist city. Routledge, pp 104–120
Pfeiffer D (2007) The dynamics of multiethnic integration: a case study of the Pico neighborhood in Santa Monica, CA, Doctoral dissertation, University of California, Los Angeles
RAND Corporation (2023). https://www.rand.org/
Santa Monica Bay Human Relations Council (2023) Vision. http://hrcsantamonica.org/vision/
Scharmer O (2009) Theory U: leading from the future as it emerges. Berrett-Koehler Publishers
Soil Food Web School (2022). https://www.soilfoodweb.com/
Van Lente E, Barry MM, Molcho M, Morgan K, Watson D, Harrington J, McGee H (2012) Measuring population mental health and social well-being. Int J Public Health 57(2):421–430. https://doi.org/10.1007/s00038-011-0317-x
Wellbeing Microgrant Program (2022). https://www.wellbeingmicrogrants.org/
White RG, Imperiale MG, Perera E (2016) The Capabilities approach: fostering contexts for enhancing mental health and wellbeing across the globe. Global Health 12:16. https://doi.org/10.1186/s12992-016-0150-3
Youthful Savings (2022). https://www.youthfulsavings.com/
Youtube (2021) Jane's Garden community 2: prepping for a thermophilic COMPOST. Retrieved November 21, 2022, from: https://www.youtube.com/watch?v=IqWwGVtf_6I
Youtube (2019) Santa Monica Mayor Gleam Davis at State of the City 2019. Retrieved April 19, 2023, from: https://www.youtube.com/watch?v=HcBOKEO6ZFs

Chapter 17
The Complementarity of POZE and AI to Optimize Quality of Life in an Urban Setting

Cornelia C. Walther

Abstract This chapter is anchored in a multidisciplinary paradigm (POZE) which describes and analyzes individual human behavior as well as social interaction. The POZE paradigm is used in two converging lines of thought. First, to analyze the components that distinguish equitable human settlements. At the level of individual behavior, the aspirational, emotional, intellectual, and social aspects of equitable cohabitation will be looked at. Complementarily, we look at the interplays among *micro* (individual), *meso* (community/institution), *macro* (country), and *meta* levels. Secondly, that multidimensional perspective is used to explore how certain applications of artificial intelligence (AI) can serve to design policy solutions that are conducive to quality of life in the urban space. Both streams of attention are pursued in view of identifying factors that serve the optimization, or a dynamic of constant improvements, of quality of life in a setting of equitable cohabitation. Based on the understanding that the individual quest for meaning and cooperation is the trigger to move from inequality to inclusion, it will be argued that human choices can transform the urban space from inequitable to fair. Without being a chapter about technology, this text looks at the potential of leveraging AI for enhanced quality of life. In short, we argue that individual aspirations (IA) and artificial intelligence (AI) can be consciously curated as a symbiosis that leads to inspired action.

Keywords AI · Complementarity · Continuum · Optimization · POZE

C. C. Walther (✉)
POZE@Ezop, Wharton School, University of Pennsylvania, Philadelphia, PA, USA
e-mail: cwalther@wharton.upenn.edu

G. H. Tonon (ed.), *Urban Inequalities*, The Urban Book Series,
https://doi.org/10.1007/978-3-031-59746-6_17

17.1 Introduction

Social optimization is an ongoing process, as the individual and collective dimensions of human existence stand in constant interplay. These dynamics can be understood and systematically addressed. The necessary switch from separation to complementarity may appear complex, yet it is possible and indispensable for individual and collective well-being, locally and globally.

The benefits that arise from respecting the responsibility that connects privilege and responsibility are mutual; the underpinning obligations are partially reflected in the collective dimension of certain human rights (Walther 2014), with a derivative in the form of social policies. In this chapter, the application of this perspective to urban policies is introduced. Ultimately, respecting the rights of everyone is in the interest of everyone. Because in a society marked by inequality everyone, including those at the top are worse off. (Wilkinson and Pickett 2009). Accepting and addressing the logic that is inherent to being human and becoming humane serves everyone involved.

Like each era before, the twenty-first century is marked by challenges and opportunities that can be summarized in a more neutral way as 'catalysts of change'. Two aspects that are different today are the massive impact of web-based (inter)action, from work over education to shopping, socialization, and leisure. That expansion of virtual activity makes social relationships at the same time extensive in quantity, while simultaneously creating a qualitative vacuum, as genuine human interaction becomes rare. In 2024, 'loneliness' was declared a health emergency by several US counties; ironically starting with the area of the Silicon Valley—where investments in technology-driven connectivity is at its peak. Never in human history have, we been as 'connected', and isolated as now. The second aspect is the amplifying effect of the web of everything that happens. The online space reflects and impacts the offline 'reality'; conversely vice versa interpersonal offline interactions are affected by online (re)actions. Nothing happens in a vacuum.

The 24/7 exposure to information leads to an inflation of sensorial inputs, which is accelerated with the recent onset of Generative AI (GenAI), which comes with the additional risk of trust erosion. When images, text, videos, and sound can be produced and shared by anyone with Internet access in a couple of clicks, 'reality' becomes a precious and rare commodity. The more sophisticated our technological tools become, the more vulnerable we are to the hijacking of our perceptions. What is real and what is fake? Authentic human connection, touch, and face-to-face time are precious to build and strengthen interpersonal bonds, which are the tissue of communities, and hence the foundation of cities that are set to support quality of life. Technology can make human bonding harder, or easier, depending on the mindset of those who design and use it, and the hybrid environment in which social interplays are taking place.

To achieve the ambitious goal of an urban setting that is genuinely inclusive, and conducive to bring out the best in all members of the community, every aspect of the urban space must be created, curated, and consciously cultivated with the ambition of equal quality of life in (the human) mind. Technology may support that journey; it can neither start nor undertake it for the human user. Our experiences are the cause and consequence of our actions and the context in which it is taking place. Individual choices are the butterfly wing that can set a hurricane in motion; this characteristic is amplified in our interconnected hybrid society. With the ever-expanding influence of social media channels and AI-powered tools, missteps made by individuals in one location have direct and indirect repercussions on the lives of others, online and offline. What we (do not) do impacts the *meso*-, *macro*-, and *meta*-dimensions (referred to as the *m4-matrix* below) which we are part of.

Even in an age where technology seems to be penetrating slowly into ever more areas of life, a meaningful existence is not a fairy tale; it is the basic configuration that humans are set up for. The central argument of this chapter is that cities can be shaped to lift individuals to fulfill their potential, alone and as part of a whole—if those who constitute the groups which they belong to, aspire to that objective and act accordingly.

From AI to policies, via public and private investments, reality is a manifestation of human choices. The two questions to be asked in view of intentional optimization are (1) what are the personal values that underpin our aspirations, and hence our choices (*micro*), and (2) how we can systematically identify and use leverage entry points to support the endeavor of maximized quality of life for a maximum of city dwellers, with impulses from the *macro* (national) and *meta* (global) policy level.

Three convergent and mutually influencing dynamics will be briefly analyzed in this chapter: (i) the POZE paradigm as a new perspective to understand the multidimensional underpinnings of human aspirations and behaviors; (ii) the emergence of AI as a tool of prosocial change; and (iii) the application of AI infused with the POZE perspective to design urban policies geared toward equitable quality life for all. These avenues assess how infrastructure and individuals could be influenced intentionally with the 'Common Good' as the shared interest—to curate an ideal context for the communities that arise from them and the people that evolve in them. The 'common good' refers to the conditions of social and economic life believed to be necessary for the prosperity and flourishing of all members of a society. It encompasses both shared benefits, such as access to essential goods and services, and collective values, such as justice, security, and environmental sustainability (Baird 2012; Mulgan 2006). It is important to note that defining the specific content of the common good is often contested and context-dependent, reflecting different values and priorities within communities (Mulgan 2006).

17.2 The Perspective. POZE and Why It Matters

In a snapshot—POZE which translates as 'inner peace' from Haitian Creole, language spoken in the country where the approach was prototyped, is an acronym that encapsulates its core parameters: a new **P**erspective to life, **O**ptimization of multidimensional interplays, the pursuit of subsequent **Z**eniths, and **E**xposure to reality as it is, and could be. Anchored in it, a twice four-dimensional prism arises which explains why individual and collective existences are not merely influenced but shaped by human aspirations, emotions, thoughts, and sensations. The causes and consequences of behavior, which is a physical manifestation of internal streams of consciousness, reflect and impact the interaction between individuals (*micro*), institutions/communities/cities (*meso*), countries/economies (*macro*), and the Planet (*meta*).

These multiple internal and intra-personal interplays should be acknowledged, accepted, analyzed, and addressed with the conscious intention of a social environment in which every person has a fair chance to thrive (Walther 2021).

Identity is a dynamic kaleidoscope that evolves over a lifetime, shaped by aspirations and actions rather than immutable past factors. Experiences play a pivotal role in shaping individuals' thoughts, emotions, and aspirations, which in turn influence their health, well-being, and behavior. These reciprocal interactions generate memories and habits, forming the fabric of personal identity. The multidisciplinary approach in the following recognizes that each person embodies multiple dimensions, reflecting the complexity of the society they inhabit (Walther 2020a). This understanding serves as the foundation for exploring the interconnectedness within both personal and collective realms.

Choices and values are influenced by perceptions, which are in turn shaped by Bayesian integration—the process of combining prior knowledge with uncertain evidence—a phenomenon deeply rooted in the intricate connection between the body and mind (Sohn et al. 2019). Negative experiences, whether chronic or acute, can narrow one's present scope of aspiration, leading to feelings of depression and diminished future ambition. Parental influence and education play crucial roles in shaping a child's worldview and future decision-making, highlighting the importance of fostering openness and inclusivity in society (Hyungryeol and Euijin 2019). However, traditionally, individuals develop within a framework of social identity markers—cultural, religious, racial, national, and socioeconomic—that often perpetuate divisions.

Understanding individual well-being as both a cause and consequence of collective well-being, four dimensions shape individuals (Fig. 17.1)—physical, emotional, cognitive, and aspirational—and four dimensions shape society (Fig. 17.2) in the *micro-*, *meso-*, *macro-* and *meta*-arenas, respectively, economic, social, environmental, and cultural (Walther 2020a, b, c). Each of these and the principles that apply in both arenas are explored next.

17.2.1 Dimensions

Human existence worldwide operates within a shared multidimensional framework, akin to navigating shifting tectonic plates. At the individual level, human experience encompasses the aspirations, emotions, thoughts, and sensations represented conceptually as soul, heart, mind, and body. Each dimension plays a distinct role in shaping our understanding of self and the world around us.

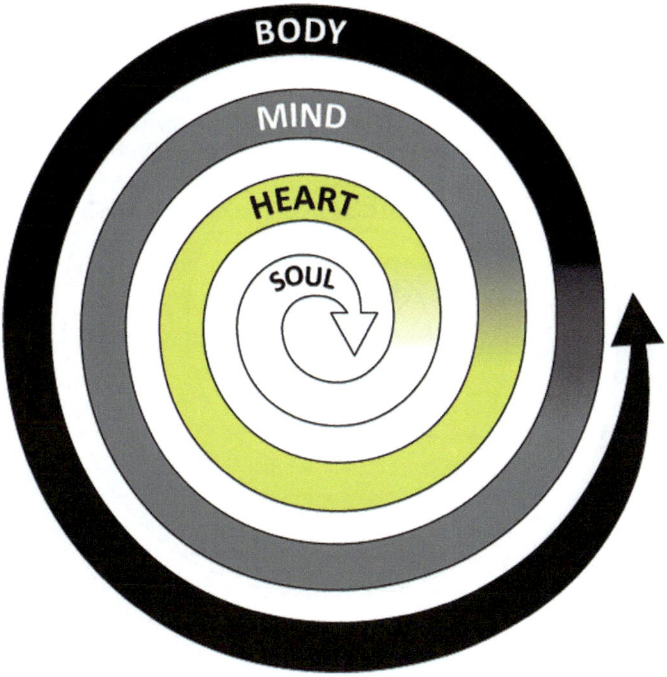

Fig. 17.1 Spiral soul to body. Interconnectedness defines our existence. From aspirations to emotions, thoughts, and actions, each element influences the next in a continuous loop. Reality is shaped by subjective perceptions, where internal dynamics dictate external experiences. Sensations feed emotions and thoughts, which in turn shape our environment and behavior. This reciprocal interaction highlights the inseparable link between internal states and external realities. *Source* Walther (2020a)

The *body* serves as the physical conduit bridging internal and external realms, akin to a receptive membrane. It functions not only as a vessel for experiencing the external environment but also as a reflection of internal processes. The body is the channel through which we express our internal world and connect with others.

The *mind* acts as a filter, analyzer, and judge of sensory inputs received through the body. Informed by past experiences stored as memories, the mind processes information, leading to decisions and actions. Understanding the interplay between sensations, thoughts, feelings, and beliefs offers a pathway to conscious influence over emotions (Kahneman 2011). Moreover, anchoring sensory experiences in the physical body provides a way to disengage from cyclical patterns of thought and emotion; rumination is stopped in its tracks. Further research into cognitive processes and the mechanisms underlying decision-making offers insights into potential avenues for conscious cognitive enhancement (Kulik and Fletcher 2016). The intricate connection between the brain and the enteric nervous system illustrates the profound influence of bodily functions on cognitive functions and emotions (Hopkins Medicine 2022). For example, disruptions in the gastrointestinal system can trigger mood changes,

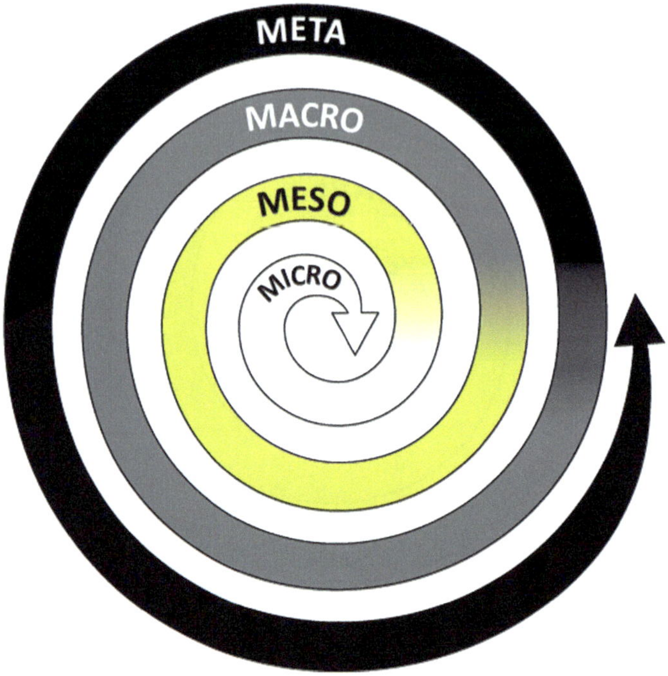

Fig. 17.2 Spiral *micro* to *meta*. Individuals constitute the *micro*-arena within a multidimensional system, shaping and experiencing the *meso*-arena through their affiliations. These communities, whether voluntary or biological, directly, or indirectly, contribute to the contours of economic, political, and cultural systems at the *macro*-arena. *Micro-*, *meso-*, and *macro*-dimensions operate within the *meta*-dimension, where each dimension influences the others while being influenced in return, within a seamless flow of change. *Source* Walther (2020a)

anxiety, and depression, highlighting the body's role in shaping perceptions and interactions; and conversely the influence that mental changes have on physiological aspects of an individual's experience.

The *heart* metaphorically represents the seat of emotions, driving human actions and decisions. Emotions serve as the underlying force guiding intellectual deliberation and motivating individuals to transition from contemplation to action (Hills and Argyle 2001). Exploring the physiological basis of emotions and their impact on behavior elucidates the ongoing relationship between affective states and cognitive processes, underscoring the significance of emotional intelligence in personal and interpersonal domains (Gigerenzer 2007). Further study of the body's mechanisms sheds light on the intricate interplay between physiological well-being and emotional states, emphasizing the importance of holistic inside-out, outside-in approaches to health (Godfrey et al. 2018).

Finally, the *soul* symbolizes the essence of humanity, reflecting the quest for meaning and purpose. As individuals embark on this journey of self-discovery, they gradually unveil their inherent potential, leading to genuine well-being and self-expression (Frankl 1946). It is used in the present context without a religious connotation. Research diving into existential psychology and the pursuit of meaning offers insights into human flourishing and the intentional cultivation of resilience in the face of adversity (Dewey 1938, 1997 reprint).

By diving deeper into the complex dynamics within each dimension and the interplays between them, individuals can gain a more nuanced understanding of themselves and their place in the world, fostering personal growth and contributing to collective transformation.

This internal multidimensionality is mirrored in the societal framework where individuals operate. In the *micro*-arena, individuals encompass four internal dimensions—aspirations, emotions, thoughts, and actions—forming the basis of their participation in broader social structures. *Meso* refers to the institutions and communities forged by individuals, encompassing familial, professional, leisure, and ideological affiliations. These serve as interfaces between *micro*- and *macro*-dimensions. *Macro* encompasses countries and economies. Lastly, the *meta-arena* encompasses the cumulative impact of *micro*-, *meso*-, and *macro*-dimensions, extending to non-anthropocentric elements like nature and supra-national institutions such as the United Nations. This interconnected system forms a comprehensive *m4-matrix*, shaping human existence and its relationship with the planet.

Comprehending the organic sophistication that shapes human perceptions of reality empowers individuals to gain deeper insights into themselves and others. This insight forms the foundation for authentic influence, a pivotal asset in driving meaningful social change. Combined with the desire for equality, such a mindset is essential for shaping and executing policies aimed at eradicating exclusion. Fostering inclusive cities, communities, and societies where everyone can thrive, starts with people as the central actor and prime beneficiary.

17.2.2 Principles

A set of fundamental principles govern both individual and collective realities: Connection, Change, Continuum, and Complementarity. By internalizing these insights through reflexive praxis, policymakers can craft iterative interventions attuned to the dynamics of a city in flux, fostering communities where all residents can flourish. A word on each to set the stage.

Connection: Nothing happens in a vacuum. Every action resonates within a vast web of interdependencies, shaping not only one's own trajectory but also influencing the paths of others. This principle highlights the profound impact of seemingly minor actions on broader societal dynamics (Gladwell 2000). Urban environments are characterized by dense networks of interactions between residents, organizations, infrastructure systems, and more. Understanding these interconnected dynamics is vital when addressing complex issues like urban inequality, isolation, or exclusion.

Change: Everything always changes. Cities, like all aspects of existence, are in a perpetual state of flux. Embracing this reality is essential for effective urban planning and development. For instance, urban areas must adapt to evolving technological advancements, demographic shifts, and environmental challenges to remain resilient and sustainable (Cairns and Krzywoszynska 2019). Cities have always been hotbeds of, technological innovation, and social transformation. Urban planning must embrace flexibility and resilience in the face of exponential change, or risk locking lower income residents into rapidly deteriorating conditions. Intentional scenario analysis will be key to navigating uncertain futures (Quay, 2010). One example of outdated city planning is the 1961 zoning code of New York City, which locks residents and businesses in the perception of social appropriateness of the city planners from the mid of past century.

Continuum: Past, present, and future are part of a continued stream of time and space. Urban landscapes are dynamic mini cosmoses which tend to preserve a certain identity overtime. Like an organism in nature, cities change over time, while preserving certain characteristics. These agile dynamic underscores the need for adaptive urban governance structures capable of responding to emergent challenges and opportunities (Evans and Karvonen 2014). Viewing urban development through a long-term, continuous lens highlights the legacy impacts of near-term decisions. For example, exclusionary zoning policies or environmental negligence often manifest in economic and health externalities passed to future generations (Rothstein 2017). However, long-term thinking also reveals opportunities for progressive change through sustained, incremental reforms over time. People's needs change, yet the foundational aspirations remain.

Complementarity: Everything is part of something else which completes it while being completed. A thriving city relies on the mutually supportive contribution of each component. Just as a vibrant ecosystem comprises a variety of species in a city each part (in)directly serves a certain function. Once cities embrace the diversity of their residents, businesses, and institutions, they flourish (Sassen 1991). Realizing the synergistic potential of diversified urban components, whether neighborhoods, demographics, or industries, relies on participatory planning (Fung 2006).

Embracing this multidimensional perspective opens the door to a pragmatic approach for large-scale transformation. As epicenters of human activity, cities may become nodes of positive change in that endeavor. No matter the size or scale, progress often begins by recognizing the fundamental human desire for meaningful and compassionate relationships, which may manifest as a proactive commitment to contributing to one's community. This symbiotic relationship benefits both the giver and receiver, echoing the ancient and time-honored principle of reciprocity observed across cultures for centuries.

Often dubbed 'the Golden Rule'. The latter is a moral maxim found in the teachings of Jesus, Hillel, Confucius, and other moral and spiritual teachers around the world, throughout centuries, continents, and cultures (Gensler 2013). It encourages mutual caring, thoughtfulness, and sympathy between people. In the bustling anonymity of urban environments, embracing the Golden Rule can foster stronger interpersonal connections, community cohesion, and collective well-being (Hanh and

Weare 2017). Researchers find that communities with high levels of reciprocity and trust tend to have lower crime rates, better health outcomes, and more robust economic growth (Putnam 2000). Practicing the Golden Rule also correlates to greater life satisfaction and emotional fulfillment at the individual level (Post 2005). Treating others as one would wish to be treated and refraining from actions one would not want done to oneself is a way of expressing one's identity in life, which is central to the personal quest for meaning (Frankl 1946). Cities can become harbors of prosocial change, and AI may be part of turning the page from exclusion to inclusion.

17.3 The Potential of AI as a Prosocial City Catalyst

Individual aspirations (IA) and artificial intelligence (AI) can be consciously curated as a symbiosis that leads to inspired action. The central nexus is the organic interplay between individual well-being and collective prosperity. One without the other is unsustainable; AI can support both, but only once the train is set in motion by human decision-makers. In the following, we look more at the possibility of tailored AI interventions in the *micro-*, *meso-*, *macro-*, and *meta*-arena. Taking the spirit of the above, we now consider the potential of AI to supercharge each collective arena in view of improved experiences for those evolve in them.

Hardware and software, policy instruments, and the programs that implement them are reflections of the aspirations and actions of human beings, be they engineers, entrepreneurs, politicians, or citizens. Those who are targeted by the deriving interventions and those who devise and implement the latter are human beings just like those who are targeted by their policies and programs. A radically multidimensional philosophy is useful in the undertaking of inclusive cities—because once we perceive all humans as living beings that evolve as the cause and consequence of twice-4 dimensions, it becomes apparent that distinctions such as color and creed, credentials, and class, are manmade and hence transient phenomena. Underneath we are the same—people who seek to survive and be happy, equipped with rights and duties to make this quest come true.

In a hybrid world where every offline action is amplified online, it matters twice to be conscious about the values that guide one's own perceptions and pursuits, which together manifest as behavior, online and off-screen. We cannot expect from the technology of tomorrow to reflect an ethical compass that the humans that design and use it today, do not follow, and manifest through their choices. AI-powered solutions hold immense potential to optimize society toward the Common Good; yet it must be configured with that intention firmly anchored in the human mind and influence the design and governance of urban spaces.

Traditionally, cities have been spaces for people to gather and connect, in view of respectively and collectively build a community that makes everyone stronger than s/he would be in isolation. Fast forward to the present day, urban spaces are often sprawling octopi where social bonds are frail, and people live in isolation. But not everything is dark.

Today's homo sapiens know more, have more, and can do more than ever before. We are better connected and (potentially) better coordinated than ever before in history. The question is whether individuals will continue to use their material and intangible potential to primarily defend their own interests or start to systematically curate an environment that is beneficial to all, and hereby secure their own long-term survival. The latter requires a holistic perspective of servicing everyone to serve the self. Cities being contained spaces as compared to the vast sphere of 'The World' they offer a laboratory to experiment with innovative approaches and serve as springboards to trigger larger-scale change.

For instance, digital social networks are already playing a major role in connecting people in the private and business context; this supercharged social networking dynamic is likely to accelerate ever faster in the years to come. Can we intentionally design it in ways that include people who do not use the internet regularly, such as the elderly or low-income families and are at risk of falling between the cracks into isolation? Social networks may propagate economic marginalization or serve to expand and catalyze economic opportunities (Granovetter, 1973). If there is solid interest from the side of public authorities, the combination of user-friendly online platforms and low-cost tools for public outreach, coupled with 24/7 refined AI-powered data analytics will allow city majors and urban authorities to stay in touch with their citizens in a regular and meaningful way. From regular opinion polls and topic-related surveys to systematically designed participatory budgeting, technology opens a wide range of opportunities for public officials to monitor the pulse of their constituencies, and to respond to their needs adequately. Will officials seize that opportunity to serve (all) their constituents better?

While the preceding exploration underscored the importance of addressing urban issues through a holistic lens, that acknowledges the interconnectedness of individual well-being, community resilience, national policies, and global sustainability efforts; the following section explores the transformative potential of integrating AI with the multidimensional POZE perspective outlined above, within the urban context. We will briefly lay out the added value that AI brings to urban spaces in efforts to improve quality of life, and the interplay among various social arenas with some examples to illustrate what could be done. We conclude with a succinct examination of the opportunity (and risk) of AI-powered data.

This is not a chapter about technology; the following examples are a tiny excerpt of the list of existing applications and potential future usages. They are laid out here to make one central argument: The question today is not whether we have enough knowledge and material means to improve quality of life for all in an environmentally sustainable way but are we willing to do so. The recommendations that conclude this chapter are an invitation to think big and beyond the status quo, to consider the option of prosocial AI in the urban space.

17.4 AI Across the *m4-Matrix*—A Brief Overview per Arena

In tandem with the type of multidisciplinary multidimensional interventions outlined above, data and AI may emerge as powerful allies when it comes to urban policy design. Through advanced data analytics techniques, AI systems can glean insights from vast repositories of demographic data, shedding light on hidden intra-urban disparities, and illuminating pathways toward equitable outcomes. From data collection to pattern recognition, AI empowers us to decipher socioeconomic dynamics and identify urban areas in (differential) need, and hence tailor fine-tuned interventions. Examples of AI-powered interventions, including some illustrations of its added value, are presented below.

Micro—AI holds promise to personalize and improve individual well-being. For example, personalized healthcare leveraging AI can empower individuals to proactively manage their health. By providing customized recommendations tailored to people's unique genetic makeup, lifestyle factors, and medical history, these AI applications enable more personalized and preventive care (Topol 2019). Beyond benefiting individuals through improved personal health outcomes, these personalized interventions can reduce burdens on healthcare systems and promote preventive community health practices overall (Krittanawong et al. 2017).

Importantly, well-designed AI solutions, be it in health care, education, or adult learning could democratize access to quality care by making it widely available, affordable, and effective for all people regardless of income level or neighborhood. By continuously gathering and analyzing data, AI systems can adapt recommendations and functionality to serve individuals' specific needs—with targeted attention toward traditionally marginalized and underserved groups (Zheng et al. 2014). AI-powered educational platforms can offer customized coaching, tutoring and lessons tailored to learners' individual abilities, pace, and preferences (Kulik and Fletcher 2016); this is particularly relevant for at-risk neighborhoods which typically face low service coverage, insufficient quality and are home to people with tight work-schedules.

Meso—Cities are the vaguest nerve of resilience; they are the cause and consequence of resilient citizens, and strong countries—*see continuum principle above.*

Individual resilience, characterized by emotional intelligence and coping skills, fuels community resilience by enabling volunteering, resource sharing, and collective problem-solving (Norris et al. 2008). Vice versa, robust community support systems, fostered by strong social networks and access to resources, bolster individual resilience (Aldrich and Meyer 2015). This synergy dynamic extends nationally, with resilient communities sharing expertise and resources which strengthens national resilience (Comfort et al. 2010), conversely national policies promoting preparedness and resource allocation empower communities and individuals (Klinenberg 2019). Thus, individual, community, and national resilience mutually reinforce each other in a virtuous cycle, that is, conducive to individual quality of life and collective well-being.

To fulfill their potential as strongholds in national crises, and supportive marinas for the individuals who reside in them, cities must be curated consciously, as unique, organically evolving eco-environments. In that regard, AI holds potential to cultivate stronger, more resilient urban centers and residential areas. For example, AI systems can analyze and aggregate data to predict and mitigate local criminality, or collective risks, like identifying areas with dense populations that are prone to natural disasters, or fragile buildings at risk of collapsing. Based on such tailored live information, the coordination of emergency responses becomes timelier, targeted, and hence effective (Meier 2015). AI-powered community health monitoring systems can track population-level health data in real-time to uncover emerging issues like disease outbreaks or environmental hazards (i.e., lead in buildings, water pollution) early on (Friedman et al. 2021). Models that are specifically designed to deal with proximity and congestion in cities allow to map trends and differences in contagion patterns between neighborhoods, but also among urban and rural areas. This allows local health authorities to promptly implement targeted interventions and efficiently (re)allocate resources to address community health needs (Friedman et al. 2021). As for most social interventions a one-size-fits all blueprint does not work; to reveal their value these models can and must be tailored specifically to the urban context- and configured with a keen consideration for local characteristics. The potential for low-cost fine-tuning is among the charms of latest large language models (LLM).

From a multisectoral perspective, integrated databases combining often disconnected data-servers with information on health, employment, housing, transportation, and education can reveal intersectional deprivations suffered by certain demographic groups in specific neighborhoods (Kitchin 2022). Advanced analysis of this composite data using AI techniques like network mapping can then elucidate pathways to systematically invest to improve people's quality of life. Such improvements can range from optimized public transit routes based on commuting patterns, to reduce travel times,—especially for residents of low-income neighborhoods (Zheng et al. 2014), via increased social worker or police deployment in certain neighborhoods at risky times, to the creation of shared community spaces such as parks or libraries that are currently left out due to outdated zoning plans. AI shows strong potential to systematically facilitate the design of inclusive community engagement. For example, AI-powered platforms were prototyped to connect residents, healthcare providers, and policymakers to co-create solutions that address on the one hand local health disparities while on the other hand promoting social cohesion (Friedman et al. 2021). It may also improve open access by blending online and offline participation for community events, hence harnessing the complementarity of technology and human interaction (Hamari et al. 2016).

Macro—AI holds promise to enable agile national urban policies. As cities are part of countries, no 'city can make it on its own'—*see complementarity principle above.* Thus, the need to complement urban planning with nation-wide urban initiatives and support (including the possibility of cross-subsidization to help poorer/smaller cities). Specifically, governments can leverage AI-powered predictive modeling and decision support systems to formulate evidence-based public policies and efficiently allocate resources across cities, dependent on the respective needs (Eubank et al.

2020). Although cities (*meso*) serve as pivotal hubs driving progress in human flourishing, national policymaking plays a vital role in providing guidance, incentives, and resources to maximize impact. National governments (*macro*) can pass legislation and allocate funding to support municipal innovations, such as tax credits for urban farms, grants for community wellness programs, or investments in AI healthcare solutions. Nations can 'prototype' model success by pioneering new policies—like congestion taxes to reduce traffic and air pollution or free public transit to increase mobility—which cities can then replicate and tailor locally.

As cities are growing, people's needs tend to evolve, and increase—see change principle above. Swelling needs and stagnating budgets are not a good match. AI can be leveraged to tailor taxes to large fortunes, in view of harmonizing the owner's contribution to the local ecoenvironment they belong to. By integrating granular, localized data, AI systems can account for regional differences and empower decentralized, dynamic decision-making to meet specific city's needs (Goldsmith and Crawford 2014). Although poverty and lack/loss of jobs is not synonymous mapping and monitoring both in real-time is useful to tailor interventions that affect people with adequate solutions. Everything is linked—*see connection principle above.* Intra and inter-city-wide data platforms combining governments, social service providers, and private sector players via crowdsourced data streams that reflect demand and offer can enable nimble, adaptive governance, and resource allocation from one city, region, or county to others, i.e., algorithms applied to detect emerging unemployment hotspots or model proposed policy impacts can inform rapid anti-poverty program implementation. Although this approach has not been used at scale, examples include the use of machine learning models to predict unemployment rates across communities based on real-time economic indicators. The predictive heat maps enabled more agile resource allocation and early crisis response programming by local government and non-profits); another way was to use quasi-experimental evaluation methods to analyze impacts of proposed welfare and unemployment insurance reforms. Findings about optimal policy design were then used to efficiently scale up pilot programs found most effective against poverty. (Cockx et al. 2018). In complementarity to this type of intervention rapid-response employment initiatives triggered by algorithmic warnings could then deploy supportive services like job retraining, transit vouchers, food assistance, and eviction protection more preemptively, intra-city and inter-city. Cross-border partnerships allow not only countries (*macro*) but also urban settlements (*meso*) to share best practices while collaborating on complex global challenges (*meta*). Networks like the C40 Cities Climate Leadership Group, comprising 97 major metropolises, coordinate ambitious climate action across geographies through knowledge exchange and aligned strategy. At the 2021 United Nations Climate Change Conference, over 1000 mayors signed a Global Mayors COVID-19 Recovery Task Force statement demanding national action to spur sustainable, healthy, and equitable urban recovery.

Meta (Global) Dimension: On a cross-border scale, AI holds meaningful potential to address urgent challenges like climate change and equitable urban development that disproportionately impact vulnerable populations worldwide. Sophisticated machine learning algorithms can rapidly analyze massive datasets—from greenhouse gas emissions to extreme weather events—to generate actionable insights informing climate mitigation and adaptation policies optimized for localized implementation (Rolnick et al. 2019). These strategies can then be shared through global municipal networks to share resources and best practices that accelerate climate resilience across cities, beyond borders.

A word on the interplay of humans (micro) and nature (meta)—Cities rely intrinsically on nature to sustain human activity. While dense urban zones often supplant ecosystems, integrating and safeguarding green space improves air quality, reduces flood risk, and enhances physical/mental health—all dimensions of well-being. Although urbanization and nature appear to be at odds, both are not only necessary for society in the twenty-first century—the former can be systematically configured to operate in synergy with the latter and hence support human well-being. AI may help in making that effort systematic. It is an urgent undertaking that serves people and planet.

Addressing the intersection of people and nature is not a luxury but an essential aspect of mental and physiological well-being and hence quality of life. Contact with nature has been conclusively linked to improved physical and psychological well-being in individuals. Spending time in natural environments lowers stress, blood pressure, and heart rate while boosting mood, cognitive function, and immunity (Frumkin et al. 2017). In particular, the presence of green space in cities provides respite from noise and crowds, opportunities for reflection and social connection, and aesthetic pleasure (Wolch et al. 2014). Urban parks, gardens, green roofs, and abundant street trees nourish human health on a neighborhood level (Van den Berg et al. 2015), but maintaining equitable access to nature requires intentional human design.

Cities, regions, and countries must implement climate-resilient design while balancing density with abundant urban tree canopy, wetland buffers, green rooftops, and ecological preservation zones. In that area AI shows promise for designing sustainable and accessible urban infrastructure. This includes increased efficiency in areas like energy consumption in public buildings or waste collection routes (Hashem et al. 2016); but also, the prioritization of accessibility for all citizens while minimizing energy consumption and carbon footprints (Angelidou et al. 2018).

Technology is not required but can ease and scale the design of cost-efficient, climate neutral, and environmentally adapted solutions. For instance, AI-powered planning tools may serve city planners to design sustainable green infrastructure and integrate it in innovative, space-saving manners throughout cities to bring the benefits of nature to residents (Jennings et al. 2016). Examples include natural play areas in place of sterile playgrounds, wildlife corridors that promote biodiversity, urban farms and forests that produce food and clean air while educating the public, and policies that protect existing ecosystems within city boundaries.

Finally, a word on international collaboration (meta)—which involves citizens (*micro*), cities (*meso*), countries (*macro*), and international initiatives and institutions (*meta*):

International partnerships, philanthropy, and knowledge sharing can accelerate progress. For example, the Rockefeller Foundation's 100 Resilient Cities initiative equipped metropolises across six continents with funding and expertise to enhance livability. And Citi Foundation and Living Cities collaborate to scale locally driven solutions that connect low-income residents with economic opportunities. Multilateral cooperation boosts the impact of city-level interventions. Additionally, transnational AI model analysis of satellite imagery can help predict conditions and direct critical aid to individuals affected by migration or conflict, including informal settlements. More broadly, AI and smart city technologies are enabling sustainable, context-appropriate urbanization worldwide through South–South and North–South knowledge transfer (Bibri and Krogstie 2017). International AI collaborations can facilitate global coordination on shared challenges like disease outbreak monitoring, early warning systems, and equitable access to scientific advances in areas like genomics and precision medicine (Vinuesa et al. 2020). Responsible cross-border data sharing can accelerate equitable urban development globally (Bibri and Krogstie 2017).

Translating the 'Golden Rule' into urban planning demands a multidimensional lens. Balancing individual needs with community cohesion requires prioritizing shared spaces, walkable neighborhoods, and equitable resource allocation. Public spaces that encourage social interaction, pedestrian friendly neighborhoods, shared community gardens, and venues for public expression all provide opportunities 'to treat others as we wish to be treated' (Jacobs 1961; Hanh and Weare 2017). But this type of sympathetic social setting does not happen by itself; it requires conscious design. Scaling up, we must consider in addition intergenerational and global impacts, ensuring sustainable development for all over time. To get a grip on this complexity AI emerges as a powerful tool, analyzing data and optimizing solutions across arenas. By embracing on the one hand the multi-perspective perspective outlined above, while on the other hand harnessing AI's potential for systematization, we can weave the 'Golden Rule' into the very fabric of our cities, creating spaces that truly embody its spirit.

In summary, AI shows immense potential to address in an organically evolving timely manner, interconnected challenges across all levels of human society. However—realizing synergistic and sustainable solutions requires inclusive governance frameworks centered on community needs (Mohamed et al. 2020), as well as responsible data collection and sharing anchored in strong ethical principles. Fully leveraging AI for social good further depends on collaborative multi-stakeholder efforts across borders.

Prosocial human ambitions combined with an AI that is configured for robust data analytics presents a formidable tool to help find solutions to improve urban quality of life for all.

AI like any technology is neutral. Like a book can be a channel for Hitlers 'Mein Kampf' or The Bible, AI can serve to expand and enhance high quality social serves or to build weapons of mass destruction. Technology is a good or bad as those who design, deliver, and use it. They are neither a magic wand nor a goal worthy of itself. What makes the value of the tool is the intention that it is used with. Fire can burn a book or cook a meal; it may kill a person or save her from being killed by the cold. AI that is used with the vocation of optimizing quality of life is a major asset for humanity. In that sense it may serve to finally make inequality and avoidable deprivation a faded memory of the past, and cities might be the entry-points to make that happen. But it will not happen by itself; nor as an automatic derivative of 'progress'. For now, most of our technological prowess serves commercial purposes. It is a choice to veer the steering wheel from money to meaning. We are all in the driver's seat in that regard.

Before we conclude, a word of special caution regarding the ethical implications of ***AI-powered data gathering and analyses***.

Especially in the context of human-centered design and the stated desire of quality of life, leveraging data analytics at *multiple interconnected* levels, must be done in a *responsible manner*. As seen earlier, once we adopt a multidimensional lens, the power of AI and big data analytics could provide critical insights to guide policies aimed at alleviating urban inequality. But it might as well backfire and offer information that jeopardizes already marginalized people and peoples further, by violating their right to privacy and dignity.

Realizing the benefits of data-driven policymaking requires grappling with real ethical risks. As algorithmic systems inherently propagate the biases of their training data, they may unconsciously exacerbate inequality due to issues from non-representative sampling to incomplete perspectives (Mehrabi et al. 2021). Thoughtful governance approaches centered on inclusivity, transparency, and human oversight are imperative. By bringing community stakeholders directly into the design process, more holistic assessments of the unintended consequences of automated decisions become possible (Morley et al. 2021).

Supposing however a rigid moral code is applied, specific areas of interest include the aggregation of disparate datasets encompassing demographics, income, education, health, and other key factors, where AI systems can discern subtle patterns of discrimination and deprivation affecting marginalized communities (Provost and Fawcett 2013). AI-powered data cleaning and preprocessing may improve the reliability of the collected information (Rahm and Do 2000). Sophisticated machine learning algorithms can then analyze these comprehensive data assets to pinpoint specific areas and vulnerable populations suffering from uneven access to opportunities (Bishop 2006; Fayyad et al. 1996). Dynamic modeling also enables policy-makers to simulate the potential impacts of various interventions, from affordable housing initiatives to targeted employment programs (Chen and Zhang 2014).

None of the outlined benefits are automatic outcomes. Every opportunity comes with a risk. The bigger the potential payout—the higher the stakes, and hurdles to overcome in reaching it. In essence, although AI and big data will increasingly inform human decision-making processes at all levels, their responsible implementation must balance quantitative precision with qualitative human values. Only, if efforts are made intentionally to improve algorithmic interpretability, it will become possible to help identify and combat embedded biases that might adversely affect marginalized communities in policy decisions (Mehrabi et al. 2021). This requires participatory frameworks where the priorities of the most vulnerable shape the engineering of technological tools from the ground up. Taking seriously such an endeavor can help cities move toward mainstreamed equity and justice for all residents.

Realizing the potential of AI-powered data relies on context-specific data governance frameworks that prioritize citizen agency, platform cooperatives, and participatory design (Johnson 2022; Van Zoonen 2020). The following illustrates what this might look like:

> Case Study: *Dunu Juzi*, a youth-led grassroots organization based in informal settlements in Nairobi, Kenya, takes a hyperlocal, participatory approach to gathering data and implementing development initiatives in low-income urban areas. It is just one an example of an organization that is taking an inclusive, community-based approach to leveraging data and technology to address urban poverty and development challenges. Some key aspects of their model include (i) working with residents of informal settlements to crowdsource hyperlocal data on issues like infrastructure, health, education, and employment using mobile phones and other technologies; (ii) using this granular, community-sourced data to uncover localized development needs and opportunities; employing machine learning on this data to predict challenges and target interventions more effectively; (iii) building grassroots partnerships with local leaders, youth groups, schools, etc., to co-design and implement context-appropriate development programs based on the data, from upskilling programs to microenterprise incubators; and (vi) focusing explicitly on amplifying existing community assets and priorities rather than taking a top-down approach. In this way, *Dunu Juzi* harnesses participatory data and inclusive governance to create locally resonant and sustainable solutions to pressing development challenges like unemployment from within communities themselves. Their model shows the potential of technology as a tool to empower residents of informal settlements to lead positive change by making decisions based on their own emic data and local knowledge. They exemplify the logic of using data analytics 'with, not on' marginalized groups, decentralized development, and urban resilience. (Kwet and Frame 2022).

Beneath the technological promise of AI lies a deeper ethical dilemma—can automated systems truly advance social justice without perpetuating historical discrimination? Even datasets that appear neutral often silently encode societal biases against minorities, the poor, and other disadvantaged groups (Mehrabi et al. 2021). Once incorporated into machine learning models, these prejudices become propagated and amplified through opaque algorithmic systems optimized solely for predictive accuracy rather than fairness or accountability (Russell 2019). Other aspects of structural power structures and lack of voice/representation may also be reproduced by AI models/apps. The very groups AI aims to empower may thus face even greater marginalization.

Technological solutionism often focuses on symptoms rather than root causes or fails to acknowledge the interconnected nature of urban challenges. However especially in the complexity of urban environments, even well-intentioned technological interventions can yield unintended negative consequences when a holistic assessment is missing (Morley et al. 2021). Failing to recognize interplays across the *m4-matrix* runs high risks of dramatic backfiring due to the butterfly effect—*connection and continuum principle*. Furthermore, initiatives driven by top-down municipal policies rather than grassroots participatory design processes often neglect the self-identified priorities of impacted communities.

To realize the promise of equitable cities, the use of AI as a tool for social progress ***must*** be guided by inclusive yet overarching ethical frameworks. Community stakeholders themselves must shape the engineering of systems intended to better their lives—sitting not just as external advisors but integrally within multidimensional design teams. Through such radical participatory praxis, the risk of perpetuating injustice through computation can give way to technology crafted with nuance, wisdom, and a shared spirit of social justice.

In sum, while AI and data analytics are powerful tools, utilizing them responsibly to create an equitable urban landscape requires comprehensive ethical foresight. From input data collection through model development, deployment, and monitoring, AI systems must be crafted based on principles of justice, diversity, and humanity-centric design (Floridi et al. 2018). This necessitates inclusive governance frameworks centered on impacted communities' needs and values.

17.5 Conclusion

(Urban) inequality is no inexorable fate, but the legacy of past decisions and circumstance that we continually renew through complacency or inaction. Yet each new dawn offers hope—a chance to reverse course toward justice. Seizing this promise, however, requires manifesting the very values we seek through four key shifts.

First, we must shift from thinking and planning with an exclusively local or international/global perspective, to a truly *glocal mindset*. Cities must become nodes of social connection, where solutions can be prototyped locally and amplified globally.

Second, we must shift from sectoral silos to an organic tapestry of private and public stakeholders. Cities must be re-designed as harbors of co-creation spanning research and practice.

Third, we must shift from disciplinary seclusion (which is not a specifically urban challenge but a caveat of research and practice in every domain of human progress) to a blend of the best that social sciences, STEM, and the arts have to offer in a shared spirit of epistemic humility.

Fourth, we must shift from the status quo as the limiting horizon of our ambitions, to a perspective of organic transformation. Every place of human cohabitation can become a place of prosocial transformation. Adopting a multidimensional perspective to live and living in cities is a first step to making cities hotspots for social zeniths, where humans are given safe space to share their best self and help humanity advances.

All the above requires candid questioning, ethical reflexivity, and the critical debate of the causes and consequences of urban inequality. Making inequity a concept of the past requires human design-centered holism. Technology can accelerate progress, without becoming its essence or sole definition. AI is a tool and may be a catalyst, but it is not a savior by its own. Progress must not outpace purpose. Automation without wisdom risks blind efficiency, or even institutionalizing injustice through opaque software architectures. Only radically participatory processes can harness computation in service of inclusive, equitable cities that are crafted to and through empowered neighborhoods.

The future remains unformed, awaiting today's choices, which are ours to make.

References

Aldrich DP, Meyer MA (2015) The resilience of place: a focus on neighborhoods and communities. In: Aldrich DP, Meyer MA (eds) People and places: the sociology of urban resilience. Oxford University Press, pp 1–20

Angelidou M, Psaltoglou A, Komninos N, Kakderi C, Tsarchopoulos P, Panori A (2018) Enhancing sustainable urban development through smart city applications. J Sci Technol Policy Manag 9(2):146–169

Bibri SE, Krogstie J (2017) Smart sustainable cities of the future: an extensive interdisciplinary literature review. Sustain Cities Soc 31:183–212

Bishop CM (2006) Pattern recognition and machine learning. Springer

Cairns S, Krzywoszynska A (2019) The changing urban landscape: interdisciplinary perspectives on cities. Springer

Chen M, Zhang X (2014) Data-intensive applications, challenges, techniques and technologies: a survey on big data. Inf Sci 275:314–347

Cockx B, Declercq K, Dejemeppe M, Inga L, Van der Linden B (2018) Scrapping the entitlement to unemployment benefits for young labor market entrants: an effective way to get them to work? Int Tax Publ Finance 25(5):1280–1308

Comfort LK, Wisner B, Cutter SL, Pulwarty RS, Hewitt K, Fordham M et al (2010) Community resilience to climate change: reducing vulnerability through collective action. J Sustain Dev 3(1):1–18

Dewey J (1938) 1997 reprint). Free Press, Experience and education

Eubank S, Eckstrand I, Lewis B, Venkatramanan S, Marathe M (2020) Evaluating the effect of public health interventions: key concepts and their importance in infectious disease modeling. Health Secur 18(2):123–131

Evans J, Karvonen A (2014) Give me a laboratory and I will lower your carbon footprint!'—urban laboratories and the governance of low-carbon futures. Int J Urban Reg Res 38(2):413–430

Fayyad U, Piatetsky-Shapiro G, Smyth P (1996) From data mining to knowledge discovery in databases. AI Mag 17(3):37–37

Frankl V (1946) Men's search for meaning. Republication by Paperback

Friedman M (2002) Capitalism and freedom (40th edition reprint of 1962). University of Chicago Press; Frumkin H, Bratman GN, Breslow SJ, Cochran B, Kahn Jr PH, Lawler JJ, Levin PS, Tandon PS, Varanasi U, Wolf KL, Wood SA (2017) Nature contact and human health: a research agenda. Environ Health Perspect 125(7):075001

Fung A (2006) Varieties of participation in complex governance. Public Adm Rev 66:66–75. https://doi.org/10.1111/j.1540-6210.2006.00667.x
Gensler HJ (2013) Ethics and the golden rule. Routledge
Gigerenzer G (2007) Gut feelings: the intelligence of the unconscious. MaxPlanck Institut. Researchgate.net
Gladwell M (2000) The tipping point: how little things can make a big difference. Little, Brown
Godfrey CJ, Aveyard P, Garnett T, Hall JW et al (2018) Meat consumption, health, and the environment. Science 361(639)
Goldsmith S, Crawford S (2014) The responsive city: engaging communities through data-smart governance. Wiley
Granovetter MS (1973) The strength of weak ties. Am J Sociol 78(6):1360–1380. https://doi.org/10.1086/225469
Hamari J, Sjöklint M, Ukkonen A (2016) The sharing economy: why people participate in collaborative consumption. J Am Soc Inf Sci 67(9):2047–2059
Hanh TN, Weare K (2017) Happy teachers change the world: a guide for cultivating mindfulness in education. Parallax Press
Hashem IA, Chang V, Anuar NB, Adewole K, Yaqoob I, Gani A, Ahmed E, Chiroma H (2016) The role of big data in smart city. Int J Inf Manag 36(5):748–758
Hills P, Argyle M (2001) Happiness, introversion–extraversion and happy introverts. Pers Individ Differ 30(4):595–608
Hopkins Medicine (2022) The brain-gut connection. https://www.hopkinsmedicine.org/health/wellness-and-prevention/the-brain-gut-connection
Hyungryeol K, Euijin L (2019) A cross-national study of the influence of parental education on intention to vote in early adolescence: the roles of adolescents' educational expectations and political socialization at home. Int J Adolesc Youth 24(1):85–101; Jacobs J (1961) The death and life of great American cities. Vintage Books
Jean N, Burke M, Xie M, Davis WM, Lobell DB, Ermon S (2016) Combining satellite imagery and machine learning to predict poverty. Science 353(6301):790–794. https://doi.org/10.1126/science.aaf7894; Jennings V, Larson L, Yun J (2016) Advancing sustainability through urban green space: cultural ecosystem services, equity, and social determinants of health. Int J Environ Res Publ Health 13(2):196
Johnson JA (2022) Global cities, local data governance: towards a situated approach. Big Data Soc 9(1):20539517221068720
Kahneman D (2011) Thinking, fast and slow. Penguin Books, London
Kitchin R (2022) Civilizing the smart city: data-driven urbanism and the critical challenges of the knowledge economy. Urban Geogr 1–8
Klinenberg E (2019) Palaces for the people: how social infrastructure can heal and unite cities. W. W. Norton & Company
Krittanawong C, Zhang HJ, Wang Z, Aydar M, Kitai T (2017) Artificial intelligence in precision cardiovascular medicine. J Am Coll Cardiol 69(21):2657–2664
Kulik A, Fletcher JD (2016) Effectiveness of intelligent tutoring systems: a meta-analytic review. Rev Educ Res 86(1):42–78
Kwet M, Frame A (2022) Digital for development: an evidence review. The GovLab
Meier P (2015) Digital humanitarians: how big data is changing the face of humanitarian response. CRC Press
Mehrabi N, Morstatter F, Saxena N, Lerman K, Galstyan A (2021) A survey on bias and fairness in machine learning. ACM Comput Surv (CSUR) 54(6):1–35. https://doi.org/10.1145/3457607
Mohamed S, Png MT, Isaac WS (2020) Decolonial AI: decolonial theory as sociotechnical foresight in artificial intelligence. Philos Technol 33(4):659–684. https://doi.org/10.1007/s13347-020-00405-8
Mulgan G (2006) Making sense of the good: and why it matters in political philosophy. Harvard University Press

Norris FH, Neria Y, Nash B, Engelhard R (2008) Community resilience as it shapes and reflects individual resilience: insights from the aftermath of the September 11th terrorist attacks. Psychol Inq 19(1):30–50
Post SG (2005) Altruism, happiness, and health: it's good to be good. Int J Behav Med 12(2):66–77
Provost F, Fawcett T (2013) Data science for business: what you need to know about data mining and data-analytic thinking. O'Reilly Media
Putnam RD (2000) Bowling alone: the collapse and revival of American community. Simon & Schuster
Quay R (2010) Anticipatory governance: a tool for climate change adaptation. J Am Plann Assoc 76(4):496–511. https://doi.org/10.1080/01944363.2010.508428
Rahm E, Do HH (2000) Data cleaning: problems and current approaches. IEEE Data Eng Bull 23(4):3–13
Rolnick D, Donti PL, Kaack LH, Kochanski K, Lacoste A, Sankaran K, Waldman-Brown A (2019) Tackling climate change with machine learning. arXiv preprint arXiv:1906.05433
Rothstein R (2017) The color of law: a forgotten history of how our government segregated America. Liveright Publishing
Sassen S (1991) The global city: New York, London. Princeton University Press, Tokyo
Sohn H, Narain D, Meirhaeghe N, Jazayeri M (2019) Bayesian computation through cortical latent dynamics. Neuron 103(5):934-947.e5. https://doi.org/10.1016/j.neuron.2019.06.012
Topol EJ (2019) Deep medicine: how artificial intelligence can make healthcare human again. Basic Books
UN Compact (2000) https://www.unglobalcompact.org/; Van den Berg M, Wendel-Vos W, van Poppel M, Kemper H, van Mechelen W, Maas J (2015) Health benefits of green spaces in the living environment: a systematic review of epidemiological studies. Urban For Urban Greening 14(4):806–816
Van Zoonen L (2020) Governing digitally integrated genomics through the city as a platform. Global Pol 11:121–130
Vinuesa R, Azizpour H, Leite I, Balaam M, Dignum V, Domisch S, Thies L (2020) The role of artificial intelligence in achieving the sustainable development goals. Nat Commun 11(1):1–10
Walther C (2014) Le Droit au Service de l'Enfant. University de Droit, Aix/Marseille UIII
Walther C (2020a) Development, humanitarian aid and social welfare. Purpose for power. Macmillan Palgrave/Springer
Walther C (2020b) Humanitarian work, social change and human behaviour. Compassion for change. Macmillan Palgrave/Springer
Walther C (2020c) Development and connection in the time of corona. Macmillan Palgrave/Springer
Walther C (2021) Technology, social change and human behaviour. Macmillan Palgrave/Springer
Wilkinson RJ, Pickett K (2009) The spirit level: why more equal societies almost always do better. In: Allen LaneWolch JR, Byrne J, Newell JP (eds) (2014) Urban green space, public health, and environmental justice: the challenge of making cities 'just green enough'. Landscape and urban planning, vol 125, pp 234–244
Zheng Y, Capra L, Wolfson O, Yang H (2014) Urban computing: concepts, methodologies, and applications. ACM Tran Intell Syst Technol (TIST) 5(3):38

GPSR Compliance

The European Union's (EU) General Product Safety Regulation (GPSR) is a set of rules that requires consumer products to be safe and our obligations to ensure this.

If you have any concerns about our products, you can contact us on ProductSafety@springernature.com

In case Publisher is established outside the EU, the EU authorized representative is:

Springer Nature Customer Service Center GmbH
Europaplatz 3
69115 Heidelberg, Germany

Batch number: 10726133

Printed by Printforce, the Netherlands